MATRIXIAL LOGIC

FORMS OF INEQUALITY

By Paul Chaplin

First edition

Published by G2 Entertainment

ISBN: 978-1-782811-71-8

Printed and bound in the UK by G2 Entertainment

PREFACE		6
CHAPTER 1	FORMS OF EQUALITY	16
CHAPTER 2	MATRIXIAL FRAMEWORK	34
CHAPTER 3	CATEGORICAL TRANSFORMATIONS	124
CHAPTER 4	FORMS OF INEQUALITY	236
CHAPTER 5	LANGUINI	299
CHAPTER 6	CHAOTIC ORDER	489
CHAPTER 7	MATRIXIAL CHRONOLOGIC	512
CHAPTER 8	MATRIXIAL ANTHROPICS	582
CHAPTER 9	COMPLETING REALITY	616
CHAPTER 10	MATRIXIAL MANIPULATIONS	663

APPENDICES:

1.	*PLAIN ENGLISH MATRIXIAL LOGIC*	699
2.	TABLE OF SYMBOLS	704
3.	LIST OF EXPERIENCES	710

PREFACE

In *I Want To Love But: Realising the Power of You,* I called this sort of communication "The Geeky Bits".

The ideas presented are intended to be accessible to the general reader, not only readers with a background in logic, philosophy or science.

Matrixial Logic ("ML") is a symbolic logic.

- You know what a "matrix" is: a set of connections. Logic is showing how things connect;
- The symbols are arbitrarily chosen correlates to linked sets of ideas.

ML is a system of symbolic logic, a tool for organising thinking about thinking, which I first developed at university. That was a long time ago, when everything was in black and white. Well, Polaroid, certainly.

I found it a useful tool in getting through a law degree at Oxford. I've since found its utility as a tool in many walks of life: from business, to personal relations.

ML is the tool which made possible the discovery of ASR[1] technology and Congruence Quotient[2] technology: both of which form the CBT core of *I Want To Love But.*

[1] Anxiety-Soothing Rhythm
[2] Five Slices

ML is a philosophical tool: a method for organising how we think. It is in principle applicable to anything that we think about.

ML does not tell you what you should think about anything. It does not prescribe the contents of thoughts, nor what conclusions you might decide to reach in your thinking about anything.

ML instead provides a set of rules for rational thought. If you wish to think rationally, you can use the rules of ML.

I do not claim that use of the rules of ML is the only way in which you can think rationally. I cannot make such a claim, because I cannot conceive of how the truth-value of such a claim could be established.

In ML terminology, I cannot think of a frame of reference in which one can solve:

$$(A) \neq ({}^{not}A) = E$$

$$>$$

$$(\text{Knowledge}) \neq ({}^{not}\text{Knowledge}) = E?$$

So, at least I'm taking my own medicine.

The bitter pill, is that ML reveals certain structures, stances, or modes of reasoning to be fallacious. That is helpful, because it allows us to gain clarity.

This is not a survey of the massive field of symbolic logic. I do not say that any of those modes of symbolic discourse are right, or wrong.[3] Nor does ML rely upon any of them.

The core method of ML can be analogised to syllogistic reasoning, but that is correlation, not cause.

Once you become familiar with using ML, you should find that you can use that tool in any of your thinking. In the course of this book, we'll look at curated applications of ML: which reveals its wide range.

I entirely accept that, in publishing this monograph, I place the premises, propositions and conclusions ("the Structure") subject to the conventions of philosophic discourse.

I try to keep a balance in this monograph between length and detail of explanation, and effective communication. The metric I use in weighing that balance, is whether I think that the reader has been given sufficient information to apply the ML tools.

Without committing the special pleading fallacy, I note the commonplace that philosophic discourse has a tendency to become burdened with demands for proof which are not observed in other disciplines.

---

[3] Although I do make clear that ML provides a perspective in which they are not agents for clarity

I therefore offer, under the principle of humility, an election which lies entirely within the sovereign power of the reader:

- If ML tools should prove to be of practical utility in reasoning, then to appreciate such use for its own value;
- To reduce such appreciation by so much as the reader is unsatisfied, for whatever reason, as to some part of the ML rules below explained;
- To place such appreciation subject to any conditionality the reader considers appropriate;
- To refuse to consider applying ML tools until the reader can acquire by that reader's standards such other information as is sufficient to satisfy such conditionality;
- To conclude, subject to further information, that as presently formulated, the ML tools have no value.

I appreciate the investment of any of your time in perusal of the remainder of this monograph.

I have found the ML tools explained in these following pages to be useful. If that be not so for you, then my hope is at least that, in determining the causes of such failure of effect, such exercise brings new information to you, which may itself be of some utility.

Paul Chaplin
2020

How to Read This Book

Turn the pages right to left, start at the top left corner and work your way across and down.

I'm having fun. And I want you, the reader to have some fun too. But, I shall make no bones about it: whether or not you are a professional logician, some of these Chapters are going to be hard going.

Indeed, labourers in that field will probably find it harder than others, because what they read is so challenging to ideas which are stamped into the DNA of classical and modern logic.

This is very much a book of parts:

Chapter 1 sets out some background. This is for the lay reader.

Part I

Thesis

Chapter 2 sets out the foundations of Matrixial Logic. There is no anticipation that much of what you read in

Chapter 2 will make much "sense" to you. Please treat Chapter 2 as if you were learning a new language, or the rules of a board game: with the merit that it has a very small alphabet and syntax.

Antithesis

Chapter 3 is one long Socratic dialogue. I use peak points from classical logic, and modern logic, to explain, by argument: (i) what problems there are[4] with those ideas; (ii) more of how ML works; (iii) why ML works; (iv) what you gain from using ML.

Synthesis

Chapter 4 is the fulcrum. Some people may not need all of Chapters 2 and 3 to follow what is being shown in Chapter 4. Even so, the previous Chapters should "come alive", once you can look back at them from the "mountain peak" of Chapter 4.

Part II

Chapter 5 is, in effect, a showcase for the practical applications of ML. In this case, in the field of linguistics. If ML can make material advances in our thinking about how language works, then that alone was worth the price of admission.

Chapter 6 is a bridge. The concepts here expand in *Secret Self*, but are of small application in the later Chapters.

Chapter 7 is another fulcrum thesis. It sets out how Matrixial Logic radically alters our views of time.

---

[4] As entirely admitted by modern logicians

Chapter 8 digests the earlier elements of Part II, providing a working model of the Human Self. This is a summary Chapter, which signposts the elaborations in *Secret Self.*

Part III

Chapter 9 uses what we have learned about ML to proffer some propositions about key concepts in western thought.

Chapter 10 provides arial footage of the city of western philosophy, following detonation of the chain-reaction thermonuclear devices, in the previous Chapters.

The Appendix sets out in 20 short Propositions, what Matrixial Logic says of itself.

You can find a handy set of Tables, as a post script. These provide a listing of the Symbols used in the book, as well as how to type them. There's also a Glossary of ML terms.

There's no separate bibliography, as it would be as long as the book itself. Throughout the Chapters, you'll find footnote references to all the key texts.

Clues

You'll find references to Clues in the text.

I'm not trying to be smart or annoying.[5] These are seeded

[5] Although I probably succeed in one, but not the other. Sorry

deliberately, to induce you to go down lines of thought. So that those other thoughts, prompted by the clues, syncopate with what's in the main text.

When you are reading, there's lots of way stations. Places where you're deliberately invited to put the book down, and have a think. The clues are one set of lay-bys on the journey.

Experiences

All the way through the book, you'll find *Experiences* that I ask you to undertake. It is really important that you follow them.

For a start, they give you an entertaining diversion from the endless lines of equations and sometimes stuffy language of logic.

More importantly, through the *Experiences*, you will get an understanding of the equation ideas, not just in your intellect, but in your emotions.

You also get to remember those ideas, in a way that no amount of blackboard repetition would.

These are experiments, in the ordinary scientific sense:

- We have a theory;
- We invent tests of that theory: operations with pre-defined outcomes, derived from the theory;
- We test the outcomes against the theory.

By doing the experiments, you the reader gain important knowledge. The way you look at the world will change. Sometimes a little: sometimes a lot.

Lines of arguments in this book are challenging to many precepts: in logic, philosophy, psychology and science. Sceptical response is understandable.

The *Experiences* provoke significant difficulties for sceptisism. As we sometimes say, following an *Experience*: well, whatever is going on here, something is going on.

Matrixial Logic provides an explanation. Now, you can seek to refuse the explanation. But you can't deny the *Experience*. Refuting the ML explanation poses some difficulty: because you only got to do the *Experience* by reason of the practical operation of ML.

If there were only *Experiences* limited in some dimension, then they might be capable of being disregarded as just an artefact of something odd. But you will undergo *Experiences* in many different areas of your interaction with yourself, and the world.

The *Experiences* are not "tricks": you can see every element of them. There are no hidden wires. You just get a simple set of written instructions: you follow them, and you decide what the result is.

The *Experiences* do not operate under hypnosis. There is no induction such as would precipitate a hypnotic state. There's just a few lines of written instructions. Indeed, you will find that you experience a widening of awareness.

These *Experiences* will show you that logic is not the opposite of emotion.[6] It is logic in action, in our life.

Logic is just as important as love, and is as universal as love. It just lasts longer, and we may one day learn to talk about it, without embarrassment.

---

[6] Sorry Commander Spock

CHAPTER 1

FORMS OF EQUALITY

Matrixial Logic is new. You can google the phrase and no results will return.

ML is a different way of doing logic. Rather than abstracting at higher orders of levels from reality, the better to free logic from the confusatory shackles of the ordinary, ML is rooted in reality.

Logic, after all, is the pursuit of reality in understanding. Logic can and should avail itself in that pursuit, of all the technical tools furnished by advanced civilisation.

But logic must guard against the mechanical fallacy: that which mistakes machinery for understanding.

As we shall survey in this Chapter, logic has, since its period of Greek innovation, sought so often to perfect the mechanical fallacy. It was only in the 20th century that First Order Logic retreated expressly from that pursuit.

With no disrespect, FOL has retreated from the battlefield of reality, and now engages only in mythicism of forms of equality. The self-recusal is understandable.

From Leibnitz and Lambert to Russell and Godel, logic

became trapped in a hall of mirrors. Haunted by problems incubated by self-reflecting. Until such a brilliant thinker as Russell could terminate his reflections with the conclusion[7] that the only meaningful word in a sentence is "the".

Logic is a scientific pursuit. Fuelled by *spectātiō* internal and external, coupled with insight and intuition, and informed by technological discovery. Then tested in theorems against reality: resulting in verification or falsification.

We appreciate that the tests of what is scientific, as adumbrated in logical positivism by the Vienna Circle,[8] cannot provide a definitive account.

One of the functions of the *Experiences*, is to allow experiments to be conducted which move beyond the circular dictates of verification and falsification.

The reader acquires new *knowledge*: a union of the subjective and the objective.[9] This goes beyond belief in the reality of a fact. It involves knowing the reality of a matter.

If it be asked how logical theorems can be tested against

---

[7] In 1959

[8] Carnap, Kuhn, Popper, primarily in this respect

[9] The only way to couch these matters at this stage in the book

reality, the course of this book should provide adequate answer.

Achieving understanding of reality can be pursued from an armchair. But adoption of such understanding by those placed contingently at the axis of social change, can alter the historic fate of nations.

History of Symbolic Logic

Symbolic logic ("SL") is commonly thought of as beginning with the Modern.

However, we must begin with Aristotle.[10] The compilation of his 6 works concerning logic, *The Organon*, left a legacy which still engages us today.

The core is syllogistic reasoning, demarcated under four different types of Categorical Sentences: universal affirmative (A), particular affirmative (I), universal negative (E) and particular negative (O). We will have cause to return to these Categorical Sentences later.

Aristotle was happy to denote place-holders for idea by letters. This is not of course the same as designating operators for functions of ideas.

Such designation, in principles, raises the question of the

[10] (384-322 BC)

relationship between the symbols, functions and reality:

Inferences

:

Equation

:

Symbols

:

Ideas

:

Reality

Modern symbolic logic expressly disavows conjunction of logical inferences, with underlying reality. That is considered to be a forlorn project of failed Higher Order Logic.

ML does not take this position. In ML, we are creating equation rules, which have *hylomorphic* relationship to reality.

> Hylomorphism, (from Greek hylē, "matter"; morphē, "form"), in philosophy, metaphysical view according to which every natural body consists of two intrinsic principles, one potential, namely, primary matter, and one actual, namely, substantial form. It was the central doctrine of Aristotle's philosophy of nature.
>
> Matter and form are parts of substances, but they are not parts that you can divide with any technology. Instead matter is formed into a substance by the form it has. According to Aristotle, matter and form are not material parts of substances. The matter is formed into the substance it is by the form it is.
>
> Surprisingly, as John MacFarlane pointed out, "the father of both formal logic and hylomorphism was not the father of logical

> hylomorphism" (MacFarlane, 2000: 255). [11]
>
> One immediate and obvious difficulty which we meet is that it is not easy to explain why Aristotle uses letters of the alphabet, like 'A', 'B', 'C', instead of concrete terms if he did not distinguish between logical form and logical matter. Here is an exemplary formulation for the first syllogism in the first figure (Barbara) from the Prior Analytics: "if A belongs to every B and B belongs to every C, it is necessary for A to belong to every C" (Pr. An., 25b37-9).[12]

There will be much more to say about this in later Chapters.

It has been noted that the history of "Euler Circles" or "Venn Diagrams" goes back to the middle ages. Margaret Baron traces[13] the ideographic concept back to Ramon Lull:[14]

> **Lull believed that, by exploring all possible combinations of certain basic principles, or categories, we could obtain knowledge of all things. To represent these principles he used circular sectors and he combined them by means of rotating circles of varying diameters. In one diagram he shows intersecting circles, thus,**

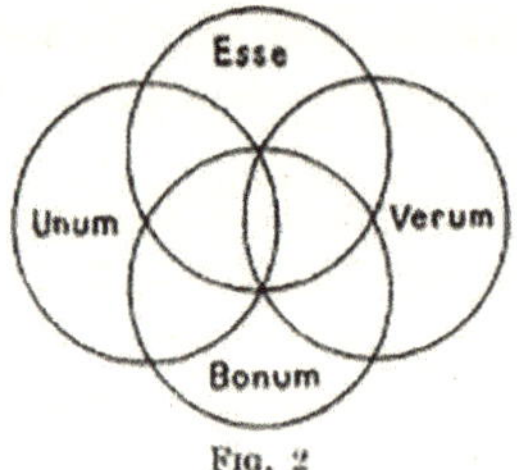

Fig. 2

[11] The Roots Of Logical Hylomorphism. Elenaragalina-Chernaya (2015)
[12] Dragalina-Chernaya op. cit
[13] Baron, Margaret E. (May 1969). "A Note on The Historical Development of Logic Diagrams". *The Mathematical Gazette*. **53** (384): 113-125. doi:10.2307/3614533. JSTOR 3614533
[14] (c. 1235-1315)

This really does look like John Venn, 600 years early. Leibnitz[15] is credited with the first invention of symbolic logic. However, his contribution is commonly understood not to have contributed to the development of SL for the next 200 years. As Margaret Baron wrote in 1969:[16]

> Nonetheless, he considered the logic of Aristotle imperfect and wanted to complete it. In so doing, he explored at some length the possibility of representing syllogistic arguments by means of geometric figures developing not only the now familiar circle diagrams attributed to Venn and Euler, but also an ingenious linear form which he considered clearer and easier to work with. The four standard categorical propositions are represented as follows:

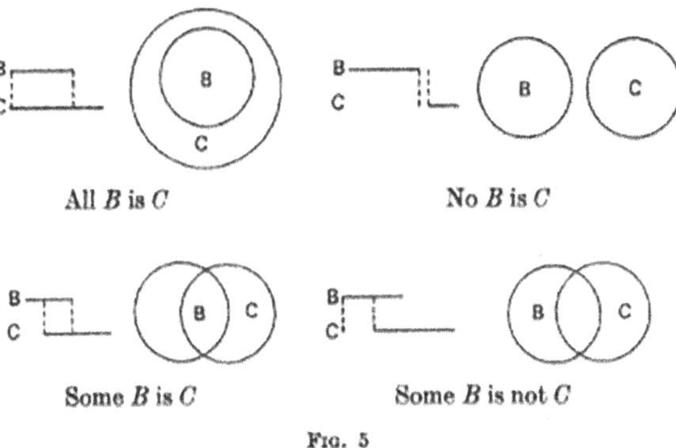

All *B* is *C* — No *B* is *C* — Some *B* is *C* — Some *B* is not *C*

Fig. 5

Boole, writing in the mid-Victorian period, nevertheless referred to Leibnitz "as if the two had shaken hands across the centuries."

[15] (1646-1716)

[16] See later

Sadly forgotten is the contribution of JH Lambert.[17] Such was his repute during his lifetime that his contemporary Kant[18] called him "der unvergleichlicher Mann." (the incomparable man). Which was the 18th century equivalent of Jimmy Page bigging up Jimi Hendrix.

His first logic was *Neues Organon oder Gedanken über die Erforschung und Bezeichnung des Wahren und dessen Unterscheidung vom Irrthum und Schein.*[19] (New Organon, or Thinking Concerning the Exploration and Designation of Truth, and of the Distinction between Error and Appearance).[20] Lambert updated his symbolic logic project over the years. Lambert was to have an influence on Venn.[21]

Ploucquet† and Lambert‡, in developing the linear form further, attempted to distinguish between certainty and uncertainty by using ordinary lines and dotted extensions. Thus, we have:

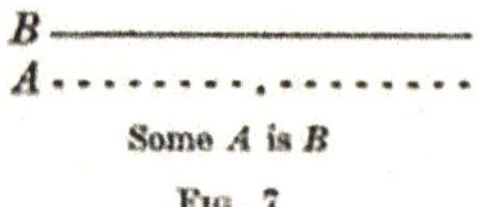

Some *A* is *B*

Fig. 7

* Couturat, L.: *Opuscules et fragments*: op. cit.: pp. 301–312, et passim.
† Ploucquet, G.: *Sammlung der Schriften welche den logischen Calcul betreffen*: Tübingen, 1773.
‡ Lambert, J. H.: *Neues Organon*: Leipzig, 1764.

[17] (1728-1777)
[18] (1724-1804)
[19] (1764)
[20] Author Trans.
[21] See later

Boole is more famous for Boolean Logic, which in its later development became in effect the basis of the binary system. Translated into circuit mechanics, this became the basis of modern computing. I type these words with this technology thanks, in original part, to Boole.

The Author must admit to a childhood fascination for George Boole (1854-1864). Like the Author, he was a son of Lincolnshire. The Mechanics Institute, at which he taught, is actually adjacent to the Lincoln Library: the locus of the Author's self-education in childhood.[22]

The Author recollects not being able to count the number of times walked past Boole's house at 3 Pottergate, about 10 minutes' walk from that childhood library.

(By Logicus - Own work, Public Domain)[23]

[22] See *Strange Days* Chapter in *I Want To Love But*. The Author (2019)
[23] https://commons.wikimedia.org/w/index.php?curid=6888771

Boole wrote:[24]

> In every discourse, whether of the mind conversing with its own thoughts, or of the individual in his intercourse with others, there is an assumed or expressed limit within which the subjects of its operation are confined.
>
> The most unfettered discourse is that in which the words we use are understood in the widest possible application, and for them the limits of discourse are co-extensive with those of the universe itself.
>
> But more usually we confine ourselves to a less spacious field. Sometimes, in discoursing of men we imply (without expressing the limitation) that it is of men only under certain circumstances and conditions that we speak, as of civilized men, or of men in the vigour of life, or of men under some other condition or relation.
>
> Now, whatever may be the extent of the field within which all the objects of our discourse are found, that field may properly be termed the universe of discourse. Furthermore, this universe of discourse is in the strictest sense the ultimate subject of the discourse.

No doubt it is sentimentality, but the Author still finds those words expressive of a beautiful idea: that the discourse of our mind is existentially in accord with the universe. Please forgive the digression. Back to the brief history of SL.

As Chris Dixon writes in *How Aristotle Created the Computer:*[25]

Boole is often described as a mathematician, but he saw

[24] *The Laws of Thought* (1854). From Cork, Ireland, rather than Lincoln
[25] https://www.theatlantic.com/technology/archive/2017/03/aristotle-computer/518697/

himself as a philosopher, following in the footsteps of Aristotle. The Laws of Thought begins with a description of his goals, to investigate the fundamental laws of the operation of the human mind:

> The design of the following treatise is to investigate the fundamental laws of those operations of the mind by which reasoning is performed; to give expression to them in the symbolical language of a Calculus, and upon this foundation to establish the science of Logic... and, finally, to collect... some probable intimations concerning the nature and constitution of the human mind.

He then pays tribute to Aristotle, the inventor of logic, and the primary influence on his own work:

> In its ancient and scholastic form, indeed, the subject of Logic stands almost exclusively associated with the great name of Aristotle. As it was presented to ancient Greece in the partly technical, partly metaphysical disquisitions of *The Organon*, such, with scarcely any essential change, it has continued to the present day.

Professor De Morgan[26] said of Doctor Boole in his *Budget of Paradoxes*:[27]

> I might legitimately have entered it among my paradoxes, or things counter to general opinion: but it is a paradox which, like that of Copernicus, excited admiration from its first appearance.
>
> That the symbolic processes of algebra, invented as tools of numerical calculation, should be competent to express every act of thought, and to furnish the grammar and dictionary of an all-containing system of logic, would not have been believed until it was proved.
>
> When Hobbes, in the time of the Commonwealth, published his 'Computation or Logique,' he had a remote glimpse of some of the points which are placed in the light of day by Mr. Boole.

---

[26] (1806-1871)
[27] Published (1915)

The unity of the forms of thought in all the applications of reason, however remotely separated, will one day be matter of notoriety and common wonder; and Boole's name will be remembered in connexion with one of the most important steps towards the attainment of this knowledge.

As Chris Dixon continues:

Boole's goal was to do for Aristotelean logic what Descartes had done for Euclidean geometry: free it from the limits of human intuition by giving it a precise algebraic notation. To give a simple example, when Aristotle wrote: All men are mortal.

Boole replaced the words "men" and "mortal" with variables, and the logical words "all" and "are" with arithmetical operators:
$x = x * y$

Which could be interpreted as "Everything in the set *x* is also in the set *y*." The *Laws of Thought* created a new scholarly field -mathematical logic - which in the following years became one of the most active areas of research for mathematicians and philosophers. Bertrand Russell called the *Laws of Thought* "the work in which pure mathematics was discovered."

The evolution of computer science from mathematical logic culminated in the 1930s, with two landmark papers: Claude Shannon's "A Symbolic Analysis of Switching and Relay Circuits," and Alan Turing's "On Computable Numbers, With an Application to the *Entscheidungsproblem*." In the history of computer science, Shannon and Turing are towering figures, but the importance of the philosophers and logicians who preceded them is frequently overlooked.

A well-known history of computer science describes Shannon's paper as "possibly the most important, and also the most noted, master's thesis of the century." Shannon wrote it as an electrical engineering student at MIT. His adviser, Vannevar Bush, built a prototype computer known as the Differential Analyzer that could rapidly calculate differential equations. The device was mostly mechanical, with subsystems controlled by electrical relays, which were organized in an ad hoc manner as there was not yet a systematic theory underlying circuit

> design. Shannon's thesis topic came about when Bush recommended he try to discover such a theory.
>
> ...Shannon's insight was that Boole's system could be mapped directly onto electrical circuits. At the time, electrical circuits had no systematic theory governing their design. Shannon realized that the right theory would be "exactly analogous to the calculus of propositions used in the symbolic study of logic."

We must mention in the same breath as Boole, William Stanley Jevons,[28] with whom Boole had a correspondence later in Boole's life.

Jevons described Boole's SL work as[29] using "dark and symbolic processes". However, Jevons gave perhaps the most succinct summary of the operational principles of Bool's SL:[30]

> Dr. Boole's remarkable investigations prove that, when once we view the proposition as an equation, all the deductions of the ancient doctrine of logic, and many more, may be arrived at by the processes of algebra. Logic is found to resemble a calculus in which there are only two numbers, o and l, and the analogy of the calculus of quality or fact and the calculus of quantity proves to be perfect.

Jevons himself explained his motivation in creating an SL,:[31]

> In this small treatise I wish to submit to the judgment of those interested in the progress of logical science a notion which has often forced itself upon my mind during the last few years. All acts of reasoning seem to me to be

---

[28] (1835–1882)

[29] *Pure Logic and Other Minor Works* (1890) §174

[30] *The substitution of similars, the true principle of reasoning, derived from a modification of Aristotle's dictum* (1869)

[31] op. cit

> different cases of one uniform process, which may perhaps be best described as the substitution of similars. This phrase clearly expresses that familiar mode in which we continually argue by analogy from like \ to like, and take one thing as a representative of another. The chief difficulty consists in showing that all the forms of the old logic, as well as the fundamental rules of mathematical reasoning, may be explained upon the same principle; and it is to this difficult task I have devoted the most attention.
>
> The new and wonderful results of the late Dr. Boole's mathematical system of Logic appear to develop themselves as most plain and evident consequences of the self-same process of substitution, when applied to the Primary Laws of Thought.
>
> Should my notion be true, a vast mass of technicalities may be swept from our logical text-books, and yet the small remaining part of logical doctrine will prove far more useful than all the learning of the Schoolmen.

We would respectfully appropriate an element of Jevon's catechism as representing our own agenda. However, we are not so bold as to claim that Matrixial Logic can serve as the imperator of "All acts of reasoning." That was not our purpose in developing it.

We do not say that it cannot so serve. But we cannot say that we have fathomed the limits of Matrixial Logic sufficiently to warrant efficacy in such boundless application.

Jevons, being made of sterner Victorian stuff, felt it competent to claim:[32]

> that these Laws are true both "in the nature of thought and things". Given that science is in the mind and not in the things, the Laws of Thought seem to be purely subjective and only verified in the observation of the

[32] https://stanford.library.sydney.edu.au/entries/william-jevons/

> external world. However, Jevons argues that it is impossible to prove the fundamental laws of logic by reasoning, since they are already presupposed by the notion of a proof. Hence, the Laws of Thought must be presupposed by science as "the prior conditions of all thought and all knowledge". Furthermore, our thoughts cannot be used as a criterion of truth, since we all know that mistakes are possible and omnipresent. Hence, we need to presuppose objective Laws of Thought in order to discriminate between correct and incorrect reasoning. It follows that Jevons regards the Laws of Thought as objective laws.

Jevons began his analysis by objecting to the non-commutative property of Aristotle's *Dictum de Omni et Nullo*: the foundation of the syllogistic system of logic.

- The principle that whatever is universally affirmed of a kind is affirmable as well for any subkind of that kind;
- The related principle that whatever is denied of a kind is likewise denied of any subkind of that kind.

Venn (the overlapping circles guy: although that work was really made famous by his successor) wrote:[33]

> To my thinking, he and Boole stand quite supreme in this subject in the way of originality; and, if the latter had knowingly built on the foundation laid by his predecessor instead of beginning anew for himself, it would be hard to say which of the two had actually done the most " (p. xxxi)

We must be indebted to Venn for his collation of historical attempts at symbolic logic.

We cannot improve on Venn's own summary. He uses the paradigm of (E) The Universal Negative.[34] As Venn says,

[33] *Symbolic Logic* (1881) p31

[34] See Chapter 3

this is an "existential proposition". Indeed, as Matrixial Logic equations show, the (E) Categorical Sentence is distinct among Aristotle's Categorical Sentences, in calling upon the null hidden form of [E][35], rather than [E] as a form of substance.[36]

406 *Historic notes.* [CHAP.

must be remembered, have been made the instruments of a more or less systematic exposition of the subject. In so far, therefore, as the notation is not entirely arbitrary—which it is in very few instances—we shall find it instructive to compare the different aspects of the same operation to which they respectively direct attention. For convenience of reference and comparison they are expressed in the same letters in each case, *S* and *P* standing respectively for the subject and predicate of the original proposition; viz. no *S* is *P*.

The analysis by which I should reach these various forms would be somewhat of the following kind:

I. In the first place we may regard the proposition as an *existential* one. In this case what it does is to deny the existence of the class of things which are both *S* and *P*.

II. We may cast the proposition into the form of an *identity*. What we then do is to make the terms of the proposition respectively *S* and not-*P*, and to identify the former with an undetermined part of the latter. The appropriate copula is then of course (=).

III. Another plan is to regard the proposition as expressing a *consequence* or *implication*: 'If *S* then not-*P*', or, 'the presence of *S* implies the absence of *P*'. This relation, of course, is not convertible, for it does not follow that the absence of *P* implies the presence of *S*. Accordingly, some kind of unsymmetrical symbol becomes appropriate to represent the copula in this case.

IV. Again, keeping more closely to the common expression of the proposition, we may regard it as expressing a relation not, as above, between *S* and not-*P*, but between *S* and *P*. This relation is convertible, and we should therefore naturally seek in this case for some symmetrical symbol.

---

[35] In effect as a Node: see Chapter 4

[36] See Chapter 2

It is notable that, in all these logic schemes, the notation expresses a single underlying mechanical principle, which is also the existential principle: A=A. The law of identity:

xx.] *Historic notes.* 407

V. Again, we may couch the proposition in conceptualist or notional phraseology, regarding S and P not as classes of things but as attributes or groups of attributes.

VI. Lastly, we may meet with nondescript attempts which aim at little more than translating the proposition as it stands, or adopting some arbitrary notation for it.

Grouping them thus, we may arrange our species as follows :—

I. Existential	1	$SP=0$	Boole.
	2	$S(0=P)$	Macfarlane.
II. Identity	3.	$S=v(1-P)$	Boole.
	4.	$S=\frac{0}{0}(1-P)$	Boole.
	5.	$S=v(X-P)$	Wundt.
	6.	$S=Sp$	Jevons.
	7.	$S=1-P-y$.	Delboeuf, Murphy.
	8.	$S=\frac{P}{\infty}$	Holland.
	9.	$S<X-P$	Drobisch.
	10.	$S<-P$	Segner.
III. Implication (non-convertible)	11.	$+S-P$	Darjes.
	12.	$S\angle\overline{P}$.	Grassmann.
	13.	$S : P'$	Maccoll.
	14.	S—$<P$	Peirce.
	15.	⊢┬ P / └ S	Frege.
IV. Common Form (convertible)	16.	$S.P$	De Morgan.
	17.	$S)(P$	Wundt.
	18.	$S>P$	Ploucquet.
	19.	$tS \parallel tP$	Bentham.
	20.	S : ━━━┿━━━ : P	Hamilton.
V. Notional	21.	$L-S\infty P$	Leibnitz.
	22.	$\frac{S}{m}=\frac{P}{n}$	Lambert.
	23.	$S>\frac{P}{n}$	Lambert.
VI. Nondescript, and Arbitrary	24.	$Sx=-P$	Maimon.
	25.	$nS-P$	Ploucquet.

Returning to Jevons, he continued:[37]

> But it is hardly too much to say that Aristotle committed the greatest and most lamentable of all mistakes in the history of science when he took this kind of proposition as the true type of all propositions, and founded thereon his system. It was by a mere fallacy of accident that he was misled ; but the fallacy once committed by a master-mind became so rooted in the minds of all succeeding logicians, by the influence of authority, that twenty centuries have thereby been rendered a blank in the history of logic. [13]

Jevons was not complaining about the law of identity. He was, instead, complaining about its allegedly too limited scope.

Jevons was seeking a mathematic logic. As he continued:

> Aristotle's dictum in accordance therewith. It may then be formulated somewhat as follows:
>
> > Whatever is known of a term may be stated of its equal or equivalent. Or in other words, Whatever is true of a thing is true of its like. [14]

We can see here in Jevons, what we were later to see with Frege, Meinod, Russell and their successors. The attempt to defuse the inequality lying at the core of Aristotle's Categorical Syllogisms.[38]

To substitute for the logical analyses of Substance, a formic logic which relies upon extrusions from pure equality.[39]

That program describes the history of symbolic logic in

---

[37] *The Substitution Of Similars, The True Principle Of Reasoning, Derived From A Modification Of Aristotle's Dictum.* (1869)

[38] See Chapter 3

[39] See Chapter 4, and the discussion of Arithmetic

the 20th century, and the expressly limited landscape of First Order Logic.

The rebuttal offered by Matrixial Logic is that: this is an enterprise at which one is bound to fail, by succeeding.

ML is not, and does not seek to be, a closed system. Any closed system defeats itself with uncomputability, the moment it seeks to apply logic to the untrammelled landscape of reality. Indeed, no closed system can even justify itself.

ML is not, then, a system in which you can simply plug values into equations and have a answer pop out, under some iteration of a law of equality.

Instead, ML deals in forms of inequality. There are definite rules as to how to invoke and manipulate equations. But ML equations illumunate: they do not effect equalising definition.

What then, is the popint of a symbolic logic which does not by its own operation define outcomes?

Let's find out by meeting Matrixial Logic.

CHAPTER 2

MATRIXIAL FRAMEWORK

Introductory

Matrixial Logic is a symbolic logic.

In Matrixial Logic: we are examining types of relationship:

(1) In E Logic: Things which are **Opposites**, and both exist;

(2) In B Logic: Things which cause or relate to other things, and which exist, but as a **Process;**

(3) In C Logic: things which do not fall within either category of relationship.

Matrixial Logic seeks to symbolise and explain objects of logical enquiry under the "3H" rubric:

- Holistic
- Heuristic
- Hylomorphic

Holistic	The whole is more than the sum of its parts. All parts meed to be understood by reference to the whole.
Heuristic	Matrices are dynamics of self-learning, by both internal and external reference.

Hylopmorphic	Forms are the forms of their contents, not merely arbitrary categories[40].

It should be added that the "hylo" is not limited to "things", but includes processes, and paramorphic relations.[41]

In this introductory Chapter, we will need to define some terms, in ways which are not entirely accurate. ML requires a ladder of understanding. Once you get to the top, the bottom rungs look quite different than when you began the ascent.

But, if you were called upon to grasp the lower rungs in accordance with perspectives provided by the topped plateau, it may be almost impossible to begin the ascent at all.

Vocabulary Note:
The word "things" is used in preference to "phenomena".

The reason is that the latter term has[42] 300 years of philosophical baggage attached to it. We do not say that such baggage has no utility within the discourses in which it has arisen.

---

[40] Naming conventions are arbitrary
[41] See Chapter 5
[42] At least

Rather, that "thing" gains in neutrality what it sacrifices in specificity.

If, in reading the former word, you prefer to substitute the latter, then we hope such substitution affects not the meaning and effect of the former in its relevant context. We must hope because we do not envisage being able to conduct such a comprehensive survey of usage of the latter, so as to be able to advance a claim to that effect.

Notation Note

You will see, in the pages that follow, endless iterations of symbols in equations, and extractions from equations. We try to maintain consistency of iteration, except where we are deliberately inconsistent so as to emphasise something.

Because Matrixial Logic involves layers, *matrices*, of concepts and operations of logic, one concept may have different reference symbols, depending where in the matrix the concept is being operated with.

The Author readily confesses an occasional slip, inconsistency, between usage. Especially in narrative. If the reader notices these, then at least what is being communicated in substance, is having effect.

We will also use a mode of narrative writing in which

we repeat the concept word, together with the symbol. Our intent is not that this should become wearisome, but so that such reiteration will allow the reader to become instantly familiar with the concept-symbol match: and also on occasion deliberately with intent to provoke thought.

Assertions

In this Chapter, we will make several unestablished context-setting assertions. In later Chapters we will unpack those assertions.

The reader is invited to reserve judgment, and to withhold agreement, conditional upon the effectiveness of such unpacking.

Exemplar Referenced to Daily Practice

An example of symbolic logic that you use every day:[43]

<u>Arithmetic</u>

$$1 + 2 = 3$$

You are so used to doing this, in your head, that you probably never think about how the bits of it are put together.

Proposition: Numbers 1–10 Relate to Each Other.

[43] The symbolic usage will be explained later

<u>Let 2 be (A) in E (the 3 form of number):</u>

$$(1) \neq (2) \; = 3$$

What the ≠ "not equals" sign does, is show that one side of the sign is not the same as the other.

We don't notice in our ordinary thinking, but that's the sign we're kind of using when we do arithmetic.

The different signs: + - x /

These are all different ways of saying that one thing is different to another thing. And when you put the two differences together, you get " = " a third thing.

The trivial level of exemplar is raised to a less trivial level by the examples:

$$(2) \neq (2) \; = 4$$

$$(2) \neq (2) \; = 0$$

$$(2) \neq (2) \; = 4$$

$$(2) \neq (2) \; = 1$$

The inequality is applied even though the integers on either side of the equality are definitionally the same.[44]

The category of ≠ inequality function is determined by

[44] Within a linear integer system

the Frame of Reference of the arithmetic operation.

Matrixial Logic is just a language. It is easier in application, as a mode of operation of information under rules, than the arithmetic done at grade school. But it is really powerful. Using this language allows us insight into matters in ways which otherwise may be unavailable.

<u>Symbolic Usage</u>

There is nothing magical nor intrinsically "right" about the symbolic usage deployed in Matrixial Logic. It is simply a short-hand for using the information encoded in the Axioms.

You can get used to it very quickly. It's no more complicated than the use of emoticons. But there are rules of emoticon usage:

☺ does not mean sad ☹ does not mean happy
in the lexicon of emoticons.

You are entirely at liberty to use ☺ to mean sad. If you explain to others that this is, for you, the graphic representation of that concept, then you can deploy that lexicon in communication with efficiency.

But it will be low efficiency, unless a practically sufficient number of people adopt the same lexicon. And you can predict confusion. But maybe confusion is your aim, in which case there is a high probability of success.

Rules

The word "Rules" signposts portended claims about how all humans see the world, and can only see the world. Stepping stones to dogmatism, and empirical faith.

That is not the intended territory. The logic of forms is not a logic that anyone has to use. They are rules of a game that you do not have to play. It may be, however, that the game is already playing you, and always does. In which case, knowing the rules looks like a good idea.

When we engage in thinking, we do so on the basis culturally contingent assumptions. These change, over place and time. Multiple assumptions can co-exist, and fail to co-exist. That is politics, then history.

Some of those assumptions seem to be deep. Wherever you peer through the clouds of politics into the lived experiences of people, you see the same assumptions being expressed in that living. And in the disciplines by which people seem to agree to understand and codify exploration of their world.

We can derive logics of form from these lived assumptions. These logics of form do not exist in some Platonic perfection somewhere. They are symptoms of organic life in thought, bounded by contingent culture.

There are forms of logic which are truly universally

accepted:[45] language, mathematics, some of the modes of scientific discourse. We do universally organise our lives around these forms of logic. This is what makes them matter.

Frame of Reference ("FoR")

FoR is multi-valent in Matrixial Logic.

This is permissible, since no analytic conclusion is sought to be drawn from any particular FoR.

FoR as merely descriptive, may be defined as:

> *a domain in which substances, moments, or ideas of them occurs, and by reference to which they assume a meaning.*

Thus, it may be descriptively useful to refer to the domain of E Logic as frame of reference, and contrast FoR which arise within it, with those which arise within the domain of B Logic.

Analysis is a method for arriving at a conceptual state of affairs which can be inferred to correlate with one of more meanings.

The modes by which meanings are inferred, and contrasted, is more than merely descriptive. It is analytic. Each FoR engages our mental appreciation in different

[45] Because they seems to work: not because they are "true" in any super-ideal sense

ways. This latter point is of importance in psychology.[46]

FoR functioning as an analytic, may be defined as:
The ascription of meaning to FoR domains by reference to each other, acting as content under a form.

Exemplar Referenced to Daily Practice
We might measure a table top in inches, a room in feet and a football field in yards.

In ML, we would refer to each as a *frame of reference*. Each FoR derives from a mode of physical measurement.

Contrasting the FoR inch with the FoR yard might generate a differential appreciation in consciousness, such that the plurality of FoRs functions as content under a form.

<u>inequality ≠</u>

Each Substance is a unique event in spacetime.
Each Moment is a unique event in spacetime.

The symbol of *inequality* ≠ expresses these Axioms.
The word "event" will come under more deliberate consideration in successive Chapters.

The inequality symbol ≠ is the vital component of Matrixial Logic.

[46] See *Secret Self*. The Author (2020)

That which is, is not any other thing which is. That is the definitional nature of what it is to be any thing.

In E Logic, we must be careful not to write (that is, to think):

$$A \neq (-A) \ldots$$

That is to take un unexpressed Form of A and to break the rules by trying to go from a Form of Content to another Content, without any intermediate step.

We do this a lot in talking to each other, or we appear to. It creates confusion, because it's not playing by the rules which we claim (by silence as to the matter)[47] that we are playing by, in making the utterance.

Matrixial Inequality

The building block of the E logic of forms is:

$$(A) \neq (nA) = \text{Form}$$

The inequality symbol is equally the building block of all other Logic forms.

If you consider the historic landscape of syllogistic philosophy, and indeed symbolic language philosophy, there are few instances of explicit inequality-based reasoning.

We certainly use exclusionary reasoning throughout our

---

[47] Or it not being obvious to our speaking partner

daily life. Also in science, moral philosophy, religion, and even in art.

Our entire vocabulary is made up of antonyms and synonyms, being different words, which we agree, in usage, to bear:

- Opposite meanings;
- The same or similar meanings.

Thus:

dog ≠ cat } Antonym[48]
dog ≠ hound } Synonym

A synonym, even a close synonym such as dog ≠ hound, is a *statement of difference.*

Matrixial Inequality is founded in the logic of differences.

The activity of Matrixial Logic is seeking the Form of each set of differences.

<u>equals =</u>

Form = the transformation of Extensions (and Moments)[49]

When we bring an Extension of Substances under a Form, we solve the inequality between the Substances in their Extensions in Form.

---

[48] Behavioural / zoological
[49] In [B] Logic

Here's a symbolic representation, to aid explanation:

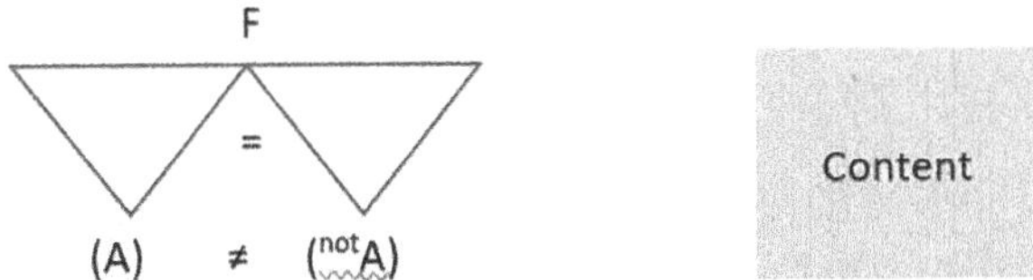

You see how the "triangles", representing each side of the inequality ≠, show the Extension of each, which Extensions mutualise under (as a) Form.

Now, you can think of the triangles merging (mutualising), which give you the = symbol.

Or, you can think of the triangles disappearing (which is generally how we do go about thinking in forms).

We can start getting all quantum or zero-point about this stuff, and explore the known unknown.

We can start making claims that Extension is the mapping of the [whatever you want to call it], which is the "real" contingency, of which the spacetime events of Substances, are merely transient.[50] Such that Form is the form not of substance, but through substance.

One can then deduce values of that avenue of theoretical

[50] Other than permanent

explanation. What's important right now, is that Matrixial Logic is not stuck in a Newtonian universe.[51]

Rules of Equality

In Matrixial method, we propose an unsolved equation, with elements disjuncted by inequality ≠, which elements then become Contents under a Form of equality = :

$$(A) \neq (^{NOT}A) = E$$

There are obvious objections at this point:

(1) *But surely this is circular*: Form is the form of its Contents and Contents are the contents of a Form.

(2) How can you independently arrive at (^{not}A), without first presupposing an E?

To which answers are given:

(1) The description of a transformation is not circular.[52]

Use of :

$$\frac{=}{\neq}$$

precludes circularity.

It is the symbolic definition of *transformation*. Transformation by reference to what? We have

[51] Not that we ever really were

[52] One atomic electron state to another; water to steam; ad infinitum

already got to that, but we will get to that.

(2) It is possible, but not necessary, to do so.

Each answer requires more information, to move beyond mere assertion. There is some difficulty in explaining these matters without recourse to specific examples. However, such recourse can serve as an intuition fallacy, so are best avoided if possible.

(1) Circularity

Let's begin with this:

- Let an object be posited (or observed) to exist;
- You perceive a <difference between>: that which is the object and that which is not;
- You can measure that perception, in any of the dimensions of spacetime, or in combination;
- If there exists a phenomenon which we categorise as "reality", which is independent of our perception, then: such object is in that reality;
- If such "reality" is spacetime, the posited object has co-ordinate extension in spacetime;
- Accordingly, the object exists in spacetime.

Now:

(a) You may assert that these domain and concepts are contained within the compound concept "object \ exist". We agree.

(b) That assertion entails the conclusion that the

function of measurement is circular: such that defining a metre, then bringing an object referentially under such definition, is circular. For example:

> The metre is defined as the length of the path travelled by light in a vacuum in 1/299 792 458 of a second. The metre was originally defined in 1793 as one ten-millionth of the distance from the equator to the North Pole along a great circle, so the Earth's circumference is approximately 40000 km. In 1799, the metre was redefined in terms of a prototype metre bar (the actual bar used was changed in 1889). In 1960, the metre was redefined in terms of a certain number of wavelengths of a certain emission line of krypton-86. The current definition was adopted in 1983 and slightly updated in 2019.[53]

(c) Whether logic compels us to agree with (b) depends upon whether the mode of measurement has reality independent of the object measured.

(d) By what means would we assess such independence?

Each answer requires more information, to move beyond the discourse of mere assertion. Fortunately, we have an entire book to assist with that enterprise.

[53] https://en.wikipedia.org/wiki/Metre

E: Opposites in Substance

Notation:
In E Logic, we denote the opposite of an (A) as:
(^{NOT}A) or (nA)
These mean exactly the same. The shorter form is simply used to save on the cumbersome keyboard process.
In spoken language, one would say: *Not A.*

What we mean by an Opposite, is something that a Thing is Not: and by which we may understand, or may define, what that Thing is.

In E Logic[54] we analyse a Substance, in relation to the opposite of that Substance.

Substance

We define *Substance* as: *that which is extended in spacetime.*[55]

Spacetime means the 4 dimensional space which is reality. In other words, the reality which is traditionally considered to be the domain of science. To put it crudely: everything which is, or the universe.

---

[54] By repetition in usage, "E Logic" become abbreviated to "E"

[55] When we get to Chapter 7, we will unwrap the concepts implicated in "spacetime"

Let's use atomic concepts to illustrate:[56]

Each atom:

- Is an event in spacetime;
- Occupies a unique co-ordinate position in spacetime.

No atom is the same as any other atom, because no atoms can occupy the same co-ordinate position.[57]

Atoms of the same state are interchangeable in spacetime.

Interchangeability is not the same as sameness. Sameness is a form, not a substance.

Nothing can exist as a sole substance. That which is sole "is" not:[58]

There will be much more to say about "spacetime" and "reality" later. However, it is irrelevant to the functioning of ML what stance a reader takes as to matters of dualism or monism.

Whether a thing is, or is not, extended in spacetime, so as to fall within the definition of Substance, subsumes itself, for the purposes of ML, under a relatively primitive level

[56] This is an aid to thinking only: don't go all quantum, yet

[57] Without change in state of the atoms

[58] We don't really mean to use the linguistic form of "being"

of analysis.

(a) A pebble is a Substance, within the definition.

(b) An atom is a Substance, within the definition.

(c) A unicorn is not a Substance, within the definition.[59]

Of these matters:

- Anyone can[60] see a pebble directly: it is appreciable in qualia[61];
- An atom is appreciable only through the use of magnifying technology;
- Neither qualia nor technology can appreciate a unicorn (but your mind can);
- Your imaginary unicorn can interact with the external world.

All readers can agree with the proposition that we appropriate information about an atom through technology, and we appropriate information about a pebble through biology.[62]

We would want to add that it is biology when enables us to appropriate that information delivered by technology.

We thus arrive at the same end point: that whether it be information as to a pebble or an atom, we appropriate

---

[59] But may function as is it were Substance: see *Languini* Chapter

[60] In principle

[61] Our senses

[62] A biological matrix

that information by means of that matrix which we are agreed upon in calling "biology".

"Biology" is itself a transitional term. The mode of transition depends upon how we choose to analyse what the components of that matrix are, and their inter-relationship.

What distinguishes: (1) the pebble and atom; from (2) the unicorn,[63] is the following:

- Category (1) things are appreciable by a plurality of persons;[64]
- Category (2) things are not.

Certainly, information about Category 2 things can be communicated by one person to another, and can be imprinted upon the universe in graphic or literary extension. Such communication and extension can function as Substance.[65]

Prior to any such communication, no unicorn is extended in spacetime. However, the unicorn and spacetime can be connected in very real ways. Chapter 7 explains how this is so.

---

[63] For present purposes

[64] Whatever one philosophically wishes to denote or analyse as the mode of such appreciation

[65] See *Languini* Chapter

Whether there are neurobiological operators which correlate to that specific idea of a unicorn formulated in the consciousness of that person may[66] be a question which is meaningful. However, the question itself accepts that a distinction is to be drawn between that which is Substance and that which is not.

Insofar as the reader may, at this point, be drawn to the conclusion that ML relies upon a dualistic account of consciousness, that is not so. As we will see over the course of this book, with its denouement in Chapter 10: ML heralds the decease of dualism. ML also repudiates materialist reductionism.

Whatever view one takes as to the operative substrate of consciousness,[67] is irrelevant to the functioning of Substance data in ML.

Therefore, of the definition of Substance as:

That which is extended in spacetime.

We may offer by way of clarificatory description:

That which is available, by means only of its existence, to appreciation by a plurality of persons.

---

[66] Or may not

[67] Ranging from "it's just a brain" to "it's separate from the brain as a thing in itself"

and we may add:

and whether such appreciation be effected
(i) directly in qualia
(ii) indirectly through technology

If any reader wishes to assume a stance of conditionality as to the above matters, then we invite the reader to revisit them in due course.

It is of the rules of the logic of forms that substance can exist only as plurality.

An exacting terminology would seems to dictate that only the word "Substances" (plural) is used in the place of that word "Substance".

You could use such terminology. But that would then create serious difficulties when you try to generate the domain of forms.

The claim, at this point in setting out the ML framework, is limited to the proposition that the definition of *Substance* given, is satisfactory under the metric proffered in the Preface.

Extension

Extension is the location in spacetime of a Substance

To be extended, is to have a co-ordinate location in spacetime.

It might be posited that the definition of extension is already contained within the definition of Substance. We disagree.

Suppose there to be a pebble. Suppose that we define Substance as *that which becomes subject to appreciation by the matrix of qualia.*

May we disregard the "hard problem of consciousness" for the purposes of this exposition[68].

Since the movement from the Scholastics to the Moderns, it would be difficult to assert that the matrix of qualia is already contained within the definition of "pebble"[69]. The property of existence is predicated upon the pebble having extension in spacetime. That is to say that existence and extension in spacetime, are the same; or that they function tautologically as equivalents.

That may be of assistance. We are not like to say that "existence" and "pebble" are the same thing, idea, or idea of thing.

The trouble with Tokenism and Reflexivism, is that without understanding of the logic(s) of forms, and the rules of illogic, which together form Matrixial Logic, the useful things they have to say, live only as assertion. That's why they inform, but don't persuade.

---

[68] On the grounds of efficiency of explanation. We will certainly return to it
[69] Unless you are Descartes, Berkely, or their latter-day idealists

Thus:

- No Substance can have existence save in extension;
- Extension is the mode by which Substance exists in spacetime

There is certainly inter-connectedness of the terms: Substance, Extension and Spacetime. However, such relationship does not necessitate circularity. In this matter, there is not circularity.[70]

The notion of "holism" is important here. Our idea of a dog is holistic. The dog is not its nose or its tail, nor even all that is in-between. We think lots of thoughts, all bound up together, in thinking of a dog.

We can usefully make distinctions between bits of those thoughts. That utility diminishes rapidly, if we ever forget that each bit is ultimately contingent on all the other bits, and vice versa. We will, sooner or later, violate the logic of forms[71].

It sounds fearfully complicated to say this, but it's actually the utter simplicity of how we do think about our world.

Substance, Singularity and Extension, are each contingent

---

[70] Although circularity of a kind can be created by positing these categories as attributes in relation to some other category. We cannot conceive what such category might be.

[71] Which each of us is "allowed" to do; but at the price of dysfunctional communication, unless we can with clarity communicate what other logic of forms we are using

upon the other.

They are like parts of the body, or parts of a car.[72] We can analyse each part on its own, down to the smallest manifestation of an event in spacetime. We can deduce connections and attributes. But all of that is redundant in meaning[73] save in relationship to the whole.

We will repeatedly encounter problems which have been created, or which fail to be solved, by reason of the *mereological fallacy*: treating only of parts of a whole; or treating of parts as if independent of the whole.

Your thought is in Form. You don't get directly from Substance to that Form.

You only get there through Extension, which is the Extension of each of those Substance events in spacetime. Unique events in spacetime are mediated by / in Extension.

Another way to put it is that Extension is what turns Substances into information. It is hardly new to say that we don't "see" Substances, we see information about Substances.[74]

---

[72] Biology and engineering are not the same, in case you were readying the blue pencil

[73] That is, so as to be consistent with the E Logic of forms

[74] Tokens, and all that

Spacetime

Spacetime is the domain of E. Until Chapter 7, by <spacetime>, we mean that set of ideas which, since acceptance of the premises of general relativity,[75] have served as the paradigm account of the universal context of reality.

Obviously, there are difficulties with that account in the quantum and cosmological dimensions of science. But let's leave those aside until we have ascended sufficient rungs on the ML ladder.

We admit that we cannot conceive of a mode of existence (nA) which is outside of spacetime (A). It may be that a form [nE] =[(Spacetime) ≠ (nSpacetime)] can be postulated.

It is correct that Singularity (Si)[76] is a hypothetical (and important) category, which does fall outside E. However, the very definition of Singularity is such that Si cannot fall under E (or B), since there is no category of (^{not}Si), or (-Si). That which is singular has no form.

We believe that this definition, with short further explication, is presently satisfactory under the metric proffered in the Preface. There will be further examination and elucidation in later Chapters.

---

[75] Say, 1905-1919

[76] See later in this Chapter, and Chapter 6

Form

Within E, there are unlimited numbers of FoRs.

We could conceptualise each FoR as an iteration of E, such that $E^1 \sim E^2 \sim E^n$. However, such conceptualisation would generally be redundant.

The actual analytic category with which we are concerned, when considering an inch, a foot a yard, is: Form.

- Any Form in E is the form of its contents.
- The contents of any Form are: Substance, in opposition, in extension

Different Forms do not transition in to E, in like manner as Substance under extension.[77]

Notation:
The Form of Substance contents (extended in spacetime and opposed in extension) is denoted as:

$$E, \text{ or } [E]$$

Thus as:

$$(A) \neq (nA) = [E]$$

A solved equation is usually written:

$$[E] = (nA) \neq (A)$$

such that the order of notation is reversed. The reversal

[77] This is an assertion of an ML rule, and I shall endeavour later to explain the basis for this rule.

serves two purposes:

- To highlight that we are considering an equation purported to have been solved;
- So that [E] may be extracted from the equation, to serve in manipulation in some further equation.

We have also found that, by reversing the order in which concepts comprising an equation, are presented to the mind, a fault-checking mechanism within the mind, is engaged.

Examples

Exemplar 1

Suppose that a box exists, such that it is smaller in size than the known universe.

What is the opposite of the inside of the box?

An answer is: the outside. It is that inside and outside, together, which define what the box is.[78]

To observe that a box exists, is to accord the attributes of Substance, related in extension in E Space, to that which together constitute the box, in that frame of reference.

What a long way round, simply to articulate the observation of a box. Yet it is by the formation of analytical tools, symbolised in ML, that we gain more

[78] As a matter of dimensionality in an FoR

efficient modes of enquiry into complicated propositions of observation.

Exemplar 2

The following example deliberately omits detail, in pursuit of present efficiency of explanation.

The purpose which informs articulation of this second example, is to convey the concept that E is not limited to the non-biological.

You, as a biological entity, are "extended" in spacetime: you have height and width and depth. You're real. You are Substance.

Each of the components of your biology, down to the atoms of which you are variably comprised through time, is Substance.

Each collection of components, such as a heart, or lung, is Substance.

Your brain is Substance. The question whether those which are products of your brain function (in conjunction with your entire physiology), is each Substance, is a matter reserved to discussion in later Chapters.

Nevertheless, the claim is made here that, what answer be proposed to that question, it makes no difference to the efficiency or utility in function of ML.

Exemplar 3

We know that 2 things are opposites because we can refer them both to something: like a background.

Imagine there are (n) white rabbits on a white background.

How many rabbits?

Can't tell.

Change the background to black: now there are 3 rabbits.

Now imagine there is only black and white in the world.

What is the background, the "Form" of rabbit counting?

We discover that the Form is black, or we suppose that there are invisible rabbits, which "exist" outside of E.

Now imagine black rabbits.

The Form of counting black rabbits is White.

This is not strictly how Form and content operates. But the example is efficient in reason at this point in explanation.

Experience: Extension

<u>Preface</u>

Throughout this book, you wil be presented with *Experiences*. You should find them entertaining and challenging.

It's by undertaking the *Experiences* that the reader will acquire understanding of the narrative and formulas. That which is sought to be explained will be more readily received in understanding, through the transformative effect upon the reader's mind, of these *Experiences.*

The *Experiences* use non-conventional forms of punctuation which are intended to assist the flow of the Instructions.

Experiences are conceived as scientific experiments. The reader is able to conduct first-hand, direct experimentation with Matrixial Logic concepts and their operation.

Empiracal validation (or falsification) through experiment is the hallmark of science.

And we are, in Matrixial Logic, doing science. Not merely theorising. Science which can actually alter your relationship with yourself and your world.

At this early stage, this will inevitably sound grandiose, and possibly slightly off-putting. The reader is asked to reserve judgment until at least the half-way mark in the book.

Please follow the Instructions. Then you can consider the Discussion analysis, and engage with the narrative which follows. Let's commence our first *Experience*.

Experience: Extension [cont]

- Be seated wherever you are reading this;
- Look around you: let's assume you're in a room;
- Look around that room from where you are sitting.

Now:

- Look to your left;
- Focus on whatever objects are there, in their placement in the room[79];
- Form a picture of what you see;
- Bring that landscape picture into your head, so that;
- When you close your eyes you have a fairly clear image of that landscape.

Now:

- Look to your right;
- Turn so that you are facing 180 degrees away from your previous left-hand view;
- Look with relaxed vision at what you see.

Now:

- While you're looking at this right-oriented perspective;
- Bring into your head the image of the left-view that you stored. Keep that image there.

---

[79] Whenever you are asked to undertake Visual work, please ensure that your object of focus is at least 2 arm-lengths away, unless directed otherwise

Now:

- While keeping that right-view image present in your head, imagine getting up and walking a few steps to your left;
- And then a few steps to the right of your chair.

Stop

<u>Discussion:</u>

(1) This was all easy[80];

(2) You seemed to feel the *Extension* of Substance around you, as you turned;

(3) As you walked, one direction seemed easy; the other direction seemed harder;

(4) This seeming was real and there are real reasons for it. For another Chapter.

<u>Content</u>

A Form is the form of its Contents

We probably don't have to do much explaining about the concept of Content.

When you are imagining those two marbles[81]: those Substances, your thought clearly has a content. But neither of the marbles is the content of your thought.

---

[80] The Experiences will become much more challenging, in later Chapters

[81] See Singularity, below

The Equation in E

$(A) \neq (^{NOT}A)$[82] $= [E]$

In a frame of reference (a "background) E,
we can "see" something and its opposite

Utilising the Box exemplar:

Proposition: A Box is Extended in Spacetime
Let Inside a Box be (A) in [E]:
$(A) \neq (nA) = [E]$
>
$(Inside) \neq (nA) = Box\ [E]$
>
$[E]\ Box = (Outside) \neq (Inside)$

Notes:

we use the ( ) brackets, just to keep the formulas tidy. They are useful when writing them in a line of text. We also use as text separators: [ ] and < > .

Also, we do not have to write that Proposition every time: it can be obvious from the text.

[82] Or (nA)

Equation Part	Explanation
Proposition: A Box is Extended in Spacetime	So that we know we are looking at Opposites
Let Box be (A) in [E]	We are looking at "Box" and trying to work out what its opposite is
(A) ≠ (nA) = [E]	The basic formula of E Space relationships
>	This symbol just means "follows" or "next
(nA)	Means "the opposite" of (A)
(Inside) ≠ (^{NOT}A)	We're looking for the Opposite of (Inside)
= Box	In relation to the "background" idea of a Box
(Inside)	So, we can see that the Opposite is (Inside)
Box = (Outside) ≠ (Inside)	We run the equation in reverse.*

* We do not have to do this. There's no "rule" about it. It just makes things tidy and easy to see, when you move from line to line.

AXIOMS in E Logic
Being is the Form of Substance under Extension
Substance is a unique event in spacetime.
To be a unique event in spacetime, is to be Extended.
Extension manifests as Form.
Any Form is the Form of its Contents.
The Content of any Form is Substance in Extension

Justification of E Logic Axioms

There is a *Frame Of Reference* ("FoR") under which these Axioms are demonstrable as true.

That FoR is the physical universe, as presented to us, and as we interact with it. Such interactions include the empirical, observable facts of reality.

We credit rule-based systems of thought[83] with the attribute of being "objective".

The rules, and predictions which can be derived from such rules, map onto a universe of "objective" reality.

---

[83] such as physics, chemistry

We take it as a datum that such universe is rationally inconceivable as having an existence which is contingent upon the observation of conscious persons.

To offer such proofs within the confines of this summary, would necessitate an entire separate book.

Exploring E Logic

When we're doing science, we try to codify our explorations and conclusions under the essential logic of forms: $E = (A) \neq (nA)$

$$\text{Water} = H_2 + O$$

The chemical formula for water, using expressions codified in the Periodic Table, is making a statement about Existence E.

That's an is statement. To be more accurate, it is a statement about the transformational conditions of an is.

H, is obviously a Substance. As is O. The + is transformational.[84] Water is a Form of : $H_2 \neq O$.

This is obviously a Vertical Perspective of Water. It takes Water not as a thing, but as the emergent Form of Content properties.

[84] The "2" is an iteration of arithmetic rule

A Lateral perspective of Water is entirely contingent upon context, which is a matter of choice. The Vertical perspective is intrinsically about excluding choice. We seek to discover the transformational conditions of an is: that targeted as the object of investigation.

Hopefully, you are by now inching away from an obvious inclination to view Matrixial Logic as dumbed-down escapee from Wittgenstein's notebooks. That is, merely a theory about language.

ML a functional set of theoretical tools and practices, for understanding the way that we tend to understand: everything, because of the way that everything is.

Transformation of Forms

Is a form of extensions of substance (A) ≠ (^{NOT}A), a Substance itself?

$$\frac{(A) \neq (^{NOT}A) = F^1}{(A) \neq (^{NOT}A) = F^2} = \quad ?$$

(1) Does a Substance lose its character as an event in spacetime, simply because we measure it?[85] If we compare two measurements,[86] two Forms, then is the resulting Form still a form of extensions in spacetime?

[85] Clue: see Chapter 7

[86] Yes: not exactly analogous: just go with to aid understanding

The answers, according to ML are:

(1) It depends on whether that measurement is the extension itself

(2) No: that is an <∑Event> \ Node.

We can by all means write:

$$(F) \neq (^{NOT}F) = S^{F}$$

And we can carry on:

$$(S^{F}) \neq (^{NOT}S^{F}) = E^{F}$$

and so on ad infinitum.

The rules of E Logic are preserved: *so long as we are prepared to acknowledge that* nothing beyond the first (S) and its correlate in inequality (^{NOT}S), is extended in spacetime.

The transformation of Forms is an important part of Matrixial Logic. We will see later how Forms become transformed, under the methodology of Nodes in E and Interference in I. These transformations will become vital in later Chapters.

The Inequality Gateway to B Logic

Once we realise the functioning of ≠ even in our ordinary everyday thinking, we become liberated in the further realisation that we use that ≠function in lots of wonderful ways.

We also become liberated from modes of thinking in equality. We come to realise that these modes are the reason why we have reached apparently intractible problems in various fields of human endeavour.

B: Processes in Negation

Notation:

In B Logic, we denote the negation of an (A) as:

(-A)

There is no need for an alternative notion (such as in E Logic), as the insertion of "-" is simple keyboard operation.

In spoken language, one would say: *Negative A*; or *Minus A; or Neg A*

From earlier:

In Matrixial Logic: we are examining types of relationship:

(1) In E Logic: Things which are **Opposites**, and both exist;

(2) In B Logic: Things which cause or relate to other things, and which exist, but as a **Process.**

Everything moves. The earth whirls through the solar system. Atoms fizzle and pop. The quantum foam, well, foams.

Each of us changes physically every moment. The last breath you took happens to a different "you" to the one you're taking now. The actual number of atoms in your body has changed. Your circulation has changed. The whole universe in which you exist, has changed. In that one moment.

In a Process, we're looking for a change in something. Something changes in itself: a flower grows taller. Or something reacts with something else: the bee pollinates the flower.

Now you wouldn't say that the flower, which has grown is the "opposite" of the slightly shorter flower, earlier today. The bee is not the "opposite" of the flower. In a way, it is part of the life process of the flower. They relate to each other in the Matrix of life.

So, when we're looking at a Process, we say:

$$(A) \neq (-A) = [\ I]$$

In a frame of reference (a "background) I,
we can "see" something and its Process

- E Space[87]
  Is about where and what things are: things that exist: things *in being*.

[87] "Space" is a convenient abbreviation and lexical equivalent of "domain".

- I Space is about how things change: things that exist as a Process: things *in becoming*.[88]

So, with I Space, we are looking at a Process:

$$(A) \neq (-A) = [I]$$

And we are looking at Moments within that Process.

Moment

We define *Moment* as: *that which occurs only in becoming.*

Moments differ by reference to the Frame of Reference:

- Walking across a street: I is walking;
- Sliding down a slide: I is sliding;
- Parachuting from a plane: I is parachuting.

They also differentiate between themselves within an FoR.

We could choose to substitute the word "Event" for Moment, with no appreciable loss of utility. The reason this is not done, is that the word "Event" is reserved for use in another ML context, where the word "Moment" would be logically unsuitable. Say you are walking a line. You start at the kerb and cross e street.

---

[88] Yes: being and becoming are intimately related. Matrixial Logic deals centrally with those very relations and the transformations between being and becoming states

- Moment 1 [we write (A)], you've stepped out;
- Moment 2 [we write (-A)], you're half way across;
- You, crossing that street, is a series of Moments (A), (-A) and so on.

Once you have the Frame of Reference (here: that street in time), you don't need to keep on writing (- -A), (- - - A), and so on. But that is what (-A) is shorthand for. It's all the moments of (A) which are *different* to any other point.

In order to stress the iteration of (A), we will later use the notation Δ^n The delta notation classically stands for "change in", and we will be using that sense a lot in later Chapters.

For example, in physics, the Entropic Second Law of Thermodynamics describes a process. That process has effect only in and through Substance. But Entropy is not Substance, it is a descriptive limitation of Moments in becoming.

We propose that the definition of *Moment* given, is satisfactory under the metric proffered in the Preface.

We reserve to later, discussion of the epistemological correlates of Substance, in relation to Moment.

Redundancy of Extension

We define *Substance* as: *that which is extended in spacetime.*

We do not define Moment by reference to extension (nor spacetime).

That which is extended cannot, in that extension, be a moment. That which is extended can become a moment, other than in that extension.

Experience: Pebble

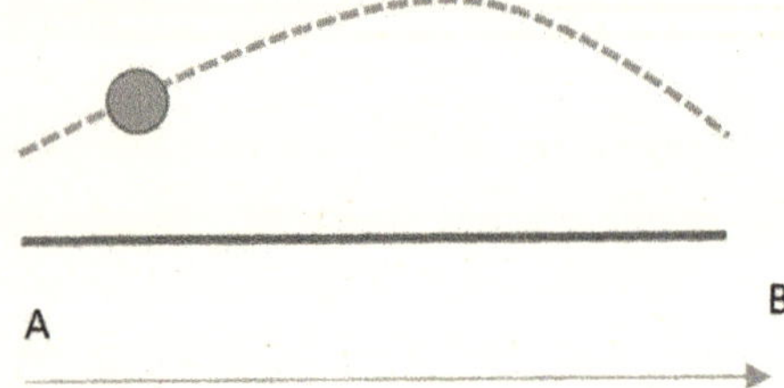

Setup:

- Just look around the room you're in;
- Turn your eyes to your left;
- Look at the space between you and the wall;
- Pick a point floating there in space and make that "A"
- Look to your right and, similarly, pick a point floating there in space and make that "B".

Now, just like the diagram, you're going to be imagining a pebble being thrown by someone (not you): and following the conventional parabola trajectory, under the ordinary influence of gravity: just as in the image.

But: don't do this with the image, do it as a projection from your head into the space in your room.

Step 1:
In your own time

- Allow the pebble to fly, slowly;
- Let the pebble take a 1-2-3 count to travel between (A) and (B);
- Repeat 2 or 3 times.

Stop

Discussion:
(1) All easy enough.

Step 2:
In your own time

- Allow the pebble to fly, slowly;
- Freeze the pebble at a spot between (A) and (B);
- Mark that spot P1;
- Let the pebble move some more and stop;
- Mark that spot P2;
- Once again, let the pebble move some more and stop;
- Mark that spot P3.

Stop

Discussion:
(1) All easy enough;

(2) Now, look at the pebble in spot P1;
(3) As you focus on the pebble in spot P1, the rest of the diagram image just melts away;
(4) there becomes just you and the pebble, at that spot P1.

To put the matter another way, that trajectory is a curated collection of moments. To be a moment necessitates that there was a past moment, and may involve a future moment. Trajectory is a transition of moments: moments under a Form of becoming.

Experience: Curve

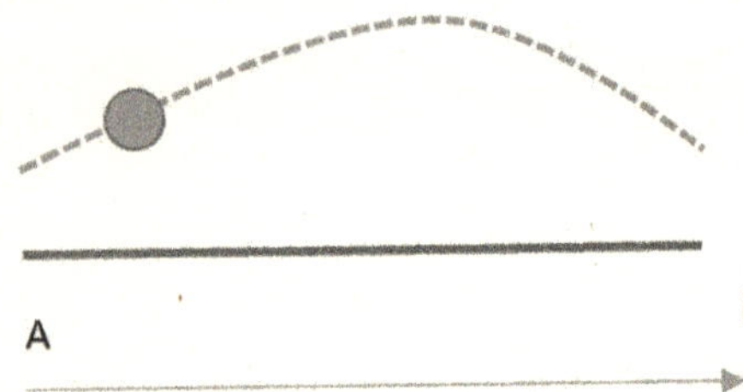

Setup:
Set up your in-the-room *Experience,* just as in *Pebble.*

Step 1:
In your own time

- Allow the pebble to fly, slowly;
- Let the pebble take a 1-2-3 count to travel between (A) and (B);
- Repeat 2 or 3 times;

Stop.

<u>Discussion</u>:

(1) All easy enough.

Step 2:

In your own time

- Allow the pebble to fly from (A) to (B);
- Try it slowly; try it fast.

Now:

- Keep the A => B trajectory line clear in your head;
- Move the pebble off the line and towards you, in the room.

Stop.

<u>Discussion</u>:

(1) It won't work;

(2) The pebble won't come to you in a straight line;

(3) You try to bend the pebble line, but then the pebble just snaps back onto its A => B trajectory line.

Step 3:

- Keep the A => B trajectory line clear in your head.

Now:

- move the A => B trajectory line towards you;
- move the A => B trajectory line away from you;

Stop.

Discussion:

(1) It won't work: in either direction;

(2) You, and the pebble, are stuck in the FoR of that A => B trajectory line.

As we invite you repeatedly to consider in this book: *whatever it is going on here, something is going on.*

Matrixial Logic explains what is going on and why. And that this is not just going on in your head: you are actually becoming entangled[89] with the conditions of reality.

What this shows us is:

- In each moment of its trajectory, there exists a pebble which is extended in spacetime: a substance extended in E. That's the pebble at spot P1;
- The trajectory itself, as Form of movement, functions in the domain of E;
- The pebble nevertheless relates to its own trajectory in time and space only as moments in that trajectory.

This is a simple *Experience* in the realm of Newtonian physics. Indeed, the pebble in an A => B trajectory line would have been familiar to an Egyptian or Athenian

[89] Clue: wait for Chapters 5 and 7

or Alexandrian natural philosopher. Or to Copernicus, Galileo, or Descartes.

Indeed, we are only one dimension and Riemann geometry[90] away from the general theory of relativity.

Already, you are seeing a fundamental difference between: [E] you and the pebble; and [I] the pebble in its trajectory.

It's not jusy that [E] and [I] are not the same: they differ in fundamental ways. The difference is so fundamental that, once you create a frame of reference in [I] you cannot actually enter that FoR as [E] so as to effect it. It is as if that FoR in [I] is occuring in some other universe, over which you have no control.

Yet, this *Experience* is all just happening in your head, isn't it? It's not "real". There is no actual pebble actually falling through an A => B trajectory parabola in your room.

Yet something is binding, limiting, controlling your actual thinking process, your Mentation, as if it was real. And in your own Mentation, it is completely real. It is as real for you as the words you are looking at, right now.

Whatever it is going on here, something is going on.

---

[90] https://en.wikipedia.org/wiki/Riemannian_geometry

Now you have your first real welcome to Matrixial Logic.

Negation

One moment is not the opposite of another, it is the negation of another.

Substance is in opposition: (A) ≠ (that which is not A).

In order that there be opposition under this inequality operation, both (A) and (nA) must exist in the same spacetime dimension.

To analogise the matter, a something cannot be the opposite of another thing, which is not there.

Moment is in negation: (A) ≠ (that which is what is no longer A).

To return to the pebble in trajectory analogy: each moment of the pebble is different to every other moment in that transition. Let's prove it.

Experience: Minus

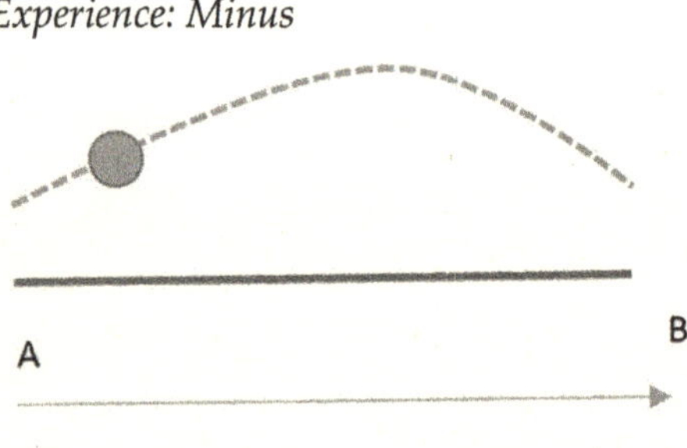

Setup:

Set up your in-the-room *Experience,* just as in *Pebble* and *Curve.*

Step 1:

In your own time

- Allow the pebble to fly, slowly;
- Let the pebble take a 1-2-3 count to travel between (A) and (B);
- Repeat 2 or 3 times.

Stop

Discussion:

(1) All easy enough.

Step 2:

In your own time

- Allow the to move along the A => B trajectory line;
- Freeze the pebble somewhere on the line.

Now

- Try to move the pebble backwards;
- Try to move the pebble back along the trajectory line: from where it is, back along the line to (A).

Stop.

Discussion:

(1) It won't go

(2) You may have tried swapping (A) and (B) around and running the trajectory in horizontal inversion

(3) But no: it still won't work. This is not an artefact of left/right eye or brain

As we said:

> a something cannot be the opposite of another thing, which is not there.
>
> Moment is in negation: (A) ≠ (that which is what is no longer A).

You created this:

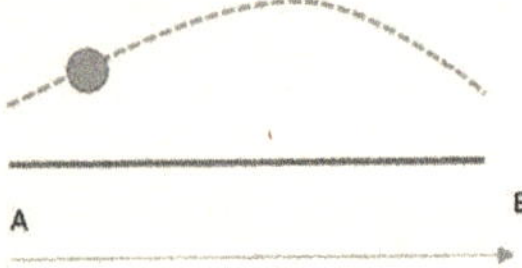

in your own head and projected it into your own room.

But you can't make a (A) Moment Δ^n which has been, come back again for that pebble which has passed through that (A) Moment Δ^n , and is now at (-A).

That Moment (-A) exists only in the negation (the extinction, annihilation) of Moment (A), and all the Moments Δ^n in between. Those in between Moments have

ceased to exist for you. They have ceased to exist.
But, there is a way out. There is a way of making Moments Δ^n work in [E]. To do that, we have to bound infinity.

Infinity ∞ : =I

Infinity∞ is the domain in which any becoming of moments manifests. Infinity is active in [E] as Information.

=I (or [I]) is the form of B Logic in ML.

Note on Notation

Although the domain of E is denoted simply as "[E]", we have developed a long habit of denoting "I" as "=I" or [I]. This is merely a keyboard and lexical convenience.

When writing of "[I]" matters, it is a nuisance to use " " or ' '. Further, when writing a sentence using '[I]' the notation can be confused with the letter I denoting the personal pronoun. In some contexts, this can prove confusing.[91]

Confusion is avoided and keyboard convenience augmented, simply be denoting the term as =I or [I].

At the risk of further confusion, but for lexical elegance, [I] is sometimes written as "B", when contrasting that with E.

E is the domain of spacetime, of things which exist. [I] is the domain of things which become. We propose that the definition of =I given, is satisfactory under the metric proffered in the Preface.

We reserve to later, discussion of the epistemological concatenation of E and =I.

---

[91] For example, when discussing matters pertaining to the Self

We mentioned a way out of [I] and into [E]. So that you can manipulate your pebble. Here it is.

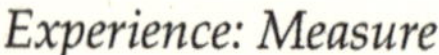

Experience: Measure

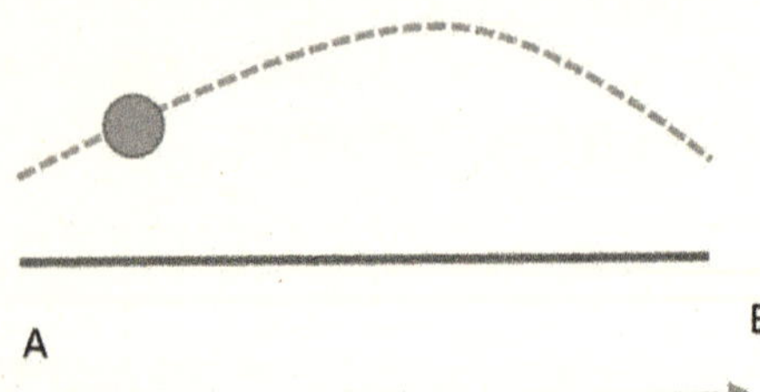

Setup:

Set up your in-the-room *Experience,* just as in *Pebble* and *Curve* and *Minus.*

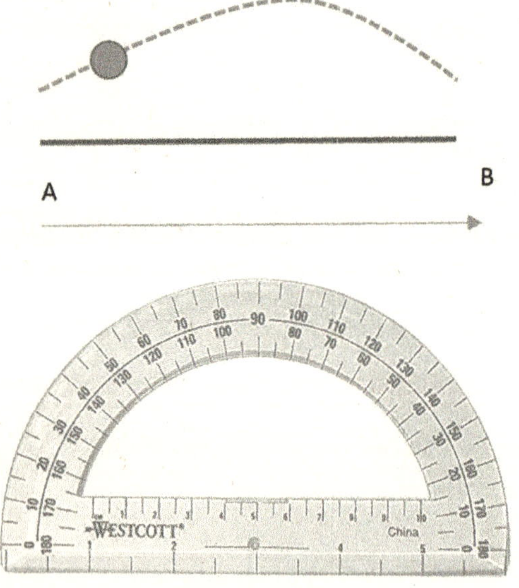

- Visualise placing this protractor over the base line and curve;
- The base line ruler is easy: visualise a series of number 0-20 along the ruler. This is your BLine;
- You see where the dotted line dots are in the pebble curve: just imagine a series of number 0-20 along that curve, and matching at each point (on a vertical drop-down) the matching number on the ruler. This is your TCurve.

Step 1:
In your own time

- Allow the pebble to move along the A => B trajectory line;
- Freeze the pebble at some 0-20 point on the line.

Now

- Try to move the pebble back along the trajectory line: from where it is, back along the line to (A).

Stop.

Discussion:

(1) Now, it's easy;

(2) The pebble hops from one BLine/TCurve point to another;

(3) Forwards, backwards; fast slow: all no problem.

Step 2:
In your own time:

- Allow the pebble to move along the A => B trajectory line;

- Freeze the pebble at some 0-20 point on the line.

Now

- Try to move the pebble back along the trajectory line: from where it is, back along the line to (A).

But:

- Try to freeze the pebble between any of the 0-20 points.

Stop

Discussion:

(1) It's stuck again;

(2) Every time you try to move the pebble so as to rest between a 0-20 BLine/TCurve point: it won't stay;

(3) The pebble just hops back to whichever BLine/TCurve point you tried to drag it from.

As we said:

> There is A way of making Moments Δ^n work in [E].
> To do that, we have to bound infinity.

We can put a boundary on an infinity $\infty$by imposing a *measurement* system. Let's reserve till later Chapters whether the measurement system is itself an infinity ∞. That allows us to escape the original ∞ infinity and move within it. We have "negated the negation", by equalising the Moments Δ^n.

But we also become stuck again. We can only navigate in

quantised steps.[92] We can't navigate in real infinity ∞.

Experience:Tangles

Setup:

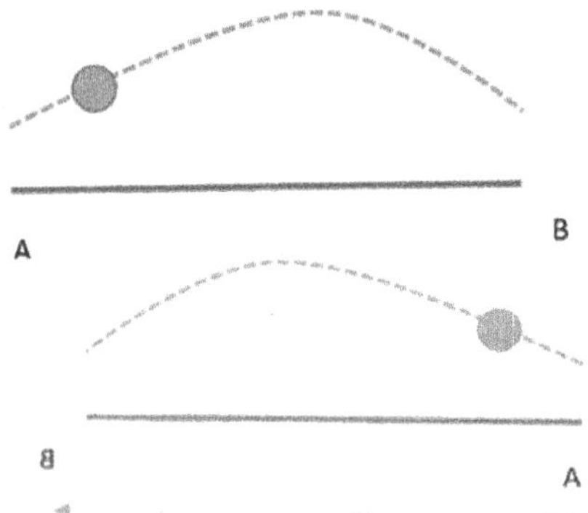

Set up your in-the-room *Experience,* just as in *Pebble* and so on.

- But now: set up another the A => B trajectory line, as the mirror of the original one
- So that you have 2 lines and 2 pebbles, in parallel
- You can have the new one in front of or behind the old one.

Step 1:

In your own time

- Allow both pebbles to move along their respective A => B trajectory lines;

[92] Clue: see Chapter 7

- Move them forwards and backwards along their respective A => B trajectory lines;
- Freeze them at the same point;
- Freeze them at different points.

Stop

Discussion:

(1) Now, it's easy;

(2) Along their respective A => B trajectory lines, each pebble enjoys complete freedom of movement.

Step 2:

- Now: place that potractor in front of you, with both lines behind: or behind both lines;
- Do the co-ordinate referencing 0-20 for the 2 BLines/ TCurves.

Now:

- Put Pebble 1 at 0 on line 1;
- Put Pebble to at 20 on line 2;
- And try to move either pebble.

Stop.

Discussion:

(1) It's stuck again

(2) Try as hard as you can, neither pebble will budge

at all

(3) It's as if there was some kind of wave function, when there were just the 2 free lines, but putting the protractor on, as frozen that wave function, so it just won't work anymore[93]

Step 3:

- Delete the idea of 0 and 20;
- Just starts with the pebbles anywhere on their separate lines, like this:

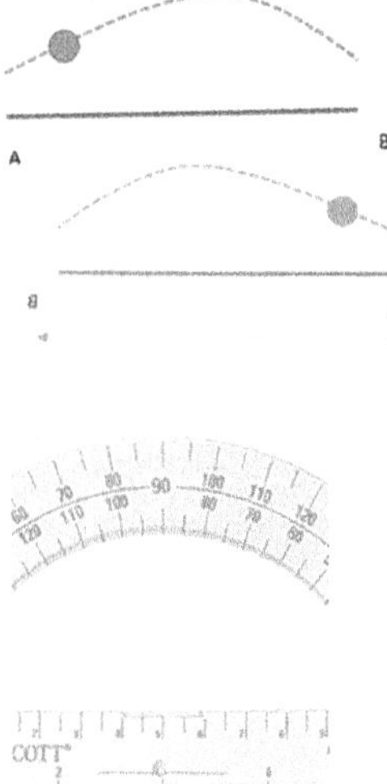

93 Clue

Now, place a part of your protractor over the 2 BLine/ TCurves.

Now:

- Try to move either pebble, however you like.

Stop.

Discussion:

(1) Now the pebbles happily move again;
(2) But only as hops from one unit to another;
(3) And it's not really a happy hopping;
(4) Once you start allowing the pebbles to move, they become random;
(5) They even start to speed up;
(6) And after a few moments, the whole thing just implodes.

Whatever it is going on here, something is going on.

For quantum mechanists, relativity cosmologists, and engineers, these *Experiences* will be even more odd. We shall say no more about that, at this point.

For now, let us conclude simply that there really is something very different about [E] and [I], and that the relationships between them bear some detailed examination: in later Chapters.

Becoming

The concept of Becoming is analytical, not teleological.

The concept is not linked to or predicated upon some grand design theory, nor any grand determinism theory.

The concept of Becoming arises from analysis of the character of transition. It is irrelevant to the concept whether or not all things are in transition from some origin, deistic or otherwise, and whether or not all transitions are determined in some way.

The concept of Becoming is formulated such that it is neutral as between these hypotheses. Things change, move and transition.[94] We claim that this datum is independent of any proposed ultimate cause or contingency of the datum.

We further claim that, were that claim to be falsified, that would not prejudice the operation of ML in [I].

Form

Within E, there is an infinity of FoRs.

That is not so within =I.

Each =I arises as a form of becoming by the moments in

[94] Intended to by synonyms

becoming which manifest in =I.

That would be circular if =I were a function of a single moment. But no moment can become in singularity. The plurality in negation of moments, which is necessary to becoming, precludes singularity: review the *Minus* and *Measure Experiences* to validate this.

Thus:

- Any Form in =I is the form of its contents.
- The contents of any =I are: Moments, in negation

Different Forms of =I do not transition further into =I. Instead, they *interfere* with each other.

Experience: Elastic
Setup:

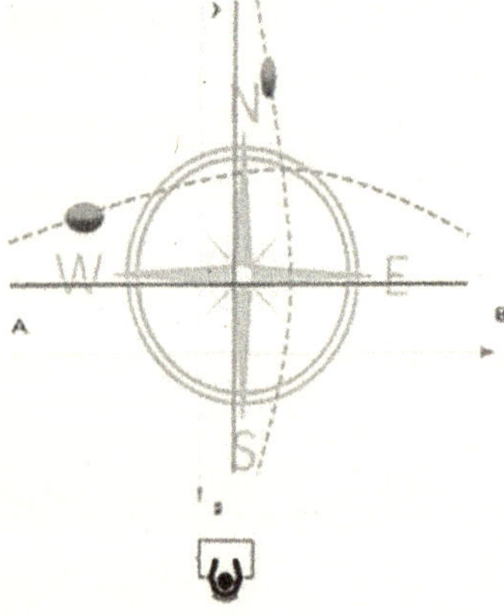

- Think of a compass;
- Think of you as the south pole;
- Set up your baselines, so that:
    - One is facing West => East (as we have done all along);
    - One is facing South => North.

Step 1:

Relax. Just blink your eyes open

- Allow both pebbles to move along their respective A => B trajectory lines;
- Just let them whizz around freely.

Stop

Discussion:

(1) This is easy;

(2) Along their respective A => B trajectory lines, each pebble enjoys complete freedom of movement: so long as you just allow them to flow.

Step 2:

Now:

- Have Pebble (A) at Point 0 in its A => B trajectory line;
- Have Pebble (B) at Point 0 in its A => B trajectory line;
- Try to move both pebbles at the same time.

Stop.

Discussion:

(1) They are stuck again;

(2) You find yourself putting huge effort into making the pebbles move: and they just won't.

Step 3:

Start as for Step 2:

- Have Pebble (A) at Point 0 in its A => B trajectory line;
- Have Pebble (B) at Point 0 in its A => B trajectory line.

But Now:

- Attach a piece of elastic between Pebble (A) and Pebble (B)
- Try to move both pebbles at the same time.

Stop.

Discussion:

(1) It takes a few moments to adjust;

(2) But now they both move fine.

We said:

> Different Forms of =I do not transition further into =I. Instead, they *interfere* with each other.

Here you see the elastic is working as an *interference device*. Sort of like an "observation".[95]

---

[95] Clue

The elastic is not any kind of infinity ∞. It is obviously limited by the changing space between the two pebbles. It is equalising the pebble movements, so that one "wave" is matched by a "trough".

Such interference functions similarly to Substance under extension.

Notation:
The Form of Moments is denoted as:

$$=I$$

Thus as:

$$(A) \neq (-A) = [I]$$

A solved equation is usually written:

$$[I] = (-A) \neq (A)$$

such that the order of notation is reversed. The reversal serves two purposes:

- To highlight that we are considering an equation purported to have been solved;
- so that =I (sometimes written as B) may be extracted from the equation, to serve in manipulation in some further equation.

We have also (again) found that, by reversing the order in which concepts comprising an equation, are presented to the mind, a fault-checking mechanism within the mind, is engaged.
[I] denotes Infinity ∞.

Examples

Exemplar 1

You are really walking.

We are interested in the information about the difference between where you were, and where you are now: in that street, by reference to time. Now, whether you want to say that the information is "real", is very much up to you. It doesn't matter whether you say it's "real" of not.

If your walking [let's call that (W)] works out that (W) ≠ (-W) sees you hit by a bus, then that seems pretty real. You could call that =I as "accident information". We reserve to later, discussion of the epistemological implications of =I.

Experience: Breathe

Step 1:

- Keep your eyes open;
- Breathe normally;
- Focus on your breathing.

Now:

- Imagine the end of your out-breath as a place [#][96] in time and space;
- Imagine you were breathing that place [#] back into you;
- And out again.

[96] This is a random symbol

Stop.

Discussion:

(1) You notice that the place [#] which began outside you, comes inside you;

(2) But then that place [#] seems to mark somewhere inside your sternum;

(3) The "real" place [#] seems to cycle out from and in to you;

(4) But there is now a ghost of the place [#] inside you. Which seems to react to the passing of your breaths.

You have experienced [I] in action. That which arises out of becoming. There is much more to say about all this, of course. Your experience will make much more sense when we come to discuss Φ in the *Languini* Chapter.

The Equation in =I

Proposition: a Breath is a Moment in I

Just to tell us what we're looking at. We don't have to write that Proposition every time: it can be obvious from the text.

Let Breath be (A) in [I]:

(A) ≠ (-A) = [I]

>

(Breath) ≠ (-A) = Breathing

>

Breathing = (Rest) ≠ (Breath)

Note: we use the ( ) brackets, just to keep the formulas tidy. They are useful when writing them in a line of text.

Equation Part	Explanation
Proposition: a Breath is a Moment in [I]	So that we know we're looking at Process
<u>Let Breath be (A) in [I]:</u>	We are looking at "Box" and trying to work out what it's opposite is
(A) ≠ (-A) = [I]	The basic formula of I Space relationships
>	This symbol just means "follows" or "next
(-A)	Means "change in" (A)
(Breath) ≠ (-A) = Breathing	We're looking for a Process of (Breath)
= Breaths	In relation to the Process idea of Breaths
(Rest) ≠ (Breath)	So, we can see that the Opposite is (Rest)
Breathing = (Rest) ≠ (Breath)	We run the equation in reverse.*

* We do not have to do this. There's no "rule" about it. It just makes things tidy and easy to see, when you move from line to line.

AXIOMS in B Logic

Infinity is the Form of becoming of Moments in Infinity.

Each Moment is unique in becoming.

To be unique in becoming, is to become in Infinity.

Infinity manifests as Form.

Any Form is the Form of its Contents.

The Content of any Form is Moments in Ininity

MATRIXIAL LOGIC

AXIOMS COMPARED

IN [E]	IN [I]
Being is the Form of Substances in Extension.	Information is the Form of Moments in Infinity.
Each Substance is a unique event in spacetime.	Each Moment is unique in becoming.
To be a unique event in spacetime, is to be Extended.	To be unique in becoming, is to Become in Infinity.
Extension manifests as Form.	Infinity manifests as Form.
Any Form is the Form of its Contents.	Any Form is the Form of its Contents.
The Content of any Form is Substance in Extension	The Content of any Form is Moments in Infinity

Opposite and Process

Can something, an (A) function both as an Opposite and in a Process?

- Not in the same frame of reference;
- We will see in the next Chapter how $(A)^n$ fly around between [E] and [I];
- But it is never the same (A): even though it may look like it is.

Names of Things

Choose a word:

[____________________________]

It needs to be a noun. It can be any noun you like, save that it must name-describe anything real in the world. Something that you can reach out and touch, right now, wherever you are.

And we mean, literally, "touch". Not metaphorically (such as "touching Jesus",[97] or "touching awareness". Play properly, or you won't learn much from this game.

OK: you have your word [______________] ?

Let's just refer to your chosen word as [*].

So, is [*] a Form or a Content under a Form?

---

[97] Even if you happen to believe that Jesus is real in the world

That is going to depend upon perspective, isn't it?

No: this is not going down the path of "it was only there because you thought of it." It matters not whether it was or was not "there" before you happened to fix your attention on it.

What we're looking at is the way that the attention is fixed.

- The Vertical Perspective: If you are considering [*] in the perspective of its properties as [*], then it's a form of Substances mutualised in Extension. You can layer those down to the atom, or up from the atom. Either way, you produce a pyramid of Forms and Contents;
- The Lateral Perspective: If you are considering [*] in the perspective of what delineates it as a Substance in Extension, then you will want to locate (NOT[*]), in whichever iterations satisfy, or inform. Satisfy, if you already know. Inform if you are experimenting.

So what are those "layers" of that "pyramid"? They are a succession of the Forms of Substance of [*]. After all, [*] is one substance: one thing: one noun. It is what it isn't, in the Lateral Perspective; and it is what isn't the process of its becoming,[98] in the Vertical Perspective.

---

[98] All of the *transformations*

Experience: Perspective

Step 1:

Have a go at visualising your chosen word [*] in:

- The Vertical Perspective

then

- The Lateral Perspective

Stop.

Step 2:

Now: do it in reverse:

- The Lateral Perspective

then

- The Vertical Perspective

Stop.

<u>Discussion</u>:

(1) You're now feeling that you're seeing [*] in a way which seems unfamiliar, yet at the same time, like it was always there.[99]

What you're doing is holding together in your awareness, a synthesis of Perspectives. It's hard to do, and it seems even harder to sustain. But we do it all the time. We will encounter more of this perspectival reframing, in later Chapters. In these initial encounters with the practice of

[99] If you're not: sorry. Or you're not really going with this *Experience*

ML, let's try this one: what's the Content of Bad?

Experience: Bad

- Think of Bad;
- Think of some examples of Bad: at least 5 examples.

Stop.

<u>Discussion:</u>

(1) What happened is that, as soon as you came to the second example, and increasingly after that, you were searching in your mind for "not good".

Bad does not exist on its own. It can only be a thing which exists, if it come under a form of negation.

So: Bad ≠ (-Bad) = (Some form of) Continuum.

So, if you want to think or say that Bad exists, you are calling into operation as a premise the form of existence: [I] = (-A) ≠ (A)

Now, Bad is an interesting iteration of [I], since (-Bad) is, definitionally, Good.

It's one of those Categories of [I], which is intrinsically paired with a pre-defined opposite.

Let's think of some things which are not so paired:

Rabbit
Ranch
Razor

What are all these? They are names of things: nouns.

Bad is not a name of anything. It is a descriptor of other things. It's an adjective, and can become a descriptor of process (badly: an adverb).

Now you start to get the juices flowing:

- A noun falls within the Form E=(A) ≠ (nA);
- But any such (A) which exists is not by mode of existence contingent upon any particular iteration of (nA).

By contrast:

- An adjective falls within the Form I= (-A) ≠ (A);
- And, any such (A) which exists, so exists only as a mode of existence contingent upon a particular iteration of (-A).

A Socratic Interlude

"But how is this materialist?"

Well sir: see that rock over there.

Yep

That is a substance extended in spacetime, yes?

Yep. Definitely.

OK: what was it when it wasn't there?

Huh?

Was that rock always there?

I guess not.

OK: what was it when it wasn't there?

Well, it was a rock, being someplace else.

OK: and before that?

Well it became a rock.

Great. And before it became a rock, was it a rock extended in spacetime?

No, obviously not.

So, you're happy that it became a rock.

Yep.

That rock, now there in spacetime, became an event in spacetime. A process of becoming, until that process crystallized as a rock event in spacetime?

Yep. I agree

A process of becoming, until that process crystallized as a rock event in spacetime?

Yep, that's cool.

Is that a materialist conception of the universe?

Yes, it must be: "event in spacetime"; "process of becoming an event in spacetime". Yep, that looks pretty materialist to me. You a physicist or something?

Not really. And materialism is monism, right?

Yes. For sure.

Great. So what is non-monist about: =I becoming an event in spacetime as the crystallization of a process of becoming anterior to =I?

Well yeah, that looks monist.

Thanks. Is it materialist?

Well, hmmm. Can I reach out and touch it, just like the rock?

Can you touch the rock? Or did you just imagine touching the rock?

OK yeah, I just imagined it. But there's stuff I can definitely actually reach out and touch.

I agree. Give me some examples.

Well, my shirt. My body.

Great. Is touching, an event in spacetime?

Yep. Definitely.

Is the sensation of touching, an event in spacetime?[100]

Yes.

How so?

Well there's touching going on. I probably exchanged some atoms with the shirt.

Love it. Where's the touching going on?

Between me and the shirt.

When was that in spacetime?

A second or two ago.

Cool. Remind me: Where's that touching going on?

Between me and the shirt.

Between what and the shirt?

Me.

So that's an experience?

Yep.

---

[100] Yes: qualia

In you?

Yep.

In spacetime?

Yep.

And if you break that experience down into how it happens...

OK...

There's a sense input becoming a feeling: of something?

Yep.

Which becomes in the continuum of what for you is appropriate to that feeling?

Mmm?

Is the shirt smooth or bobbly?

Bobbly.

So, your sensation translates as a feeling in the continuum of bobbly-ness?

OK, yeah I see. Yeah.

And that forms a feeling of bobbly-ness?

Yep.

Which you can access as a thought: "that's bobbly".

Yep.

Was that an event in spacetime?

Well yeah. It must be. What else could it be?

Are events in spacetime materialist?

Definitely.

Great. Last thing: did you really feel that feeling?

Yeah, of course I did. I know I did.

Is the real, real?

Yeah, obviously.

So now a challenge:

- At what point in the narrative did we switch from E Logic to B Logic?
- Or was it the same logic (whether E or B) all the way through?
- Or was it both all the way through?

Forms as Substances

Please re-read the section on *Transformation of Forms*.

We said:

> The rules of E Logic are preserved: *so long as we are prepared to acknowledge that* nothing beyond the first (S) and its correlate in inequality (nS), is extended in spacetime.

Here's your challenge:

Is there a difference between:

- *Treating as if it were a Substance,* a Form of Forms (F) ≠ (nF) in E Logic; and
- =I being as a Substance

We will leave you to turn that one around, while we open the next doorway.

Singularity

Singularity is an important conceptual category in ML.

We denote Singularity concepts as: [Si], or sometimes in narrative text (Si).

An example is God. What is it about God? Just one. God is separate from the "real" universe. You can think about that idea of God, but you can't touch or see God: unless you're a prophet.

Here, we're just using that idea of "God" to show you what a Singularity concept can look like.

However, it is not like the concept of zero "0", which constitutes an element in a linear co-ordinate system. Singularity functions only reflexively.

1+1=3, is a mistake in applying the rules of arithmetic. In being mistaken, it can helpfully illuminate the rules. By all means, the equation could be articulated as a manifestation of some other set of arithmetic rules. You can claim that the equation is permitted within the set of arithmetic rules, of which it is supposed to be a manifestation. But you are then either lying, or deluded.

Substance and Moment are that which is real under the logic of forms, and only those (and their iterations).

Experience: Marble

Step 1:

- Clear your mind;
- Think of a blank space; a nothing space;
- Just nothing.

Now:

- Imagine a single, solitary marble[101].

Stop.

<u>Discussion</u>:

(1) Already you are imagining a marble in the context of something else. You can't help it. It is how our minds are born to work.

Step 2:

- Have a rummage around what you can think about that solitary marble you are imagining.

Stop.

<u>Discussion</u>:

(1) You see that each time you try, you have to resist the inclination to imagine something else. As you successfully resist, the marble seems to disappear. The

---

[101] imagining an atom is a strain on anyone's imagination

less you resist the more "real" the marble appears.

Step 3:

Now:

- Imagine two marbles.

Stop

(1) The feeling of strain disappears, doesn't it. With those two marbles,[102] you feel like you could do anything. Well, you can play marbles, for a start.

That's why Singularity is so useful as the idea of an idea which you can't actually think *in*: you can only think *of*.

C: Chaos

So, we now have two Categories of Matrixial Logic: E Logic (the logic of being in Existence); and B Logic (the logic of becoming in Infinity).

We have seen how they are distinct, but separated. They are modes of Logic, not different universes.

We have seen how they operate together, not just in parallel, but each becoming in reality contingent upon the other. That leaves C Logic: the logic of chaos in order.[103]

[102] Substance as plurality

[103] That's putting it romantically, rather than scientifically

We live in a universe of C Logic. We live our experiences in C Logic.

C Logic instantiates where the biological, the social and the individual meet. Which is always, all the time, everywhere.

Chaos is that space in which no substance or moment solves under a form.

Now, we could get smart, and from the outset suggest that C Logic is the "real deal" and that E & B Logics are merely the appearances of C Logic.

That would make sense: that domain of the unknowable until it is known, and the unbeing until it has become.

What, though, can we possibly say about not_is and non_become?

A moment's reflection tells us that we can say everything about not_being and non_become, which "everything" is reflexive of in_being and has_become.
So, we are arriving at Forms of Forms: C.

$$C = (\text{Is}) \neq ({}^{\text{not}}\text{Is})$$
$$C = (\text{Become}) \neq ({}^{\text{not}}\,\text{Become})$$

How can C be both? How dualist is that?Well, how can light be a wave and a particle? How materialist is that?

C Logic is the universal logic of Potential.

So, we have "half" a formula:

$$(A) \neq (Z^n)$$

More formally, we can write:

$$\frac{(A) \neq (Z^n)}{C}$$

where the denominator merely expresses what we already know from writing $\neq (Z^n)$. In other words, that may be iterations of an (A) going on, but we have no idea what (if anything) one (A) has to do with (Z^n).

We find content for C by exclusion. There are some terms which just won't solve in E or B. Those, by definition, are C terms.

C terms are things that we know we don't know.[104] We can hypothesise that there are things which we don't know that we don't know. Since any such hypothetical things cannot enter any account of knowledge,[105] we don't need a category of logic for them.

---

[104] In the sense that we can't subsume them under a FoR

[105] Which is what philosophy is

Which C is which and how can we tell? The answer is that we cannot tell until C/C has resolved itself into order: whether that be Extension or Becoming.

So what's the point of C? Well, what's the point of light? Or dark matter?

A universe filled only with Being and Becoming would be a pretty poor one. Indeed, it's impossible to think of such a universe.

Experience: Nothing

Step 1:

- Try to think of nothing, for a few moments;
- We don't mean, tune out. Instead, think of nothingness, a void.

Now:

- Think of something.

Now:

- Try to think of the something, without the nothing.

Stop.

<u>Discussion</u>:

(1) It doesn't work, does it?

(2) Once you have conjured thought of a void, whatever else you think of, seems to reference in

some way that void.

Step 2:

- Allow yourself to notice your thoughts and feelings;
- Just sink into your awareness, and relax. This isn't meditation. Forget about Buddhist mountains. Just allow your normal, everyday thoughts and feelings to be around.

Stop.

Discussion:

(1) You notice there's something there, at the edge of your feelings and thoughts.

(2) You can't really call it something. It's like a presence. Kind of like a ghost which is there, but not there.

(3) And if you reach out with your thoughts and feelings to "touch" it, you seem to be drawn "out" of yourself into what seems like infinity.

Step 3:

- Allow yourself to notice your thoughts and feelings;
- Notice that *something there,* at the edge of your feelings and thoughts.

Now:

- Think of that Void place.

Stop.

Discussion:

(1) You noticed something there, at the edge of your feelings and thoughts

(2) As you thought of the Void place, that *something* disappeared.

Welcome to the domain of C Logic.

We instantiate with C, by means of Singularity Bridges.

Singularity Bridges

What (Si) can provide is a *transformation of inequalities.*

For ease of writing, we call such a transformation a *Singularity Bridge,* or just a "Bridge".
In building a Bridge:

- We use an unsolved formula as a *Function 1*

$$(A) \neq (Z^n) \neq Si$$

- We take another unsolved formula as *Function 2*
- We call the difference between these two ≠ Functions a *Disjunct*
- We perform a transformation of the ≠ *Functions,* such that:

$$[(A)] \neq [(A^n)] \neq Si$$

We can read $[(A)] \neq [(A^n)] \neq Si$: that is, the two iteration of ≠ can be equivalents: as disjuncts. Each ≠ can cancel out the other. We need to find the Bridge which can effect such cancellation. That is to formulate a *Transformation of Inequality Functions.*

We denote an Si Bridge like this:

$\dfrac{\neq}{\neq}$	=	?
Disjunct	Bridge	Form

That Bridge must be a Singularity: that which is, but only outside E and B of our universe; and which is the boundary condition of E and B of our universe: the event horizon which touches both sides.

The Form may be in E or B.

- If a Bridge arises in E, then we denote that as:

  } External

- If a Bridge arises in B, then we denote that as:

  } Internal

We then write an Si Bridge, using the notation:[106]

[Si: Martian≠/Venusian≠]	= Alien =	[Being]	} External
Disjunct	*Bridge*	*Form*	*Domain*

If we want to propose a link an External with an Internal domain, we denote with: |

[106] The contents in this example are frivolous, and the equation is not accurately solved

So, for example:[107]

[Si: Martian ≠/Venusian≠] = Alien [Being] } External

|

[Si: Fear ≠/ Joy≠] = Stress [Becoming] } Internal

Note that the | symbol does not denote a *formula,* but a propositional connection between formulas.

Syllogism

Those familiar with syllogistic reasoning will have noticed that the formulas in E and B follow the syllogistic pattern:

Syllogism	**Formulae**	
Major Premise P1	(A)	(A)
Minor Premise P2	(nA)	(-A)
Conclusion	[E]	[I]

We are indeed employing syllogistic reasoning in locating (nA) and (-A); then in formulating the Form, which the conclusion / solution.

This may well lead you to the view that Matrixial Logic has been designed as a way of assisting us to see more clearly how it is we think about matters.

We will see much more about this in the next Chapter.

[107] The contents in this example are frivolous, and the equation is not accurately solved

Uniform Terminology

This is the uniform terminology we use in Matrixial Logic:

Logic Type	Attribute	Method	Space
E	Existence	solve equations in	[E]
B	Becoming	solve equations in	[I]
C	Chaos	solve equations in	[C]

E Logic is the logic of extensions of Substance. That's the derivation of "E". C Logic is chaos logic: so the C denomination makes obvious sense.

You can see that B Logic is the odd one out. We solve B Logic equations in I, not B.

This is simply a stylistic choice, to avoid confusion. If we used "B" as the solution space, it would be easy to become confused between "being" and "becoming". If we used "I" to denote the Logic Type, it would also get confusing.

That said, we always write [I] equations as:

$$(A) \neq (-A) = [I]$$

There is also a deeper philosophical reason for the E/E : B/I distinction. Frames of reference work differently in E Logic and B Logic. Extensions are iterations of E. Moments are not iterations of B.[108]

[108] For later discussion

Concluding Remarks

We arrive at our starting point. Our concern with data coordinate in spacetime.

It is a commonplace not to examine closely the categorical assumptions with which we approach "scientific" thinking.

Part of the problem lies with the inherited architecture of 2,500[109], which lies, not as a cityscape, but as sedimentary layers of human history.

This is not blameworthy. We cannot help but be where and when we are. But if we reject the tools of archaeology, our new architecture will lie and fall upon infirm foundations.

The compelling concern of those who feel compelled to be so concerned, is to deny the reality of data. To make the mistake that data and insensate objects are of the same kind of something. To seek to apply what appears to be the logic of objects[110] to what appears to be the logic of data[111], and in so doing, to violate the very premises of object logic.

Such violation is only possible at all, of course, because the violators are operating the logic of data, in the first place.

---

[109] Or even longer, depending how you trace the history of modern thought

[110] E Logic

[111] B Logic

What also makes it possible, is that forms of data logic act as Substances in co-ordinate spacetime.[112]

Here, we bring unity in diversity. It cannot be made to matter whether forms of data logic:

- *are* Substances in co-ordinate spacetime; or
- *act as* Substances in co-ordinate spacetime.

because the only forum in which any mattering is said to have meaning, is that in which Forms *act as* Substances.

We can endorse the claim that objective knowledge about the world is gained with most certainty through the laws of science.

The rules of E Logic cannot be used to suborn the rules of B Logic, without violating the rules of E Logic. It may be that this ML framework review has already provided the reader with some insight into the applications of ML.

Although that was not the purpose of this Chapter, if that be its effect, then subject to the rubric of efficient explanation, the work has served beyond its purpose.

Next, we turn to examine using Matrixial Logic to look at classical and modern logic.

---

[112] So, we answer the challenge set out in *Forms as Substances*

CHAPTER 3

CATEGORICAL TRANSFORMATIONS

Logic still engages with the framework and rules laid down by Aristotle. Using the Aristotelian Categorical Sentences allows us to compare transitions[113] into ML equations. These begin to provide us with our first insights into the internal logic of the ML equations rules.

We can then, through a brief survey of certain historic schema of logic, illuminate elements of matrixial logic.

As Holmes famously remarked of Watson: [114]

> "It may be that you are not yourself luminous, but that you are a conductor of light. Some people without possessing genius have a remarkable power of stimulating it.

This is the ML viewpoint of classical and modern logic.

---

[113] "Transitions" is used, in place of the more obvious "transformations", as this latter term is reserved for technical use in ML

[114] *Hound of the Baskervilles.* Conan-Doyle (1901). Kindly disregard the refencences to "genius". Of possible passing interest: the Author lived in Conan-Doyle's former lodge house in Crowsley (near Henley) in the 1990's

ML Transitions of the 4 Categorical Sentences

There are four types of Aristotelian categorical sentences ("ACS"):[115]

(1) (A) Universal Affirmative
(2) (E) Universal Negative
(3) (I) Particular Affirmative
(4) (O) Particular Negative

The following Table references the OEIO forms of Categorical Sentence to the appropriate ML equation equivalent:

Aristotle Categorical Sentence Transitions to ML

	C S	ML Equation	Type
A	Every A is B	(A) ≠ (-A) = B[I]	Categorical Universal
E	No A is B	(A) ≠ (B) = N[E]	Reflexive Universal
I	Some A is B	(A) ≠ (nA) = B[E]	Reflexive Opposition
O	Some A is Not B	(A) ≠ (nA) = B/^{n}B*	Unreflexive Opposition

Table 3.1

* ACS (O) is a 'bookmark', to be considered later.

Marginal Note:

We can, for completeness complete the octet of ACS:

Singular:	Callias is just.	Callias is not just.
Indefinite:	(A) human is just.	(A) human is not just.

Since the ML transitions are so obvious, the reader is invited to essay them.

[115] Using the abbreviations invented by the medieval scholars

We are here engaged in the limited exercise of demonstrating that ML equations can function in place of ACS.

However, this is in provides valuable information. The ACS equivalence:

- Illustrates the operation of ML equations;
- Demonstrates ML equation consistency;
- Provides the foundation for showing the transition of ML equations in parallel to the domain of Aristotelian categorical syllogisms.[116]

Let's first of all locate a perspective which makes sense of the difference between ML equations of the Affirmative, in [I] and the Negative, in [E].

Experience: Landscapes
Step 1:

- Look out onto a landscape: urban or rural;
- Just relax into the view: see it however you see it.

Now:

- Think to yourself the words "all is me"

Stop.

[116] See later in this Chapter.

Step 2:

- Look out onto a landscape: urban or rural;
- Just relax into the view: see it however you see it.

Now:

- Imagine the landscape looking at you;
- Imagine the landscape saying "all is us".

Stop.

Discussion:

(1) Step 1 felt fine. Quite natural.

(2) Step 2 felt alien. If you put some energy into the Step 2 projection, you may have felt quite uncomfortable.

You have just experienced a difference between:

A	Every A is B
E	No A is B

We proceed, to expand *Table 3.1* as follows.

ML Equations

{A} Every A is B: AaB
Every Donkey is Animal
(A) ≠ (-A) = I
>
(A) ≠ (-A) = B[I]
≈
(Donkey) ≠ (-Donkey^n) = B[I]

{E} No A is B: AeB
No Donkey is Horse
(A) ≠ (nA) = E
>
(A) ≠ (B) = C[E]
≈
(Donkey) ≠ (Horse) = Null [E]

{I} Some A is B / Not All A is B: AiB
Some Donkey is Brown
(A) ≠ (nA) = E
≈
(All Donkey*) ≠ (nAll Donkey*) = Brown [E]

* Note: the 'All' is definitionally inferred in the Category "Donkey"

{O} Some A is not B / Not All A is not B: AoB
Some Donkey is not Brown
(A) ≠ (nA) = B/^{n}B[E]
≈
(All Donkey*) ≠ (nAll Donkey*) = ^{n}Brown/Brown [E]

We can then provide explanation as to the derivation of ML equations.

The Proofs of course depend upon assumptions derived from the ML Axioms stated in the *Framework* Chapter.

Explanatory

{A} A belongs to every B (Every A is B)
(Every) Donkey is Animal / All Donkeys are Animals
(A) ≠ (-A) = I
>
(A) ≠ (-A) = B[I]
≈
(Donkey) ≠ (-Every Donkey^n) = Animal[I]

Proofs:
<u>(A) (in plurality of iteration) [≠ (-A^n)] is stated to hold iteration; as (B), which is a non-iterative term.</u>

[P1]
(1) Any FoR of iterate plurality must be in I.
(2) No (A) or [≠ (-A^n)] can function as Form.
(3) Ergo: B must solve as Form in [I]

[P2]
(1) (A) in iteration (-A^n) is stated to limit B
(2) Ergo: B cannot:
(2.1) iterate (A) ; (2.2) iterate (-A^n)
(3) Ergo: B must function as domain of (A) and (-A^n)

[P3]
Were it stated that One Single A is B, then:
(1) That would state an impossibility; because
(1.1) nothing can be in spacetime at the same time and in the same dimension as any other thing; otherwise
(1.2) two things would be in the same extension as one thing, which is a violation of the laws of spacetime.
(It is established that one thing can iterate in 2 different extensions in spacetime).[117]
(2) The Single A would of necessity be a Singularity (Si).
(2.1) No Si has any FoR.
(2,2) QED: no FoR could exist in which (one) A is B.

[P4]
(1) It is required that no (A) subsist in categorical opposition to any other (-A^n).
(2) Only an (A) which subsists in categorical opposition "≠" to any other (nA) can subsist under a Form of E.
(2A) No E can be an FoR for (-A^n).
(3) Ergo: (-A^n) must solve in I.

[P5]
(1) It is required that (A^n) solve with B,
(2) Chaos does not admit any solution.
(3) Ergo: (-A^n) cannot solve in C

[117] https://phys.org/news/2015-01-atoms.html

Discussion:

1. "Every" or "All" dictates a succession of (A)-like iterations ($-A^n$).
2. It is not necessary to transform (Man) into (Men), since (Men) by definition constitutes as the iterated Moments of (Man).
3. What all ($-A^n$) have in common is their FoR in B.

Explanatory

{E} A belongs to no B (No A is B)
No Donkey is Horse / Donkeys are Not Horses
(A) ≠ (nA) = E
>
(A) ≠ (B) = C[E]
≈
(Donkeys) ≠ (Horses) = Null [E]

Proofs:
[P1]
(A) (in universal plurality) is stated to oppose [≠(nA)] (in universal plurality); as (B)
(1) Any FoR of opposition must be in E
(2) No (A) or [≠(nA)] can function as Form
(3) Ergo: that which is neither (A) nor (B) must solve as Form in [E]
[P2]
(1) "No Donkey" [(A)] is a compound of:
(1.1) [There exists a (donkey)]
(1.2) ≠ [There exists an iterated plurality of ($donkey^n$)]
(1.3) = Donkeys[I] : stated as [(A)]
(2) "A Horse" [(B)] is a compound of:
(2.1) [There exists a (horse)]
(2.2) ≠ [There exists an iterated plurality of ($horse^n$)]
(2.3) = Horses [I] : stated as [(B)]
(3) The statement requires that 2 Forms be opposed, as if each were Substance.
(4) Thus: [(A)] ≠ [(B)] = E
(5) E cannot be comprised in either Substance, so must be the form of their extensions.

(6)	C is accordingly the present temporal Form of those extensions.
[P2A]	
Suppose it be stated: No Donkeys were Horses.	
(1)	Repeat [P2](1)-(5) above.
(2)	C is accordingly the past temporal Form of those extensions
[P2B]	
Suppose it be stated: No Donkeys will be Horses	
(1)	Repeat [P2](1)-(5) above.
(2)	C is accordingly the future temporal Form of those extensions
P3]	
(1)	It is not required that any (A) iterate in (A^n).
(2)	Only an (A) which iterates in (A^n) can subsist under a Form of I.
(2A)	No I can be an FoR for (nA).
(3)	Ergo: (nA) must solve in E.

Discussion:

1. Given the Solution of the Universal Affirmative in =I, it seems counter-intuitive that the Universal Negative solves in E.
2. However, the intuition arises from the appellation "Universal".
   (1) *No Man is a Horse* is only a universal reflexively.
   (2) Contrast "Every Man", which is a universal by reference to its own iterations (Man^n).
3. It might therefore be said that:
   (1) Categorical Sentence (E) is *Reflexive Universal,* whereas
   (2) Categorical Sentence (A) is a *Categorical Universal.*

4. It will be seen that Categorical Sentence (E) does not quite comfortable sit on a single ML equation

line. **That's because this seemingly innocent formula is actually a Node <∑E>.**

5. We will be meeting Nodes later, and in great detail.

Explanatory

{I} Some A does belong to B (Some A is B)

Some Donkeys are Brown

(A) ≠ (nA) = E

>

(A) ≠ (nA) = B[E]

≈

(Some Donkey) ≠ (nSome Donkey) = Brown [E]

Proofs:

[P1]

(A) is stated to oppose [≠(nA)] as a function of (B)

(1) Any FoR of opposition must be in E

(2) No (A) or [≠(nA)] can function as Form

(3) Ergo: that which is neither (A) nor (nA) must solve as Form in [E]

[P2]

(1) (A) is precluded from iteration (-A), because there is a mode of (A) which exists in opposition to (A).

(2) There being no iteration, the equation cannot solve in (=I).

(3) Ergo: the equation must solve in (E).

Discussion:

1. This is an inequality of simple opposition; such opposition (nA) obviously functioning as form in (E).
2. That which limits (A) is the Form, rather than categorical iteration ($-A^n$).

Plurality Note

1. For discriminatory accuracy, it is entirely plausible to write:

$$(A^n) \neq (nA^n) = B[E]$$

≈

(Some Donkey^s) ≠ (nSome Donkey^s) = Brown [E]

2. As will be seen later, this additional notation, for plurality, becomes very important in dealing with the Ackrill modified O.[118]

Explanatory

{O} Some A does not belong to B (Some A is not B)

Some Donkey is not Brown

(A) ≠ (nA) ≠ nE

E is an indeterminate (contingent) form

≈

(All Donkey*) ≠ (nAll Donkey*) = ^{n}Brown/Brown [E]

Proofs:

(A) is stated to oppose [≠(nA)] as a function of (nB)

[P1]

(1) Any FoR of opposition must be in E

(2) No (A) or [≠(nA)] can function as Form

(3) A Form which is ^{n} is indeterminate. It can only function as an oppositional reflexive in itself

(4) Ergo: that which is neither (A) nor (nA) must solve as Indeterminate Form in [E]

[P2]

(1) (A) is precluded from iteration (-A) as its own Substance.

(2) There being no iteration, the equation cannot solve in (=I).

(3) Ergo: the equation must solve in (E).

---

118 See later in this Chapter

Discussion:

1. The (O) Sentence has always been regarded as the "odd one out".
2. Matrixial Logic explains why this is so: it is because what is being used as Form is itself indeterminate.
3. ML Rules dictate that any thing stated as ^{NOT} must itself be composite in a form.
4. Thus, while (O) does solve in ML as: = ^{n}Brown/Brown [E]; that is functionally equivalent to writing ≠ ^{n}Brown.
5. In other words, the equation cannot be solved as determinate without more information.

Error in Statement

The reader will see later in this Chapter how discovery by Aackrill of the correct formulation of (O) changes matters considerably.

Experience: Category

Step 1:

- Look around your room;
- Look at an object and a wall: 2 "things".

Now:

- Focus on the 2 things;
- Focus on the space between them;
- Try to formulate a way in which all of 1 thing are all of the other thing.

Stop.

Discussion:

(1) You felt yourself running through options for the "all is" form.

(2) You ended up finding some idea, of which both things are examples, like: made of atoms; mine; in the house

(3) The idea you created is a Form: and you made the particular characteristics of the things disappear. They became iterations of that Form.

(4) So: you created a form of Infinity [I], of which these very different things, became example Moments.

Evidence that ML Equations correctly transition ACS

The modern logic possibility matrix provides a useful validation of the ML equation transitions of the 4 Categorical Sentences.

	ACS	ML Equation	Type
A	Every A is B	(A) ≠ (-A) = B[I]	Categorical Universal
	Possibility Matrix		Some B are A Some A are B All B being A is a possibility Some B are not A is a possibility
E	No A is B	(A) ≠ (B) = N[E]	Reflexive Universal
	Possibility Matrix		No B is A Some A are not B Some B are not A
I	Some A is B	(A) ≠ (nA) = B[E]	Reflexive Opposition

	Possibility Matrix		Some B are A All B being A is a possibility All A being B is a possibility Some A are not B is a possibility Some B are not A is a possibility
O	Some A is Not B*	$(A) \neq (nA) = B/{}^{n}B$	Unreflexive Opposition
	Possibility Matrix		All B being A is a possibility No A is B is a possibility No B is A is a possibility Some B are not A is a possibility Some A are B is a possibility Some B are A is a possibility

Table 3.2

* Note: the reader should come back to this Table and apply the ML equation in (O), following the Ackrill translation.[119]

The reader is invited to apply each iteration in each Possibility Matrix to each ML equation.

Choosing an example at random:

A	Every A is B	$(A) \neq (-A) = B[I]$

<u>Possibility Matrix: Some B are not A is a possibility</u>

$$>$$

$$(A) \neq (-A) = B[I]$$

$$\geq$$

$$(-A)^n \Leftrightarrow \{B\}^n$$

[119] See later in this Chapter

There is ambiguity about exactly what "some" means here. It may be that "some" cannot be reached by iteration within an FoR, but has to be achieved by a TransFormation.[120]

Notation:

The new, unfamiliar symbols:

⇔ { } n

Symb	Meaning / Use
These ML symbols denote operation within an FoR	
>	Shows we are moving from an FoR to operations within an FoR
n	Shows we are iterating, or one might say, "pluralising"
	This is a "shadow" iteration within an FoR. It is an inherently unstable (conditional) iteration. So the operation cannot be notated by the classic =
{ }	These brackets contrast with the usual [ ]. They notate that the Form is being operated "internally". It notates that this is no longer a real Form, but a "shadow" form.[121]

Preliminary Implications of ACS/ ML Transitions

At this early point in the exposition, it is to be expected that a remarkable feature of ACS / ML transitions has become apparent.

---

[120] See later in this Chapter

[121] Note how this re-appears in *Languini* Chapter

The Divergent Frames of Reference

Aristotle Categorical Sentence Transitions to ML

	C S	ML Equation	Type
A	Every A is B	(A) ≠ (-A) = B[I]	Categorical Universal
E	No A is B	(A) ≠ (B) = N[E]	Reflexive Universal
I	Some A is B	(A) ≠ (nA) = B[E]	Reflexive Opposition
O	Some A is Not B	(A) ≠ (nA) = B/^{n}B	Unreflexive Opposition

ACS (E) and (I) solve in [E] equations.

Yet, ACS (A), the universal affirmative, solves only in [=I] equation. This idiosyncratic iteration is, from the perspective of classical predicate logic, is a strange result.

We will see later in the Chapter how the equation solution for ACS (I), which is the sub-alternate of ACS (A), actually transitions.

At this preliminary stage, it could be thought that the difference in transitions is merely an artefact of a classification operating by unreasoned fiat.

We will see that is not so, when we come to consider ML equation transitions of categorical syllogisms.

We can perhaps best begin the exploration by considering ACS (I), which effects the most simple ML transition.

I Some A is B (A) ≠ (nA) = B[E] Reflexive Opposition

The possibility matrix allows:
Some B are A
All B being A is a possibility
All A being B is a possibility
Some A are not B is a possibility
Some B are not A is a possibility

These possibilities all occupy the disjunct space between subject and predicate in ACS (I). It can be said that the ML equation "maps that space". It is *matrixial.*

It is immediately attractive to consider that we are entering the terrain of Leibniz's "Intensional" Algebra of Concepts.

Although the genius of Leibniz secures respect,[122] we most quickly and obviously part company at the theorem:

> "*Theor. 6.* Universalis Affirmativa et Universalis negativa sibi opponuntur contrari`e".

Leibniz's work, is worthy of respect for what it does teach us. However, the deliberate precision (in context) of the formulation *sibi opponuntur*[123], gives rise to no doubt that his logic was not Matrixial. We will return to Leibniz later in the chapter.

[122] And noting Leibnitz' influence on Boole: see *Forms of* Inequality Chapter
[123] Oppose themselves; in themselves are opposed

The Particular Affirmative in Inequality

The engaged reader will have noted that ACS (I), the Particular Affirmative: *some (A) is (B),* produces in ML transition:

(A) ≠ (nA) = B[E] Reflexive Opposition

To return to the rather gnomic statement under the Discussion rubric:

2. That which limits (a) is the Form, rather than categorical iteration ($-a^n$).

We are dealing with the categorical statement (I):

Some A is B.

We ask: what is the *Frame of Reference* of this proposition? Let us examine the elements. In doing so, we will need to take care with our existential vocabulary. Doing so produces rather ungainly combinations of words.

The reader's perspective on what follows should be greatly enhanced by the time that the reader has digested the *Languini* Chapter. The reader is invited to re-read the following section: what appears obscure now, should then appear illuminating by its self-evident quality.

For clarity of writing only, and inferring nothing from the denotation, let us write:

A as (A) B as (B) is as [is]

Now proceeding:

[1]

Hypothesise (A):

(1) If we wish to predicate anything of an (A), then hypothesis (nA) or (-A) is necessitated.

(2) We wish to predicate of (A) *existence*: that (A) [is].

(3) hypothesis (nA) or (-A) is necessitated.

Objection: Be it contended that premise (1) is a false, or undemonstrated premise.

Response:

[A]

a. If (A) can have the property or predicate[124] of existence without any frame of reference, then by what means is appropriated in thought:[125] (i) the (A); or (ii) the idea (A).

b. That rhetorical question is answered thus: by no means can (A) be appropriated in thought if there be no FoR.

c. It then be contended that such FoR is furnished in the act of thought itself.

d. We agree.[126]

[B]

e. To think a thought entails not any requirement that such thought be referenced to any that which is not thought.

f. Insofar as we can conceive of entirely empty thought, that is agreed. However, insofar as thought is of or about, or in any relation to any content of though, then, we return to proposition [A](c/d).

[C]

g. To contend that the thinker and the thought (which has some content) relate in hylomorphic isolation[127] is self-contradictory, since that contention renders both thinker and thought subject to a frame of reference.

h. It be contended that the referent thinker can serve as (B) .

[124] If the reader considers those to be materially different in this context

[125] If the reader considers (i) and (ii) to be materially different in this context

[126] If this be not agreed, then by what description or analysis is it stated that one can think a thought without thinking it?

[127] As its own "self-contained" frame of reference

i. Answered: let it be assumed[128] that this be so. Then, by what manner of analysis can one propose "Some (A) is (B), following extraction of (A) from (B)?

j. That rhetorical question is answered thus: in accordance with the reader's own postulates: by no means.

[D]

k. It now be contended that (A) can be appropriated in thought without reference to (nA) or (-A).

l. It be responded, then: by what means of ideation is (A) extracted from the universe, or the void?[129]

m. That rhetorical question is answered thus: by no means, without simultaneously ideating the reference of (A) as (nA) or (-A).

n. It now be contended that the universe or void can serve as referent.

o. We agree. [130] Such universe or void is, by the reader's own admission, (nA) or (-A).[131] We will see practical demonstration of this rubric in the *Languini* Chapter.

[E]

p. It be contended that the referent universe or void can serve as (B) *ipso solo.*

q. Answered: let it be assumed[132] that this be so. Then, by what manner of analysis can one propose "Some (A) is (B), following extraction of (A) from (B)?

r. That rhetorical question is answered thus: in accordance with the reader's own postulates: by no means.

[F]

s. It be contended that the referent (B) can serve as the referent for (A).

t. Answered: let it be assumed[133] that this be so. Then, by what manner of analysis can one propose "Some (A) is (B)", when the hypothesis of (A) is a priori premises upon (B)?

---

[128] Only for present purposes

[129] If the reader considers those to be materially different in this context

[130] Only for present purposes

[131] If the nomenclature be not assented to, let the phrase "other than (A)" serve arbitrally

[132] Only for present purposes

[133] Only for present purposes

u. That rhetorical question is answered thus: in accordance with the reader's own postulates: by no means.

[G]

v. By what means, then, can "is" be postulated of (A) if there be no state of affairs in which (A) is not?

w. If it be contended that "is" can serve as the frame of reference for (A), that is to repeat the contention already thrice advanced and rebutted.

[H]

x. It is thus demonstrated that: (A) without existence cannot be comprised in thought.

y. No ideation of any frame of reference for (A) is possible, other than (A)'s opposition (nA), or negation (-A)

z. Q.E.D. If we wish to predicate anything of an (A), then hypothesis (nA) or (-A) is necessitated.

[2]

Hypothesise (B):

We have established that:

the opposition (nA) or negation (-A) of (A) is necessarily called to thought by thinking [*(A) is*]

(1) (A) ≠ [(nA) or (-A)], is potential to a Frame of Reference

(2) (B) is called to be hypothesised in some relation to (A)

(3) Hypothesised (B) provides a Frame of Reference in respect of [(A) /(nA) /(-A)].

Objection: Be it contended that (3) is false, or undemonstrated.

Response:

[A]

a. It be contended that (B) could call forth its own referents [(nB) / (-B)].

b. Agreed. That is not an objection to the Conclusion.

[B]

c. Thought could, in principle, call forth an infinity of (a), (B), $(x)^n$.

d. Agreed.That is not an objection to the Conclusion.

[C]

e. The premise of Objections [A] and [B] is that (A) is not in the domain of the instruction to hypothesise (B). That premise is false.

f. Accordingly, the hypothetically infinite domain of self-referent[134] thoughts is delimited by the referent in respect of which instruction is given: that being the instruction which calls forth[135] the hypothecate (B)

[3]

Hypothesise Selection Amongst a Plurality of $(A)^n$ as a function of (B):

(1) Hypothesised (B) provides a Frame of Reference in respect of [(A) / (nA) /(-A)].

(2) Bounded plurality ("some") precludes infinity.

(3) The Frame of Reference in (B) is of [(A) / (nA)].

Objection: Be it contended that (2) is false, or undemonstrated..

Response:

[A]

a. Infinity cannot bind any plurality of it sown iterations.

b. That which cannot bound without cannot bound within = tautology

c. Bounded plurality ("some") precludes infinity

[4]

The opposite of (A) "is" (nA):

(1) Bounded plurality ("some") precludes infinity

(2) A is not the opposite of A: tautology.[136]

(3) Only Substance [(A) ≠ (nA)] presents otherwise than under infinity ∞.

Objection: Be it contended that (3) is false, or undemonstrated..

Response:

[A]

a. Let is be proposed that some other phenomenon (z) function as the opposite of (A). You may designate (z) as

[134] In opposition or negation

[135] In this case

[136] But nor is A identical to A

just that, or elephant, or by any denomination.

b. Yet any such as you have chosen *functions* in opposition to (A).

c. Such function is arbitrary.

d. Such function is indiscriminate as to (A).

e. Moreover, you seek to violate the condition of existence of (A): that (A) not be a singularity.

f. Thus, you propose an infinity ∞ of singularities.

g. Therefore, of (A), that which is hypothesised as other than (^{NOT}A) constitutes as ∞

[B]

h. As previously stated:

For discriminatory accuracy, it is entirely plausible to write:

$$(A^n) \neq (nA^n) = B[E]$$

i. The ^{n} serves to denote plurality "*Some*" of that which serves as the same Substance in Frame of Reference B.

[C]

j. As a cautionary note, were "*Some*" to be replaced by a definite number (an integer), then the contingency in E would "collapse" into an infinity ∞ bounded by [I].[137]

[5]

Plurality of $(A)^n$ subsists only as (A) ≠ (nA):

(1) Of (A), that which is hypothesised as other than (^{NOT}A) constitutes as Moment in [I]

(2) Only Substance [(A) ≠ (nA)] presents otherwise than under [I].

(3) To predicate [is] of (A) is necessarily to predicate (nA).

Objection: Be it contended that (3) is false, or undemonstrated.

Response:

[A]

a. It was demonstrated that, that which is (A) in plurality cannot subsist in an infinity of iterated $(A)^n$.

b. Ergo, that which is "some" must consist of non-iterations of (A).

[B]

---

[137] Refer to Proposition [3]; and see later in this Chapter

c. That which is "some" must oppose that which is not "some".

d. Ergo [(A) ≠ (nA] in the domain "some".

[C]

e. Were no (B) hypothesised, then some other frame of reference would be needed such that "some" (A) could exist.

[6]

(A) ≠ (nA):

(1) The notation ≠ functions as affirmation of inequivalence of one side of the symbol to the other.

(2) The notation ≠ functions as denial of inequivalence of one side of the symbol to the other.

(3) Any other symbol (not otherwise in use in ML) could serve in place of this symbol.

(4) The functions of such symbol would remain the same.

(5) It is not thought that these propositions, being arbitrary denomial choices, require proof, by syllogism, or otherwise.

[7]

"Some" is undefined:

(1) The meaning of "some" is: that plurality which is not bound by numerical quantification nor limit.

(2) It is not thought that this proposition requires proof, by syllogism, or otherwise.

[8]

(B) functions as frame of reference

To predicate [is] of (A) is necessarily to predicate (nA).

(1) [(A) ≠ (nA)] requires a Frame of Reference.

(2) B has been mandated as referent to discrimination amongst $[(A) \neq (nA)]^n$.

(3) (B) provides a Frame of Reference for $[(A) \neq (nA)]^n$.

[9]

(B) functions as frame of reference for discrimination between $[(A) \neq (nA)]^n$:

(1) [Some] $[(A) \neq (nA)]^n$ is without discrimination.

(2) The direction that [Some] $[(A) \neq (nA)]^n$ be of B introduces a discrimination function amongst that [Some].

(3) (A) ≠ (nA) = B [E]

[10]
[Some A is B] ≈ [(A) ≠ (nA) =B]

It is admitted to be beyond the capacity of the Author to:
(1) hypothecate every conceivable objection
(2) answer each objection hypothecated above such that no further objection can be hypothecated.

The course of any public discussion of these presents will afford ample opportunity to rectify these inherent limitations of this communication medium and method.

The = Function
It is necessary to adopt a somewhat discursive approach to discussion of this subject.

In classical, predicate and propositional logic, and in arithmetic and algebra, the = symbol functions as the conclusion of an *argument*.

Putting the matter another way, the function is a statement of equivalence or indeed identity.

Giving the reader a break from the antagonisms, equations and syllogisms, let us take a historical moment:[138]

[138] https://en.wikipedia.org/wiki/Equals_sign

The etymology of the word "equal" is from the Latin word "*æqualis»* as meaning «uniform», «identical», or «equal», from *aequus* («level», «even», or «just»).

14.ᴢ.—+—.15.ꝭ═══════71.ꝭ.

The first use of an equals sign, equivalent to $14x+15=71$ in modern notation. From The Whetstone of Witte (1557) by Robert Recorde.

Howbeit, for easie alteratiõ of *equations*. I will propounde a fewe exãples, bicause the extraction of their rootes, maie the more aptly bee wroughte. And to auoide the tediouse repetition of these woordes : is equalle to : I will sette as I doe often in woorke use, a paire of paralleles, or Gemowe lines of one lengthe, thus: ═══, bicause noe .2. thynges, can be moare equalle. And now marke these nombers.

Recorde's introduction of "="

The = symbol, now universally accepted in mathematics for equality, was first recorded by Welsh mathematician Robert Recorde in *The Whetstone of Witte* (1557). The original form of the symbol was much wider than the present form. In his book Recorde explains his design of the "Gemowe lines" (meaning *twin* lines, from the Latin *gemellus*[1]

> *And to auoide the tediou e repetition of the e woordes : is equalle to : I will ette as I doe often in woorke v e, a paire of paralleles, or Gemowe lines of one lengthe, thus: =, bicau e noe .2. thynges, can be moare equalle.*

> *And to avoid the tedious repetition of these words: is equal to: I will set as I do often in work use, a pair of parallels, or Gemowe lines of one length, thus: =, because no 2 things, can be more equal.*

"The symbol = was not immediately popular. The symbol || was used by some and æ (or œ), from the Latin word *aequalis* meaning equal, was widely used into the 1700s" (*History of Mathematics*, University of St Andrews).

In classical and contemporary logic[139] Aequalis has a passive function. Aequalis is the road sign at the end of journey, signifying destination.

In such as in $x + y = z$: the aequalis functions merely as the conjunction of the 2 sides. It is the symbolic expression of tautology.

Thus, in:

$$x + y = z$$

$$>$$

$$z - y = x$$

the aequalis is demonstrating a conjunction relation between assortments of these 3 iterations, under rules.

> As a marginal note, the attentive reader may be thinking that [=z] is there functioning as the form of an inequality. It is indeed, but that algebraic equation is actually a truncated expression of:
>
> $$(x) \neq (-x) = [I]_1$$
>
> $$\Sigma\ ([I]_1 \neq [I]_2) \neq (\textstyle\int +) = z\ [I]$$
>
> $$(y) \neq (-y) = [I]_2$$

In Matrixial Logic, = does not act merely as conjunction. Rather = has an active function:

- the disjunct [(A) ≠ (nA)] or [(A) ≠ (-A)], radiates a field of

[139] Mathematics and so on

potential FoRs

- the [=E] or [=I] function collapses that potential into a specific FoR

A way of stating this, which captures some of the reality of the matter, is that the = [E] / [I] is engaging an FoR, relative to our perception.

Beyond Equals

Experience: Roomed

Step 1:

Imagine an empty room, with uniformly blank surfaces, around the size of a squash court:

- Consider what you "see" in that empty room.[140]
  You see the walls, floor, ceiling.

Now:

- Focus, and try to see the "space" in, filling the room.

Stop.

Discussion:

(1) As you intensify the focus, you begin to invent imaginary flecks of dust, or atoms: seeking to invest the emptiness with some substance(s).

Step 2:

[140] Say a squash court, without markings

Now, imagine a line, like a laser light in museum security system. It must cross the room, in a straight line

- Now imagine 2 lines;
- Now multiply the lines;
- Keep multiplying them.

Stop.

Discussion:

(1) You can do the first few lines, easily.

(2) But, as you multiply the lines, eventually they start to bend: they become spaghetti.

(3) If you kept going long enough with the multiplication, you find the light lines turning into a homogeneous liquid, just filling up the room space.

(4) That works for a while, but as the lines radiate in a *field of potential FoRs,* they become jumbled and entangled, and eventually lose their coherence completely.

Step 3:

Now, imagine a ball in the room.[141] A basketball size ball.

- Now, project a laser light out from the ball;
- Keep projecting more laser lights from the ball.

Stop.

---

[141] Floating free, or located anywhere

Discussion:

(1) Now it is easy.

(2) You picture lines, connecting the ball to the walls, ceiling and so on.

(3) The lines stay straight. They remain coherent.

(4) After a while it is like the ball just will not let you insert any more lines.

Step 4:

Now, imagine a single atom in the room. A tiny point, which is visible only to your mind's eye.

- Now, project a laser light out from the atom;
- Keep projecting more laser lights from the atom.

Stop.

Discussion:

(1) This works just like Step 3.

(2) It seems to make no difference what size the object is: just so long as there is an A (the object) and a B (the room).

Conclusions:

1. The empty room is blank canvas.[142] The only FoR is the bounded room space.
2. There is a *potential intersection* of FoR's, but the potential is unfulfilled.

[142] Please mark this idea: we will return to it

3. Coherent 'light' cannot maintain its coherence, when it is merely a product of that single FoR. In other words, coherence cannot sustain in unbounded infinity ∞.
4. The introduction of the Ball (that which is not empty space), *collapses that potential into a specific intersection of FoRs.*
5. Now ∞ becomes bounded by the intersection of FoR's.
6. When you use the atom as the "anchor" for this 2nd FoR, that works too. In fact, you find that populating the room with coherent, straight laser lights to be the easiest exercise.
7. The atomic laser light shows that the FoR intersection which bounds ∞ is not a box, or container: it is a *point*. It could be the tiniest imaginable point in the quantum universe. So long as it is there.[143]

Suppose Aristotle now wished you, in this fact-specific context, to give effect to Categorical Sentence (I): *Some A is B*.

What would you do?

You would, of course add Balls to the RoomSpace [E].

Other ways of articulating the [=E] or [=I] function:

[143] Yes, we have just meta-explained the "big bang"

(1) Graphically:

$$(A) \neq (nA)\ [= E]$$

(2) Matrixially:

(A)	≠ ∫ [= E]	(nA)

But (3) What you cannot[144] write is:

$$(A) = [= E] = (nA)$$

The Roomspace clearly does not *aequalis* the Ball or the nBall.

Yet another manner of articulating the [=E] or [=I] function is as follows:

- Aequalis flattens: reduces to a unidimensional plane;
- [=E] or [=I] matriculates: expresses a multidimensional dynamic.

The aequalis is a device which, when used other than as matriculation, renders potentially unsound any cumulation of functions.

That is not so in a single plane of reference, such as arithmetic, or geometry. There, each integer and

[144] Operating the rules of ML, or indeed of classical logic

subdivision operates under a rule of infinity ∞, from which is abstracted for computational convenience, the formulae of iteration.[145]

Matrixial TransFormation

Back to that Room. We stated that:

> The FoR intersection which bounds ∞ is not a box, or container: it is a *point*.

This provides us with our first exemplar:

Matrixial transFormation ("MtF")

$\underline{(A) \neq (nA) = E}$

$>$

(Atom) ≠ (nAtom) = RoomSpace [E]

≠ [~] $\sum FoR_2 = [I]$

(Light) ≠ (nLight) = RoomSpace [E]

$\geq$

[(Atom^{NOT}Light)] ≠ [- (Light^{NOT}Atom)] = Point [I]

$>$

$\underline{(A) \neq (-A) = I}$

Notation:

We have explained the equivalence of n≈^{NOT}

[145] We will see much more about this in Chapter 4

Symbol	**Meaning / Use**
These are symbols used in ML TransFormation	
[~]	Intersection: indicates a transforming point between 2 frames of reference.
∑	In logic and mathematics, this sign usually means "sum". In ML, it means "aggregation"; "taken together as". Its use with [~] is almost redundant. However, when we are typing lines of ML equation,s the ∑ figure is a useful "reminder" that we are now dealing with combined FoRs.
FoR_2	The subscript shows us the number of Forms that we are aggregating. The subscript number can be anything from $_2$ to $_n$

Discussion:

(1) There will be much more to say about Matrixial transFormation (MtF) in this and later Chapters.

(2) In this MtF, we see how 2 types of Substance in E (being) become transformed into 2 Moments in =I (becoming).

This gives us some ML Principles:

(1) Substances relate to Spacetime ◊.

(2) Moments relate to Infinity ∞.

(3) ◊ Spacetime is Actuality.

(4) ∞ Infinity is Potential.

(5) An intersection of: ∑ ◊ [~] ◊; > produces ≥ a point in ∞ Infinity.

(6) An ∞ Infinity point collapses Potential.

What is missing from this account is the *plenum* in which such interaction of [E] Forms can have actual realisation:

contrasted with mere equations written on a page.

We will come to that in the next Chapter. There, we will discover how this aggregate figure of Forms becomes comprised in a ***Node***.

Actual, Potential and Infinity

A paradigm of infinity is mathematics: the endless series of integers and the space between integers and the ideas of what fills those spaces.

The work of Cantor[146] showed us: $\sum FoR_2$ (◊ [~] ◊) > Point [I]. It demonstrated how we can create sets from infinity, and thus bound infinity.

What Cantor, and following him, Russell[147] could "see, but not observe"[148] was the dimorphic character of FoRs.

Scene and Elements

In our classical way of thinking about the Room, Ball/ Atom and Lights ("the Elements" and "the Scene"), we would apply number theory.

We would treat each of these Elements being manifestations or expressions in a single dimension: thus to count them. This is not "wrong". As a matter of physics, and common

---

146 (1845-1918)

147 And Whitehead

148 Sherlock Holmes again: *A Scandal in Bohemia*. Conan-Doyle (1896)

sense, were that Scene to be observed in real life, we could measure each Element of the Scene.

We could apply, or derive formulae from Scene Elements and their interactions. Indeed, we can apply, or derive from this Scene formulae of:

- Newton's laws of motion[149]
- General relativity
- Quantum mechanics

A Scene could hardly be more "real" in spacetime than that.

The above point is made so as to disassociate the reader from any idea that ML is only referable to what is going on inside the reader's head.

The reader requested to take it on trust that what the reader Experiences in the Scene is the same as any other reader. By way of proof from the negative, were that not so, then each reader would, at this point, be holding a different book.

The use of the world "Model" is ambiguous. It also raises a host of questions about the means of model construction and external reference.

That said, may we pray in aid the use of that word, temporarily, as a step on the ladder of explanation.

---

[149] If we allow the ball / atom to move or bounce, or we allowed 2 balls/ atoms

Let us, therefore, temporarily, assume a 1-1 correspondence between the thoughts experienced by the reader,[150] and Experience Model: and "reality" in the Scene out there.

Unidimensional Scenarios

We would usually treat all Elements of the Scene as being part of a physically whole reality. We are used to thinking of the Elements as being extrusions from an underlying reality of potential.

There are good reasons for this. They are explored in detail in the following Chapters.

ML Transitions of Categorical Syllogisms

Using the Categorical Syllogisms allows us to compare transitions into ML equations. These provide further insights into the internal logic of the rules of ML equations.

The Author wishes to acknowledge to the reader indebtedness to Terence Parsons, for his scholarship.[151]

This next section applies transitions of *valid categorical syllogisms,*[152] gratefully extracted from his ground-breaking work on medieval logical scholarship.

---

[150] "In" mind or brain: whichever you prefer

[151] *Articulating Medieval Logic* Parsons, Terence (2014)

[152] Op.cit p. 16

Aristotle Categorical Sentence Transitions to ML

	C S	ML Equation	Type
A	Every A is B	(A) ≠ (-A) = B[I]	Categorical Universal
E	No A is B	(A) ≠ (B) = N[E]	Reflexive Universal
I	Some A is B	(A) ≠ (nA) = B[E]	Reflexive Opposition
O	Some A is Not B	(A) ≠ (nA) = B/^{n}B	Unreflexive Opposition

We have now seen a demonstration of the above ACS transitions to ML equations.

We now examine transitions of Aristotelian Categorical Syllogisms ("ACSY").

ACSY Barbara

We take the first figure and mood Barbara: the syllogism of 3 universal affirmatives (In ML "Categorical Universals"):

Figure 1 (ch. 4) Barbara (26a1)	ML Equation
Every M is P	(M) ≠ (-M) = P[I]
Every S is M	(S) ≠ (-S) = M[I]
∴ Every S is P	(S) ≠ (-S) = P[I]

Table 3.3

We then iterate:

1. The ACSY in tabular form
2. The ML transitions
3. A Hybrid of both.

The Hybrid is intended to illustrate parallax cross-reference of both domains of concepts, as a guide to understanding. Here we use: Soldier \ Man \Pink

	ARISTOTLE					
P1				Su/E-Man	is	Pr / Pink
P2		Su/E-Soldier	is	Pr / Manly	is	Pr / Pink
C	∴	Su/E-Soldier	is	Su/E-Man	is	Pr / Pink
	R	1	2	3	4	5

	MATRIXIAL LOGIC					
P1				(M) ≠ (-M)	=	P[I]
P2		(S) ≠ (-S)	=	M[I]	=	P[I]
C	∴	(S) ≠ (-S)	=	(M) ≠ (-M)	=	P[I]
	R	1	2	3	4	5

	HYBRID					
P1				Every Iteration of (Man) = Manly [Form]	is	Pink [I]
P2		Every Iteration of (S)	is	Manly [Form]	is	Pink [I]
C	∴	Every Iteration of (S)	is	Every Iteration of (Man) = Manly [Form]	is	Pink [I]
	R	1	2	3	4	5

Table 3.4

The orange typeface in Hybrid shows the implicit articulation of:

(1) P1/R4 in P2/R4

(2) P1/R5 in P2/R5

(3) P1/R1 in C/R3

(4) P1/R4 in C/R4

It is notable that such implicit articulations belong to the original ACSY.

This is the table with those implicit articulations removed, and the blanks numbered in co-ordination with the numbered list above:

	MATRIXIAL LOGIC					
P1				(M) ≠ (-M)	=	P[I]
P2		(S) ≠ (-S)	=	M[I]	(1)	(2)
C		(S) ≠ (-S)	=	(3)	(4)	P[I]
	R	1	2	3	4	5

Table 3.5

What may be obvious, is that ML equations immediately reveal elision from Subject (P1 / R3) to Predicate (P2 / R3).

In the ordinary language Greek of the original composition, it was by no means necessarily obvious from any particular word whether in assumed place as Subject or Predicate, and in like manner, in scholastic Latin.[153]

The reduction of words to symbols in First-Order Logic ("FOL") has made that original opacity the more obscure. Indeed, FOL continues the paradigm introduced by Leibnitz and Lambert[154] of seeking to empty Subject and

---

[153] Not being a Greek scholar, my reading of Aristotle is exclusively in English and bad Latin: so I must take this by acceptance of authority

[154] See Chapter 1

Predicate of content, so as to be able to manipulate them under a single dimension of articulation. By contrast, ML equations, being hylomorphic,[155] inevitably reveal the FoR in which a content and form is operational, thus identifying the function within an equation of each term.

Implicit Domain Recognition

This almost trivial transition exercise reveals something important about the Barbara ACSY, and, by extension, all ACSY: the *implicit domain.*

> Every M is P
> Every S is M
> ∴ Every S is P

So far, so 2,300 years ago. Now, notice the change in M from S subject (in Major Premise) to P predicate (in Minor Premise).

The Barbara ASCY looks as if it has this linear form:

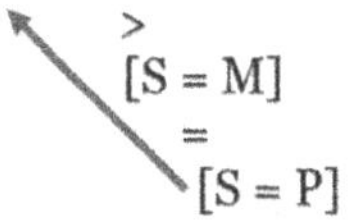

>
[S = M]
=
[S = P]

or:

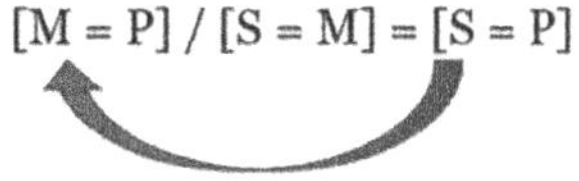

[M = P] / [S = M] = [S = P]

[155] See *Framework* Chapter

Indeed, it does, in a dimension where there is no distinction drawn between a concept having *undifferentiated potential both as S and P*.

Yet the question then arises: what are the conceptual[156] requirements of a term which can function with *undifferentiated potential both as S and P*?

That question leads in succession to asking whether there exist concepts which do not have *undifferentiated potential*.

Perhaps there are concepts:

- Which can only fulfil an S term
- Which can only fulfil a P term
- Which can fulfil either, but only under conditions[157]

We will return to this important matter.

Now, suppose we take a classic verbalisation of Barbara:

P1	Every Man	is	Mortal
P2	Every Soldier	is	a Man[158]
C	Every Soldier	is	Mortal

The rules of logic, accepted for 2.3 millenia, tell us that this syllogism is correct. Any argument which follows this form is true, subject only to the premises having factual truth value. In Barbara, there remain the same:

[156] Or linguistic, if that is an important distinction here

[157] And if so, of what consist such conditions

[158] Obviously not factually true; but an irrelevant datum here

- P1/P and C/P
- P2/S and CS

In other words, there is no change in domain of S Soldier and P Mortal. Each is acting as an [I] Form of iterated contents.[159]

Now contrast Man, which acts both as:

- S Man in P1
- P Man in P2

In ordinary language, and in language substitution by FOL, the *undifferentiated potential both as S and P,* of Man, is not a matter of interest.

This matter is obscured by the functional equivalence, in ordinary language, of these alternate terms:

P1	Every Man	is	Mortal
	All Men	are	
	Men		are
P2	Every Soldier	is	a Man
	All Soldiers	are	Men
C ∴	Every Soldier	is	Mortal
	All Soldiers	are	

Since the universal necessarily includes the plural form of

159 Tautology

the singular denomination of that universal, the meaning of the term, in syllogistic syntax, is unaltered by such alternative usage.

It is to be noted that we are not here making any claim that such alternative terms usage is incorrect. It is a different point: the syntactical structure, with its equivalent alternates, further obscures the *different implicit domains inherent in the syllogism.*

The ML equation for a Categorical Universal (a Universal Affirmative, in ASC) is:

$$(A) \neq (-A) = B[I]$$

We stated of the Categorical Universal that:

(A) (in plurality of iteration) [≠ (-A^n)] is stated to hold iteration; as (B), which is a non-iterative term.	
[P1]	
(1)	Any FoR of iterate plurality must be in I.
(2)	No (A) or [≠ (-A^n)] can function as Form.
(3)	Ergo: B must solve as Form in [I]
	[P2]
(1)	(A) in iteration (-A^n) is stated to limit B
(2)	Ergo: B cannot:
(2.1)	iterate (A) ; (2.2) iterate (-A^n)
(3)	Ergo: B must function as domain of (A) and (-A^n)

We now return to answer the inquiry:

Perhaps there are concepts:

- Which can only fulfil an S term
- Which can only fulfil a P term
- Which can fulfil either, but only under conditions

What we find is that the answer lies not in any particular concept or word. It is in the framework in which the concept or word is used.

Man and Men are being used syllogistically as unidemensional actors. What that use disguises, is that there are two completely different framework concepts in use here:

- M as Substance;
- M as Node.

M is being used multi-dimensionally. The syllogism works only because one use of M crosses the matrix.[160]

Let's look at Table 3.5 again:

	MATRIXIAL LOGIC					
P1				(M) ≠ (-M)	=	P[I]
P2		(S) ≠ (-S)	=	M[I]	(1)	(2)
C		(S) ≠ (-S)	=	(3)	(4)	P[I]
	R	1	2	3	4	5

Table 3.5

The (S) term is conditioned, placed in a framework, by its two expressions of form in [I]: as Man and Pink.

The term (P) is conditioned by its action as form of inequality of iterations Δ^n.

---

[160] See Chapter 5

But here, we have a problem:

P1	(M) ≠ (-M)
P2	M[I]

We can presume that (M) ≠ (-M) because we think we already know that the form is [Men] [I].

But there is no necessary reason that this be so. Socrates is a man \ Plato is a man. This in itself does not give us any categorial form "Men". We could be saying: Socrates is a man \ Not a donkey.

The clarity provided by Matrixial Logic shows us that (M) is not a primary premise. Indeed, it is not a premise at all.

It is supposedly premised that: <All Men are Pink>

But we do not yet know that there is a form of <Men>. We attain that form only through the second premise expression: <All Soldiers are Men>. But that is a categorisation of <Soldier>, not <Men>.

In the major premise, <Men> is functioning arbitrarily. It is only in the minor or second premise that <Men> acquires a Form as <Man> in [I].

Yet, we are obliged to read that secondarily acquired Form back into the first premise, in order that the first premise function in a logical statement at all.

The first premise cannot in logic properly function as a major premise at all. We need a pre-premise which specifies <Men> as the iteration of a universal <Man>. We need to create (M) either in opposition (nM) or negation (-M).

Until we have done that, the term <Men> is empty. The first premise should properly read:

<Some undefined things> are <Pink>

Once we notice this, the Barbara syllogism fails. It becomes a mere collection of assertions, which do not have mutual support in a unidimensional framework.

At the core of ACSY is the use of an undefined, and indeed empty set: <Some undefined things> are <Pink> .

The necessary consequence is that any attempt to utilise syllogistic reasoning is riven by inherent paradox. For readers unfamiliar with the history of western logic, let it be explained that this is a problem equivalent to finding out that information can travel faster than the speed of light.[161]

For those brought up in Aristotelian logic, this is all quite disturbing. The familiar becomes foreign. We realise, with a crash, that there is a deep fault line in Barbara.

But if Barbara is flawed, then the entire edifice of

[161] The physics community is still struggling, understandably, with this one

Aristotelian logic in the ACSY, is flawed.

The entirety of the western logical philosophy project[162] collapses.

Leibnitz Intensional Interpretation

We must first recognise the great debt of gratitude owed, not only to Leibnitz himself, but to his modern emmanuensis Wolfgang Lenzen. Without his seminal 2016 paper,[163] the labours of this brief critique would have been rendered longer, and less well-formed.

We confess to some ambivalence as to the significance of Leibnitz in the foundational landscape of ML. We might perhaps offer a metaphorical note:

> in complex narrows of alleyways, constructed in the epoch of mighty monarchs, upon foundations of the ancients, Herr Leibnitz devised an umbrella, *intensioned* to guard one from the unwelcome conjunction of water and the law of gravity. The apparatus served less function than hoped: the spokes being uncanvassed, and the instruction manual dictating an inverted clasp of the ferule.

To put it another way, Leibnitz was right in the wrong way, and wrong in the right way.

[162] And the Eastern, which uses a 5 term syllogism: where 2 of the terms iterate 2 of the other 3 terms

[163] Lenzen, W. Leibniz's Logic and the "Cube of Opposition". *Log. Univers.* 10, 171–189 (2016). https://doi.org/10.1007/s11787-016-0143-2

Part Upon the Square

We begin, of course, with the familiar Square of Opposition:

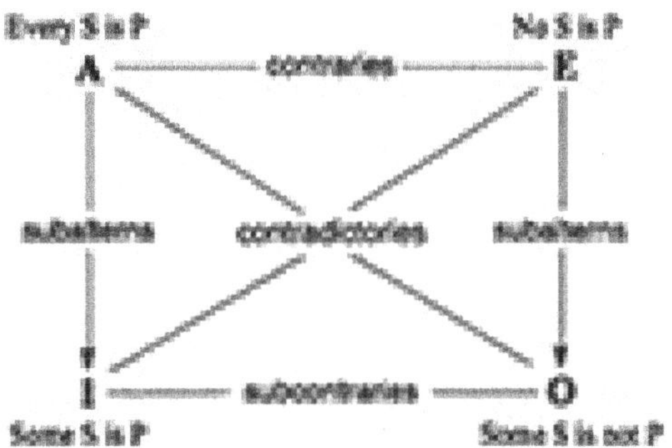

As Lenzen states:

> Third, Leibniz admits the unrestricted use of conceptual negation. In contrast to propositional negation, ¬ , concept negation ("Not-A") is formalized as ~A. Hence, in particular, one can form both the inconsistent concept A~A ("A Not-A") and its "tautological" counterpart ~(A~A) ("Not-(ANot-A)".

We may contrast Aristotle:[164]

> The proposed categories have, then, been adequately dealt with. We must next explain the various senses in which the term 'opposite' is used. Things are said to be opposed in four senses: (i) as correlatives to one another, (ii) as contraries to one another, (iii) as privatives to positives, (iv) as affirmatives to negatives.
>
> Let me sketch my meaning in outline. An instance of the use of the word 'opposite' with reference to correlatives is afforded by the expressions 'double' and 'half'; with reference to contraries by 'bad' and 'good'. Opposites in the sense of 'privatives' and 'positives' are' blindness' and 'sight'; in the sense of affirmatives and negatives, the propositions 'he

[164] *Categories* 10, Logic (Organon). Trans EM Edgehill

> sits', 'he does not sit'.

In the Aristotelian lexicon, therefore, the "contraries" of Square of Opposition are Class (iv) "opposites": *affirmatives to negatives.*

> But in the case of affirmation and negation, whether the subject exists or not, one is always false and the other true. For manifestly, if Socrates exists, one of the two propositions 'Socrates is ill', 'Socrates is not ill', is true, and the other false. This is likewise the case if he does not exist; for if he does not exist, to say that he is ill is false, to say that he is not ill is true.
>
> Thus it is in the case of those opposites only, which are opposite in the sense in which the term is used with reference to affirmation and negation, that the rule holds good, that one of the pair must be true and the other false.

Now, disregarding for the moment the Russell problematica[165] of *non-referring expressions* (negative existentials), we already see in Aristotle the non-hylomorphic conception of Class (iv) opposition.[166]

It can thus be seen that the seed-germ of the Leibnitzian concept of *intensional opposition* is contained here. At this point, it is superfluous labour to do other than cite Lenzen further:

> As regards the relation of conceptual containment, A ∈ B, it is important to observe that Leibniz's formulation 'A contains B' expresses the so-called *intensional* view of concepts as *ideas*, while we here want to develop an *extensional* interpretation in terms of *sets of individuals*, *viz*. the sets of all indi viduals that fall under the concepts A and B, respectively.

[165] *On Denoting* (1905), and its proposed solutio

[166] It accepted, of coursem that Aristotle's Logic is generally considered to proceed as a paradigm of hylomorphism

> Leibniz explained the mutual relationship between the "intensional" and the extensional point of view in the following passage of the *New Essays on Human understanding*:
>
> > "The common manner of statement concerns individuals, whereas Aristotle's refers rather to ideas or universals. For when I say *Every man is an animal* I mean that all the men are included amongst all the animals; but at the same time I mean that the idea of animal is included in the idea of man. 'Animal' comprises more individuals than 'man' does, but 'man' comprises more ideas or more attributes: one has more instances, the other more degrees of reality; one has the greater extension, the other the greater intension" (Book IV, ch. XVII; cf. [1] VI, 1, p. 486; my translation).
>
> ... For Leibniz, the *extension* of a predicate A is not just the set of all *existing* individuals that (happen to) fall under concept A, but rather the set of all *possible* individuals that have that property.

From the Matrixial Logic perspective, this juxtaposition of the intensional and etensional qualities of things as ideas, is both tantalising and frustrating.

At this point in the exposition, we offer a *Conceptual Short-Cut*:

The problematica of *non-referring expressions* is itself an "empty set". That is to say, the problem itself arises only from the attempt to interpret subject/predicate relations in a framework of non-hylomorphic extension.

The attempt, that is, to construct valid forms of reasoning unidimensionally: which results in the foundational paradox considered in the last section.

We see the illusory problem arising in the 13th century, with

William of Shyreswood's syncategorematic treatment of 'is' as equivocal, in that it can indicate either: (1) actual being (*esse actuale*); or (2) habitual being (*esse habituale*).[167]

Shyreswood's syncategorematic project has at least the merit of duo-dimensional thought.[168] That non-hylomorphic extension framework itself arises from the attempt to render Class (iv) Opposition as a uni-dimensional set of linear co-ordinates: which is the Boolean project, as systematised by Venn.

That uni-dimensionality arises as a functional necessity of seeking to place opposition relationships under a single arithmetic plane of summed equality, reflexive of a superstructural assumption of equality.

Now, to be fair, that assumption is justified by the Aristotelian advocacy of *intensional equality*.[169]

In reading Leibnitz, we should be cautious in attributing to his calculus a foundational *inequality*.

---

[167] c.1240. As a biographical delight, William was treasurer of Lincoln Cathedral in 1254. Just missing Boole's Pottergate house by 600 years, and around 100 yards from the Chapter House (construction commenced 1220)

[168] Characteristic of the Mythic: see *Modern Man In Search Of A Soul. Jung* 1933

[169] As a marginal note, it is perhaps striking how *modern* is the manner in which Aristotle treats of intensional equality, contrasted with the 3 other categories of opposition

On the contrary, the Leibnitzian project,[170] was to rescue subject/predicate relations from the medieval scholastic confusion immanent in the individualisation (and particularisation) of terms.[171]

Such confabulation of the 4 classes of opposition is exemplified by *Introductiones in Logicam.*[172]

> Rule [II] The sign 'every' or 'all' requires that there be at least three appellata.[173]

This might be called duo-dimensional non-hylomorphic extension-ism: the very opposite of what Leibnitz was trying to achieve.

By attempting to restoring *intensional equality* to Aristotle, Leibnitz was able to rise above (or better perhaps) parallel to it: with a quantifying calculus.

However, the further the direction of travel in that parallel, the more Leibnitz removed himself from the conception of Matrixial Logic.[174]

We see that Leibnitz struggles to combine unilinear negation:

---

[170] As the Author visualises it

[171] A pressing concern in 1659

[172] William of Shyreswood

[173] Kretzman 1968, p. 23

[174] Which, though bad for Modern philosophy, granted the merit of this labour to the Author

> *A non-A* contradictorium est.

with multi-linear negation:

> *Possibile* est quod non continet contradictorium seu *A non-A*[175]

As Lenzen states:

> The next element of the algebra of concepts - and, by the way, one with which Leibniz had notorious di culties - is *negation*. Leibniz usually expressed the negation of a concept by means of the same word he also used to express propositional negation, *viz.* 'non'. Especially throughout, the statement that one concept, A, contains the negation of another concept, B, is expressed as 'A est non B', while the related phrase 'A non est B' has to be understood as the mere propositional negation of 'A contains B'.

This, one trusts, suffices to place Leibniz not merely some distance away from Matrixial Logic, but in quite another dimension.

We see that Leibniz assumes the law of identity:[176]

> Conj 1 $A \in BC \leftrightarrow A \in B \wedge A \in C$
> ("That A contains B and A contains C is the same as that A contains BC"(§35)

Which leads Leibniz to declare:

> Neg 2 $A \neq \sim A$
> ("A proposition false in itself is 'A coincides with not-A")', (§11)

As Lenzen continues (Author emphasis):

[175] Where the term does not hold within itself contradiction
[176] Of which more later

> On the one hand, as Leibniz himself proved in the important fragment "Primaria Calculi Logici fundamenta" of August 1st, 1690, Neg 4 follows from Neg 2 , since if one would have, for certain concepts A and B, both A = B and A = ~B, then *because of the transitivity of identity* one would obtain B = ~B in contradiction to Neg 2.
>
> On the other hand, Neg 2 is conversely obtained from Neg 4 by considering just the special case where B = A. Hence the fundamental axiom of identity, A = A, entails that A = ~A!

The reader can see that Leibnitz was here in the brink of a great discovery: Matrixial Logic just under 400 years early. Yet his motivation to rescue the aequalis from the depredations of meieval scholars, [177] led him down the path of erecting defences around aequalis.

It might be said that Leibnitz's *intensional* project had, as its fundamental but unrealised foundation, the mutation of the aequalis from a passive, to an active function.

Sadly, aequalis is not fitted to serve that function. As Lenzen continues:

> The intended extensional interpretation of ~A is just the set-theoretical complement of the extension of A, because each individual which fails to fall under concept A eo ipso falls under the negative concept ~A:

$$\phi(\sim A)=\overline{\phi(A)}.$$

We thus arrive at the apotheosis of a journey beginning c350 BC in Athens, taking a ramble in medieval

[177] As Leibnitz saw it

scholarship, assuming a right angle in 1650,[178] becoming straightened again into Boolean set theory: the new paradigm of extensionalism, and then revitalising as the zombie problematica of Russell's *non-referring expressions.* [179] And all for a lack of ≠ inequality.

Not Fair and Square

The paradigm equations of ML are by now familiar:

$$(A) \neq (nA) = E$$

$$(A) \neq (-A) = I$$

The Universal Affirmative solves in I, whereas the Universal Negative solves in E. To any reader schooled in FOL, this result is counter-intuitive.

We must again grant indebted acknowledgement to the work of Terence Parsons. His teaching paper on this matter[180] is so compelling in its clarity, that it would be idle effort for the Author to attempt the like labour in his own words (Author emphasis added):

Most contemporary logic texts symbolize the traditional forms as follows:

Every *S* is *P*	$\forall x(Sx \rightarrow Px)$
No *S* is *P*	$\forall x(Sx \rightarrow \neg Px)$
Some *S* is *P*	$\exists x(Sx \,\&\, Px)$
Some *S* is not *P*	$\exists x(Sx \,\&\, \neg Px)$

---

[178] Not quite, but allowing potetical licence in dating

[179] *On Denoting* (1905), and its proposed solution: see later in this Chapter

[180] Parsons, Terence, "The Traditional Square of Opposition", *The Stanford Encyclopedia of Philosophy* (Summer 2017 Edition), Edward N. Zalta (ed.), URL = <https://plato.stanford.edu/archives/sum2017/entries/square/>.

If this symbolization is adopted along with standard views about the logic of connectives and quantifiers, the relations embodied in the traditional square mostly disappear. The modern diagram looks like this:

THE MODERN REVISED SQUARE:

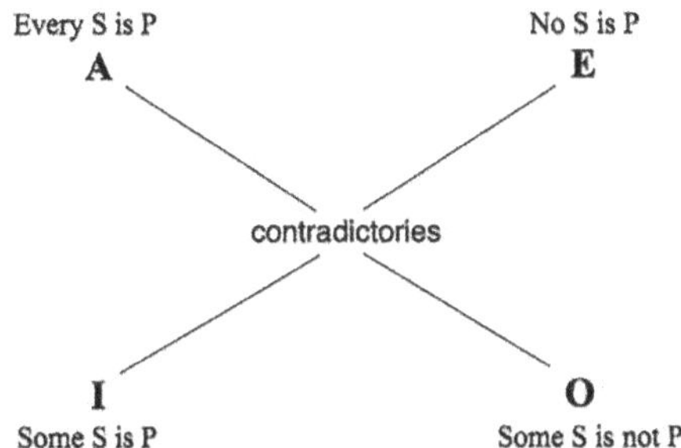

Every modern logic text must address the apparent implausibility of letting 'Every *S* is *P*' be true when there are no *S*s. The common defense of this is usually that this is a logical notation devised for purposes of logic, and it does not claim to capture every nuance of the natural language forms that the symbols resemble.

So perhaps '$\forall x(Sx \rightarrow Px)$' does fail to do complete justice to ordinary usage of 'Every *S* is *P*', but this is not a problem with the logic. If you think that 'Every *S* is *P*' requires for its truth that there be *S*s, then you can have that result simply and easily: just represent the recalcitrant uses of 'Every *S* is *P*' in symbolic notation by adding an extra conjunct to the symbolization, like this: $\forall x(Sx \rightarrow Px)$ & $\exists xSx$.

Why does the traditional square need revising at all? The argument is a simple one:[2]

[2]This argument is given e.g. in Kneale & Kneale 1962, 55–60.

Suppose that '*S*' is an empty term; it is true of nothing. Then the **I** form: 'Some *S* is *P*' is false. But then its contradictory **E** form: 'No *S* is *P*' must be true. But then the subaltern O form: 'Some *S* is not *P*' must be true. But that is wrong, since there aren't any *S*s.

The puzzle about this argument is why the doctrine of the traditional square was maintained for well over 20 centuries in the face of this consideration. Were 20 centuries of logicians so obtuse as not to have noticed this apparently fatal flaw? Or is there some other explanation?

One possibility is that logicians previous to the 20th century must have thought that no terms are empty. You see this view referred to frequently as one that others held.[3] But with a few very special exceptions (discussed below) I have been unable to find anyone who held such a view before the nineteenth century. Many authors do not discuss empty terms, but those who do typically take their presence for granted. Explicitly rejecting empty terms was never a mainstream option, even in the nineteenth century.

Another possibility is that the particular **I** form might be true when its subject is empty. This was a common view concerning *indefinite propositions* when they are read generically, such as 'A dodo is a bird', which (arguably) can be true now without there being any dodos now, because being a bird is part of the essence of being a dodo. But the truth of such indefinite propositions with empty subjects does not bear on the forms of propositions that occur in the square. For although the indefinite 'A dodo ate my lunch' might be held to be equivalent to the particular proposition 'Some dodo ate my lunch', generic indefinites like 'A dodo is a bird', are quite different, and their semantics does not bear on the quantified sentences in the square of opposition.

In fact, the traditional doctrine of [SQUARE] is completely coherent in the presence of empty terms. ***This is because on the traditional interpretation, the O form lacks existential import.*** The **O** form is (vacuously) true if its subject term is empty, not false, and thus the logical interrelations of [SQUARE] are unobjectionable. In what follows, I trace the development of this view.

...

2.2 Aristotle's Formulation of the O Form

Ackrill's translation contains something a bit unexpected: **Aristotle's articulation of the O form is *not* the familiar 'Some *S* is not *P*' or one of its variants; it is rather 'Not every *S* is *P*'.**

With this wording, Aristotle's doctrine automatically escapes the modern criticism. (This holds for his views throughout *De Interpretatione*.) For assume again that '*S*' is an empty term, and suppose that this makes the **I** form 'Some *S* is *P*' false. Its contradictory, the **E** form: 'No *S* is *P*', is thus true, and this entails the **O** form in Aristotle's formulation: 'Not every *S* is *P*', which must therefore be true.

When the **O** form was worded 'Some *S* is not *P*' this bothered us, but with it worded 'Not every *S* is *P*' it seems plainly right.

Recall that we are granting that 'Every *S* is *P*' has existential import, and so if '*S*' is empty the **A** form must be false. But then 'Not every *S* is *P*' *should* be true, as Aristotle's square requires.

On this view *affirmatives* have existential import, and *negatives* do not - a point that became elevated to a general principle in late medieval times.

The ancients thus did not see the incoherence of the square as formulated by Aristotle because there was no incoherence to see.

[6] Ockham (SL I.72): "In affirmative propositions a term is always asserted to supposit for something. Thus, if it supposits for nothing the proposition is false. However, in negative propositions the assertion is either that the term does not supposit for something or that it supposits for something of which the predicate is truly denied. Thus a negative proposition has two causes of truth." Loux 1974, 206.

For such a lengthy quotation, the reader may be forgiven for asking "so what?"

Infinity is not Extension

For the answer, let us return to the traditional ACS:

Aristotle Categorical Sentence Transitions to ML

	C S	ML Equation	Type
A	Every A is B	(A) ≠ (-A) = B[I]	Categorical Universal
E	No A is B	(A) ≠ (B) = N[E]	Reflexive Universal
I	Some A is B	(A) ≠ (nA) = B[E]	Reflexive Opposition
O	Some A is Not B	(A) ≠ (nA) = B/^{n}B	Unreflexive Opposition

For readers "feeling" the equations of ML, ACS (O) may hitherto have felt "ugly". Certainly, the incompleteness of B / ^{n}B seems odd.

Now, let O be translated as *Not every A is B*:

$$\text{A} \qquad (A) \neq (-A) = B[I]$$

$$>$$

$$\text{O} \qquad (A^{n}) \neq (-A^{n}) = B[I]$$

Now the ML forms of inequality shine. *The Universal Affirmative and the Universal Negative now partake in the same frame of reference.*

Ackrill/Aristotle Categorical Sentence Transitions to ML

	C S	ML Equation	Type
A	Every A is B	$(A) \neq (-A) = B[I]$	Categorical Universal
E	No A is B	$(A) \neq (B) = N[E]$	Reflexive Universal
I	Some A is B	$(A^{n}) \neq (nA^{n}) = B[E]$	Reflexive Opposition
O	Not Every A is B	$(A^{n}) \neq (-A^{n}) = B[I]$	Reflexive Negation

Table 3.6

For readers who enquire why the Author did not merely clarify this intrigue at the outset, the answer is:

- ACS (O) did not figure significantly in the text so far. and accordingly had little prejudicial effect upon explication;
- The history of logic has proceeded, as Parsons says, under the illusion of the mistranslated (O);
- Having taken the reader through the analyses thus far in

this Chapter, the significance of the Ackrill (O) translation is vastly enhanced.

This now presents the opportunity to revisit the empty set conundrum, and present new solutions.

We can now present the Matrixial Logic equivalent of the ACYS Square:

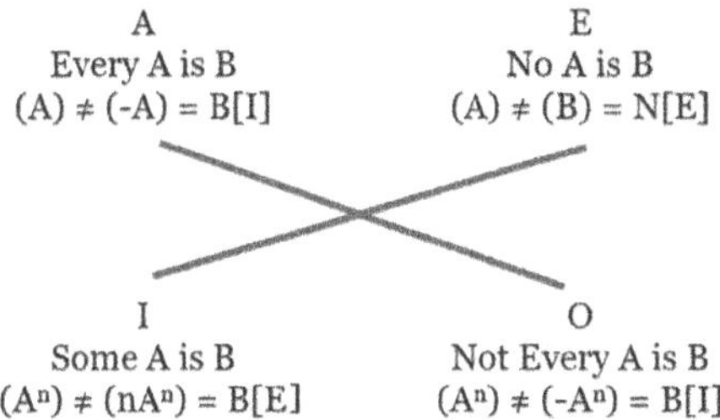

This mode of presentation instantly reveals the hylomorphic nature of the different [E] and [I] forms of inequality.

It now, perhaps, becomes obvious why we said, at the conclusion of Chapter 1:

> We can see here in Jevons, what we were later to see with Frege, Meinod, Russell and their successors. The attempt to defuse the inequality lying at the core of Aristotle's Categorical Syllogisms.[181]
>
> *To substitute for the logical analyses of Substance, a formic logic which relies upon extrusions from pure equality.*[182]
>
> That program describes the history of symbolic logic in the 20th century, and the expressly limited landscape of First Order Logic.

---

[181] See Chapter 3

[182] See Chapter 4, and the discussion of Arithmetic

The rebuttal offered by Matrixial Logic is that: this is an enterprise at which one is bound to fail, by succeeding.

Applying ML Equations in Logic Laws

For lay readers, some context is important:

The **laws of thought** are fundamental axiomatic rules upon which rational discourse itself is often considered to be based. The formulation and clarification of such rules have a long tradition in the history of philosophy and logic. Generally they are taken as laws that guide and underlie everyone's thinking, thoughts, expressions, discussions, etc. However, such classical ideas are often questioned or rejected in more recent developments, such as intuitionistic logic, dialetheism and fuzzy logic.

According to the 1999 *Cambridge Dictionary of Philosophy*,[1] laws of thought are laws by which or in accordance with which valid thought proceeds, or that justify valid inference, or to which all valid deduction is reducible. Laws of thought are rules that apply without exception to any subject matter of thought, etc.; sometimes they are said to be the object of logic

The term, rarely used in exactly the same sense by different authors, has long been associated with three equally ambiguous expressions: the law of identity (ID), the law of contradiction (or non-contradiction; NC), and the law of excluded middle (EM).

Sometimes, these three expressions are taken as propositions of formal ontology having the widest possible subject matter, propositions that apply to entities as such: (ID), everything is (i.e., is identical to) itself; (NC) no thing having a given quality also has the negative of that quality (e.g., no even number is non-even); (EM) every thing either has a given quality or has the negative of that quality (e.g., every number is either even or non-even).

Equally common in older works is the use of these expressions for principles of metalogic about propositions: (ID) every proposition implies itself; (NC) no proposition is both true and false; (EM) every proposition is either true or false.[183]

The 3 primary laws of logic, in more detail, are:

(1) Identity

All things are equal to themselves: P=P

Something is what it is, and it is not what it is not

A statement cannot remain the same and change its truth value

Law of Identity:

X = X (expressed as an equality relation);

X =/= ~X (expressed as an inequality relation).

Therefore: Law of Identity (LOI) states that

"Something (X) is what it is {X = X},

and

It is not what it is not {X =/= X}."

Thus:

Let I1 : = {X=X}

Let I2: = {X=/= ~X]

LOI = [X=X]^[X=/=~X] ; = the conjunction ("and") of I1 and I2; that is,

Law of Identity: LOI = I1 AND I2.

(2) Non-Contradiction

No proposition can both be and not be (the case).

Nothing can both be and not be (i.e. something cannot both be and not be): ~(X^~X); where (~) is negation ("not") and (^) is logical conjunction ("and").

In the same aspect, no statement is both true and false

A proposition X and its negation ~X cannot both be true (at the same time, in the same sense, simultaneously): ~(X ^ ~X);

[183] https://en.wikipedia.org/wiki/Law_of_thought

It is not the case that X and ~X are true together, which logically excludes contradictions.

Implications of LNC (Law of Non-Contradiction):

- Contradictory propositions (or contradictories) cannot both be true.
- At least one of the contradictories is false.
- Both of the contradictories are allowed to be false together, but not true together.
- LNC does NOT logically exclude the option that X and ~X are both FALSE together: it can be the case that neither X is true nor ~X is true.

(3) The Excluded Middle

All propositions either are or are not (the case).

Every statement is either true or false.

Something either is or is not; where "or" is to be understood as an inclusive disjunction (allowing for the option that something both is and is not).

Alternatively stated: A proposition X is true or its its negation ~X is true; which can be reformulated as saying a proposition X is either true, or false, or both, but cannot be neither.

That is, LEM implies that it cannot be the case that neither X is true nor ~ is true; equivalently stated as: A proposition X and its negation ~ X cannot both be false together.

This last statement means that the "neither-nor" option which states that "Neither X (is true) NOR ~X (is true)" is excluded by the law of excluded middle

(LEM): X V ~X

"Either a proposition X is true or its negation ~X is true"

= is equivalent to +
"It is not the case that neither X is true nor ~X is true"
because the negation or complement of "neither-nor" (joint denial) means "either-or" (inclusive disjunction).

The law of excluded middle can be reformulated as follows:
X and ~X cannot both be false together:

It is not the case that:
'NEITHER X (is true) NOR ~X (is true), where X = an arbitrary proposition obeying bivalence: X XOR ~X by the definition of a proposition.

We then have the **9 Modes of Argument**:

(1) Modus Ponens (M.P.)
-If P then Q
-P
-Therefore Q

(2) Modus Tollens (M.T.)
-If P then Q
-Not Q
-Therefore not P

(3) Hypothetical Syllogism (H.S.)
-If P then Q
-If Q then R
_Therefore if P then R

(4) Disjunctive Syllogism (D.S.)
-P or Q
-Not P
-Therefore Q

(5) Conjunction (Conj.)
-P
-Q
_Therefore P and Q

(6) Constructive Dilemma (C.D.)
-(If P then Q) and (If R then S)
-P or R
-Therefore Q or S

(7) Simplification (Simp.)
-P and Q
-Therefore P

(8) Absorption (Abs.)
-If P then Q
-Therefore If P (P or Q)

(9) Addition (Add.)
-P
-Therefore P or Q

These modes all follow from the law of identity. They are extrusions from a single plane of equality.

Modern Predicate Calculus

With Russell, we arrive at the foundations of 20th century analytic / predicate philosophy.

Aristotle had sought[184] to remove the problem through non-hylomorphic abstraction. Medieval scholars simply shoved the problem through the door marked "syncategoremic". Leibnitz was clearly troubled by the dimorphic attributes of the concept of "existence", and sought to solve its problems by adopting the intensional approach.
Boole and his mid-Victorian successors returned to

[184] In Class (iv) Opposition

Aristotle, but with a new mechanistic unimorphism, (intuitively) given an intensional patina by the substances of number, then geometric form.[185]

At the turn of the 19th century, that patina was discarded. We arrived at modern symbolic logic, ushered in through the work of Russell and Whitehead.

Modern predicate calculas takes as its starting point the *mereological fallacy,* which consists of carving up a unimorphic conception of "is" into 3 sub-domains, but without recognition that the sub-domains each instantiate a completely different form of inequality:

Cicero is wise	Px	(C) ≠ (nC) = Wise [E]
Cicero is Tully	x = y	(T) ≠ (nT) = C [E]
Cicero is	x	(C) ≠ (-C) = Is [I]

Before advancing another step, a question arises which seems[186] not to have occurred to Russell and Whitehead, nor indeed to any who confer analytic integrity upon the calculus:

> *If we may sub-divide a domain of existence into 3 separate FoR's, by what means are those sub-divisions related to the whole: save through some 4th domain of existence?*

The answer cannot be sought in A=A. The law of identity

[185] The Venn Diagram

[186] The Author admits inability, through ignorance of all that has been said on the matter, to be more conclusive

is no kind of law because it offers no foundation for prescription. It is merely an invented tautology, and has been since its invention.

Moreover, the tripartite sub-division of the domain of existence in which A=A is nothing other than a violation of the law of identity.

It is the creation of a foundational empty term and empty set: which then allows you to go anywhere and do anything.

That is precisely why it works: until it doesn't.

In the first volume of *Principia Mathematica,* Russell and Whitehead wrote:[187]

> Thus corresponding to any propositional function *x*, there is a range, or collection, of values, consisting of all the propositions (true or false) which can be obtained by giving every possible determination to *x* in in *x*. A value of *x* for which *x* is true will be said to "satisfy" *x*. Now in respect to the truth or falsehood of propositions of this range three important cases must be noted and symbolised.
>
> These cases are given by three propositions of which at least one must be true. Either (1) all propositions of the range are true, or (2) some propositions of the range are true, or (3) no proposition of the range is true.
>
> The statement (1) is symbolised by "(*x*) *x*," and (2) is symbolised by "(*x*). *x*." No definition is given of these two symbols, which accordingly embody two new primitive ideas in our system.
> The symbol "(*x*). *x*" may be read as " *x* always," or " *x* is always

[187] (1910) p15-16. Author emphasis

true," or " x is true for all possible values of x."

The symbol "(x). x" may be read "there exists an x for which x is true" or "there exists an x satisfying x," and thus conforms to the natural form of the expression of thought.

At this point, that violation of Matrrixial Logic rules, which is immanent in the Russell/Whitehead model ("RW Model"), becomes appararent.

It is declared, as a fundamental axiom of the model that:

[where] (1) all propositions of the range are true
is symbolised by "(x) x,"
or " x is true for all possible values of x."

[where] (2) some propositions of the range are true
is symbolised by "(x). x."
there exists an x for which x is true

To explain the error:

(1) The UA (Categorical Universal) equation is, of course:

$$(A) \neq (-A) = B[I]$$

$$\approx$$

$$[I]B = (-A) \neq (A)$$

(2) The PA (Reflexive Opposition) is:

$$(A) \neq (nA) = B\ [E]$$

$$\approx$$

$$[E]B = (nA) \neq (A)$$

(3) It is not legitimate, within the rules of ML, to effect

admixtures of the *contents* of Informational [I] and Existential [E] forms, without the use of express Node terms <∑E> which exhibit crossing the matrix.

(4) Yet, that is exactly what the ACS and ACSY effect.

(5) ML reveals the non-hylomorphic principle of Aristotelian Class (iv) opposition.[188]

(6) The RW Model acts as if φ and x (predicate and subject) are independent atomic units (Atomism").

(7) As if each were, indeed, a singularity. A singularity which has a noetic[189] "existence" beyond the event horizon which is created by the act of collation and determination in truth equations.

(8) In short, the RW Model treats φ and x as if each functions in like manner under both Forms of Inequality.

(9) That is a fundamental error.

(10) The consequences of that error are thus pre-ordained, by actions in the logic of the symbolic forms adhered to it, and through which it is thereafter expressed.

(11) Accordingly, without reading further,[190] it is entirely predictable that the RW Model will encounter difficult with the ACSY Square, the moment that it attempts to deal with φ x relations which involve

---

[188] *Categories* 10, Logic

[189] Or ethereal

[190] Although one always should

forms of forms, singularities outside the square,[191] or chaos relations.

(12) It is to be noted that Atomism, without regard to form of inequality, functions inevitably as deism.[192]

We now take more time to examine the model proposed in *On Denoting*.[193]

> **My theory, briefly, is as follows. I take the notion of the *variable* as fundamental; I use "C (*x*)" to mean a proposition[2] in which *x* is a constituent, where *x*, the variable, is essentially and wholly undetermined. Then we can consider the two notions "C (*x*) is always true" and "C (*x*) is sometimes true".[3] Then *everything* and *nothing* and *something* (which are the most primitive of denoting phrases) are to be interpreted as follows:—**
> **C (everything) means "C (*x*) is always true";**
> **C (nothing) means "'C (*x*) is false' is always true";**
> **C (something) means "It is false that 'C (*x*) is false' is always true".[4]**
> **Here the notion "C (*x*) is always true" is taken as ultimate and indefinable, and the others are defined by means of it.**
>
> [3] The second of these can be defined by means of the first, if we take it to mean, "It is not true that 'C (*x*) is false' is always true".

Taking our ACS table:

Ackrill/Aristotle Categorical Sentence Transitions to ML

	C S	ML Equation	Type
A	Every A is B	(A) ≠ (-A) = B[I]	Categorical Universal
E	No A is B	(A) ≠ (B) = N[E]	Reflexive Universal
I	Some A is B	$(A^n) \neq (nA^n) = B[E]$	Reflexive Opposition
O	Not Every A is B	$(A^n) \neq (-A^n) = B[I]$	Reflexive Negation

Table 3.6

[191] The present king of France is bald

[192] The anthropomorphic or otherwise alienated mode of appearance is a mere cultural contingency

[193] *Mind*, New Series, V14, No.56 (Oct 1905) p479-493

and fitting it into Russell's schema: we see that Russell is effecting a wholesale collapse of the ACS categories.

"Truth" is no longer hylomorphic to anything, except whatever constituents Russell arbitrarily selects. This is to turn integral order into directed chaos.

The result is atomic croutons, floating around in context-dependent soup. You may indeed arrange the croutons as you please. They will always bear whatever relationship to the soup you chose to "Denote". Because it is you who are tilting the soup bowl.

It is, in other words, a subjectively hylomorphic enterprise, in which all of the atoms are arranged according to Russell's subjective frame of reference. Therefore, necessarily, it always "works".

Except when it doesn't:

> **Consider next the proposition "all men are mortal". This proposition[1] is really hypothetical and states that *if* anything is a man, it is mortal. That is, it states that if x is a man, x is mortal, whatever x may be. Hence, substituting 'x is human' for 'x is a man,' we find :—**
> **"All men are mortal" means "'If x is human, x is mortal' is always true".**

So far, so good. Aristotles Class (iv) mode of opposition, in operation.

But then:

> This is what is expressed in symbolic logic by saying that "all men are mortal" means "'x is human' implies 'x is mortal' for all values of x". More generally, we say:—
> "C (all men)" means "'If x is human, then C (x) is true' is always true".
> Similarly.
> "C (no men)" means "'If x is human, then C (x) is false' is always true".
> "C (some men)" will mean the same as "C (a man),"[1] and
> "C (a man)" means "It is false that 'C (x) and x is human' is always false".

Which leads Russell to the oops moment:

> "C (every man)" will mean the same as "C (all men)".

Here, we see clearly, the problem in mixing your (E) "all men" with your (I) "every man".

Put the matter this way: Russell wanted to escape what he saw as the difficulty that non-hylomorphic predicates which are domained in (E), do not have a discrimination function for when (A) is a nullity. Russell was trying to escape the dilemma of the foundational empty term and empty set.

That is, of course because (x)=(x): which constrains "C", by making (x) independent of the FoR. Out of the ACS frying pan, into the soup.

Where (A) ≠ (nA) solves only in [E] and (A) ≠ (-A) solves only in [I]: then no such problem arises in the first place.

The Glabrous Gallic Majesty

The agenda of *On Denoting* was stated as:

> **The course of my argument will be as follows. I shall begin by stating the theory I intend to advocate;[1] I shall then discuss the theories of Frege and Meinong, showing why neither of them satisfies me; then I shall give the grounds in favour of my theory; and finally I shall briefly indicate the philosophical consequences of my theory.**

The result of the methodology pronounced under the motivation of that agenda was:

> If "the author of *Waverley*" meant anything other than "Scott", "Scott is the author of *Waverley*" would be false, which it is not. If "the author of *Waverley*" meant "Scott", "Scott is the author of *Waverley*" would be a tautology, which it is not. Therefore, "the author of *Waverley*" means neither "Scott" nor anything else – i.e. "the author of *Waverley*" means nothing, Q.E.D. (1959, 85)

That result hardly appears to have met the rubric announced at the start of the Russell project.

Under the rubrics of Matrixial Logic, no problem arises from the question of literary authorship asked by George IV.

To see why this is so, we first refer back to the ACS Transitions table:

Ackrill/Aristotle Categorical Sentence Transitions to ML

	C S	ML Equation	Type
A	Every A is B	$(A) \neq (-A) = B[I]$	Categorical Universal
E	No A is B	$(A) \neq (B) = N[E]$	Reflexive Universal
I	Some A is B	$(A^n) \neq (nA^n) = B[E]$	Reflexive Opposition
O	Not Every A is B	$(A^n) \neq (-A^n) = B[I]$	Reflexive Negation

Table 3.6

We then apply the ML transitions to the sentence elements:

Scott is	[an author]	*Waverly*	*the Author of*
(S) ≠ (-S) = is [I]	(A) ≠ (nA) = X[E]	(A) ≠ (-A) = W[I]	(Of) ≠ (nOf) = Is [E]
Every "Scott" is[194] a form (of person[195]) in the process of life in its becoming	There exist some people X who are authors	Waverly is a form of the process of authorship (A)	There exists a crytallised authorship
(S) ≠ (-S) = is [I] (1)		(A) ≠ (-A) = W[I] (2)	
	(A) ≠ (nA) = X [E] (1)		(Of) ≠ (nOf) = Is [E] (2)

Thus:

[I](1) transforms Nodally with [E](1) □[I](2) transforms Nodally with [E](2)

Note: The blue and cream 2-fold layering is not an essential function in the infographic. The layering is simply provided to aid dimorphic thinking.

You could simply say "Scott identifies with Waverly", or "Waverly is of Scott": because there you are matching [I] becomings [FoRs].

It is because you want to attribute a Process to a Substance, that you require a Nodal transformation of each Process.

It is only because Russell was constrained by how he saw his atomic system that this astonishing rigmarole is required.

---

[194] Or was or will be: as in classical and quantum mechanics, time's arrow is irrelevant

[195] Real or hypothetical: you can specify a further condition of reality

When we view the Matrixial Logic equation forms, the "Denotation" construction becomes very simple:

- There are Authors of things (books): which is a Substance relationship in [E];
- There are forms of Authorship (a Process) in [I];
- We use names to attribute [I] to [E] (and vice versa.

As we will see in Chapter 5, this is simply a Node,[196] interacting with Moments Δ^n in an infinity ∞.

Not content with atomic crouton soup, the Theory of Denotation finds itself need two soup bowls: (i) for *primary occurrences*; (ii) for *secondary occurrences*:

> The distinction of primary and secondary occurrences also enables us to deal with the question whether the present King of France is bald or not bald, and generally with the logical status of denoting phrases that denote nothing. If "C" is a denoting phrase, say, "the term having the property F," then
>
> "C has the property ϕ" means "one and only one term has the property F, and that one has the property ϕ".[1]
>
> If now the property F belongs to no terms, or to several, it follows that "C has the property ϕ" is false for *all* values of ϕ. Thus "the present King of France is bald" is certainly false; and "the present King of France is not bald" is false if it means
>
> "There is an entity which is now King of France and is not bald,"
>
> but is true if it means
>
> "It is false that there is an entity which is now King of France and is bald".
>
> That is, "the King of France is not bald" is false if the occurrence of "the King of France" is *primary*, and true if it is *secondary*. Thus all propositions in which "the King of France" has a primary occurrence are false; the denials of such propositions are true, but in them "the King of France" has a secondary occurrence. Thus we escape the conclusion that the King of France has a wig.

---

[196] The aggregation of Content Forms in [E]

From the perspective of Matrixial Logic method, it is somewhat perpexing as to why an "empty set" prompts any attitude of complex thought.

Experience: Invisible

Setup:

- Look around your room;
- Imagine an invisible paperclip in your head (not easy).

Step 1:

- Now imagine the invisible paperclip there, in the room;
- Just hanging in the air.

<u>Discussion</u>:

(1) We know from our Chapter 2 *Experiences,* that this is a lot easier.

(2) Instead of trying to imagine a solitary marble, you are thinking: (Room) ≠ (Invisible Paperclip).

(3) But what is the Form of that inequality?

Step 2:

- Now imagine that the invisible paperclip <u>was</u> there, in the room, 50 years ago: but is not there now.

<u>Discussion</u>:

(1) This is easy. You go into a historical perpective: as if you were visiting your room as part of an old castle, in the past.

(2) You can imagine all sorts of historical artefacts that have since disappeared: like suits of armour, swords, and so on.

(3) The invisible paperclip is just like those ancient swords: it was then, but isn't now.

(4) So it's now easy to see that the form of inequality: [(Room) ≠ (Invisible Paperclip)] = Time [I]

Step 3:

- Now imagine that the invisible paperclip will be there, in the room, 50 years in the future: but is not here now.

Discussion:

(1) Again, it's is easy. You go into a reverse-historical perpective: as if you would be visiting your room as part of a spaceship, in the future;

(2) You can imagine all sorts of sci-fi artefacts that have not yet appeared: like transporter machines, ray guns, and so on;

(3) The invisible paperclip is just like those sci-fi ray guns: it isn't now, but will be;

(4) This further shows that the form of inequality: [(Room) ≠ (Invisible Paperclip)] = Time [I]

But, we have a problem with this sort of Time [I]. We could say: (Present) ≠ (-Present) = Time [I]. But then how do we account for different modes of (-Present): past, future, long ago, near future and so on? There's no mode of discrimination within the equation.

We are therefore led to reason that Time [I] is not an accurate mode of Form. Instead <time> must be some result of Nodes. Thus, we have different Nodes giving us <∑past> \ <∑present> \ <∑future.

We don't need to worry about how those Nodes work at the moment. That will come in Chapter 7.

What we can see is that <the present king of france is bald>, is simply a propoistion about an invisible paperclip. Clearly there was a <present king of france>. All we are doing in Russell's paradox is making the active function of <time> in "is" work unidimensionally: as if "is" and "was" are merely participles of the verb "to be", rather than complex multi-dimensional constructs.

If there really were an empty set problem in human thought, then out imaginative life would be severely disabled.

- For those who wish to consider the pastoral life of unicorns, disappointment awaits. For all that might be said, or even thought, or unicorns, is *false*;
- Consideration of the future is barred to us. We can conceive of the future, but only a domain with null truth value;
- Indeed, although we may hold to a reasonable degree of scientific certainty that unicorns do not exist, and be sure that the future has not yet occurred, what truths can we speak of that data for which we have less than certain

> empirical knowledge? The answer, in Russell's system, must be none.

And that self-immolating scepticism is exactly the final destination of atomic soup:[197]

> One interesting result of the above theory of denoting is this: when there is anything with which we do not have immediate acquaintance, but only definition by denoting phrases, then the propositions in which this thing is introduced by means of a denoting phrase do not really contain this thing as a constituent, but contain instead the constituents expressed by the several words of the denoting phrase. Thus in every proposition that we can apprehend (*i.e.* not only in those whose truth or falsehood we can judge of, but in all that we can think about), all the constituents are really entities with which we have immediate acquaintance. Now such things as matter (in the sense in which matter occurs in physics) and the minds of other people are known to us only by denoting phrases, *i.e.*, we are not *acquainted* with them, but we know them as what has such and such properties. Hence, although we can form propositional functions C (x) which must hold of such and such a material particle, or of So-and-so's mind, yet we are not acquainted with the propositions which affirm these things that we know must be true, because we cannot apprehend the actual entities concerned. What we know is "So-and-so has a mind which has such and such properties" but we do not know "A has such and such properties," where A *is* the mind in question. In such a case, we know the properties of a thing without having acquaintance with the thing itself, and without, consequently, knowing any single proposition of which the thing itself is a constituent.

If it is thought that *The present King of France is bald* presents some kind of problem, then the solution is utter simplicity:

[197] op. cit. p492-493

The present	*(Kf)*	*is Bald*
(A) ≠ (nA) = X[E]	(Kf) ≠ (-Kf) = Is [I]	(Ba) ≠ (nBa) = Is [E]
There exists Substance co-ordinate in spacetime, in the present	(Kf) is a form of the process of life in its becoming	There exists a gla-brous state of being
(A) ≠ (nA) = X[E] (1)		(Of) ≠ (nOf) = Is [E] (2)
	(S) ≠ (-S) = Is [I] (1)	

Solution:

- (Kf) ≠ (-Kf) = Is [I]
- there is no reciprocal in Iteration or Substance
- because (Kf) cannot transform, but (-Kf) can.

The Author views construction of the above table as a matter of utter simplicity, applying the methods of ML. It can become so to the reader.

So to become, it is necessary only that the reader apply the following rubrics:

- A datum which consists of Substance in co-ordinate spacetime, has its existence in FoR [E]
- A datum which consists of a life in being, or otherwise to say, that which is organic,[198] has its existence in FoR [I]
- Any universal proposition has its FoR in [I]
- Any other proposition has its FoR in [E]

[198] And is being considered in its own life path

At this stage of the exposition, questions as to why this is so, why this should be so, and how it can be that this is so, are reserved to a later Chapter.

There is a sudden scraping back of academic chairs. Particularly those belonging to 20th century behaviouralists, and monists of every description: *but this is dualism!*

We reply that it is most patently not. To measure inches and miles, is to use two different frames of reference. To which differentiation, the most Cromwellian monist would not ascribe the arch-sin of dualism.

To play association football by rules different to American football, is not obviously to adopt a dualist method in setting rules.

In other words, discrimination is not dualism, although dualism appears to[199] require discrimination.

What ought to be dealt with now, is the matter that: this is not a contrast of "subjective" and "objective" view.

If one insists on using this terminology, then such user would[200] be bound to grant that both FoRs are objective:

---

[199] But in logic does not

[200] It is supposed

(1) It is not that we are trying, by subjective transmutation, to place ourselves as an accompanying homonculus within the mind of Scott, or His Gallic Majesty, so as to pose questions from within some personal mindscape frame of reference.

(2) It is no different to asking questions such as: was that book written by Aristotle; is that animal a unicorn; did Harry Potter vanquish Valdemort; and so on.

(3) If it be objected that one cannot reliably distinguish that which is organic from that which is not, then the answer is two-fold:

(4) First: you who think or communicate a matter are the sovereign of the FoR in which you choose to think or communicate. Should you wish to grant the qualities of organic life to a brick, you are at liberty to do so. You are able to discriminate between being and becoming in the ordering of your own thoughts. Such discrimination is not, in the ordinary discourse of your thought, challenging. Should it become so, in respect of any matter, you can inspect it.

(5) Secondly: the domain of science is relied upon by all to effect the articulation of such discrimination.

(6) One is minded to notice the Cairns-Smith theory of crystalline abiogenisis,[201] as illustration of

---

[201] Genetic Takeover and the Mineral Origins of Life (1982, repub 1987)

> a scientific idea of abridgement of the being / becoming dichotomy. However, even that theory is proposed as a mechanism for the transformation of that which is in being to that which becomes.

Those academic chairs are now flung over, and perhaps drinks spilled: *but this is not even dualism, it is being-becoming Platonic idealism!*

Good people of the faculty, calm yourselves.

May we briefly refresh our acquaintance with Platonic idealism:

Polytheistic

Being: Demiurge >Produces *ideas/forms*[202] in perfection:

<bridge>

Becoming: mortals perceive *substance*

And for good measure, Aristotelian dualism:

Monotheistic

Singularity: Unmoved mover / Logos

<bridge of Forms>

substances > mortals perceive.

Then, using the same summary schema, Matrixial Logic:

Materialist Atheism

mortals perceive > forms < Substance E
Life I

[202] Jesus and the angels, et al

At this stage, we now submit, for the purpose of efficient explanation, this infogram of question and answer:[203]

#	Question	Answer
(1)	By "mortals" you just poetically mean human beings	Yes: as did Aristotle
(2)	Do you include also animals	No
(3)	Why not	See later Q &A
(4)	By "perceive" do you mean only Qualia?	No
(5)	But Qualia are included	Yes: so far as that is considered by any to be a useful concept
(6)	Does "perceive" include the tools of perception	Meaning what?
(7)	Modes of measurement, computation and so on	Yes: but we must exercise caution in what FoR we use in these cases
(8)	Do forms exist	The question is ambiguous as to its FoR
(9)	Are there a sorts of form like colour	Yes
(10)	And do they have a mode of existence in ML	Yes
(11)	Are there sorts of form like music	Yes
(12)	And do they have a mode of existence in ML	Yes
(13)	Are there sorts of form such as ideas	Yes
(14)	And do they have a mode of existence in ML	Yes
(15)	Can we only perceive through forms	No

[203] So that Socrates also gets a look in

(16)	So, we can perceive otherwise than through forms	Yes
(17)	When, or in what domain do we perceive through forms	When we engage in self-aware thought
(18)	When, or in what domain do we perceive otherwise than through forms	When we engage with reality otherwise than in self-aware thought: a different mode of perception
(19)	Does self-aware thought include dreams	Yes
(20)	So dreaming thoughts are the same as waking thoughts	No
(21)	What is the difference	Dreaming thoughts are unmediated by organic sensibility
(22)	Did god or some other singularity grant us the capacity for perception	No
(23)	Did god or some other singularity grant us these forms	No
(24)	Did god or some other singularity grant Substance or Life	The emotional / political answer is No. The logical answer is that the question is nonsensical
(25)	Do Substances exist	The question is ambiguous as to its FoR
(26)	Are there a sorts of substance like rocks: inorganic things	Yes
(27)	And do they have a mode of existence in ML	Yes
(28)	How do we know that	Because your [I] can become an [S] for you,[204] and the forms of [S] are how you specify the attributes of knowledge

[204] A very crude summary of the more nuanced process explored in Chapters 4 and 5

(29)	Is that not circular	No. It is the opposite of circular. It is dimorphic, which precludes circularity.
(30)	Are there a sorts of substance like trees: organic things	Yes
(31)	How do we know that	Same answer as (28)
(32)	How do we know the difference between the organic and inorganic	We learn the difference behaviourally from the moment of birth
(33)	Is that a claim of genetics	That depends upon what you wish to state as the domain of genetics
(34)	Is that a claim of neurobiology	Insofar as neurobiology is proposed to describe deterministic outcomes: No
(35)	Is that a claim of behaviourism	Insofar as we may form analytic hypotheses, based on observation: Yes
(36)	Is it a claim that this difference in our perception of the organic/inorganic, is necessary	The question is ambiguous.
(37)	Is that difference a teleological necessity	No: see (15) and (16)
(38)	Is that difference a behavioural necessity	Observation cannot produce a universal: an [E] cannot produce an [I]
(39)	Is that difference, manifested in behaviour, universally observed	Yes: unless we observe any exceptions
(40)	Is this a claim that we perceive the organic and the inorganic differently?	We, being organic ourselves: Yes
(41)	These differentiated forms of perception of different types of entity	Yes...
(42)	Do these perceive reality	They are our reality
(43)	So they are Kantian a priori spectacles	No: we reject to concept of A Priori
(44)	But there is an objective reality	There is for each of us, Yes

(45)	Meaning, we can only perceive objective reality, subjectively	No: meaning that the invention of a purported FoR in which both of these persepctives can be delineated, in a single FoR, is a mereological mistake
(46)	Look: if there were no human apes on planet earth, would planet earth still exist. Yes or No?	Under the merological fallacy ("Mf") which is the foundational premise of the question: Yes
(47)	Is there a reality separate from our subjective experience of reality	There is no such separate reality as "our" subjective experience
(48)	Is there a reality separate from any individual's subjective experience of reality	Yes
(49)	If that individual died, would that separate reality continue?	Yes: but not for that individual.
(50)	If every individual ceased to be, would that separate reality continue?	Under Mf, Yes: but not for any such individual. Which contradictions evidences that there is indeed an Mf
(51)	Are these differentiated forms of perception something which is a priori in something like a Kantian sense	No: and in nothing like a Kantian sense
(52)	By what means do they arise in us / for us	See (35): in and as our developmental path from birth
(53)	Could we effect thinking otherwise than in / through / by / as these ML forms	Any FoR created by use of Mf can give us any answer to that
(54)	Same question, otherwise than under an Mf based FoR	No means of knowledge perceptible as thought
(55)	Per (16), we can perceive otherwise than though forms	Yes
(56)	So, how we perceive is not limited to ML forms	Agreed

(57)	For example, our sensory experience of pain sensation is not limited to ML forms	No means of knowledge perceptible as thought
(58)	So this is dualism, as a line is drawn between thinking and feeling	No: without the feeling, there is no thinking
(59)	But there is a difference	There are "factual" differences between all states of affairs
(60)	But there exist a domain in which ML FoRs have effect, and a domain in which they don't	This question relies upon the Mf
(61)	You mean to say that we think and feel in different ways	Yes
(62)	It is a human being who is doing the thinking and feeling	Yes
(63)	And because it is a human being (and not hypothetically some other entity), we notice that these activities are conducted differently	Yes
(64)	Some other entity: an animal, or an alien, or a fiction, might conduct these activities differently than how we do	Yes: under the rubric that, without committing the Mf, we can so discuss these matters, with respect to such entities
(65)	So your claim is that human individuals think and feel in ways that are categorically different, not because of anything a priori...	Agreed
(66)	Rather that this... observed	Yes, and then analysed
(67)	...categorical difference between thinking and feeling, is exactly what it is to be human	Yes
(68)	Is this capacity for categorical difference purely individual	No: time requires space, and vice versa
(69)	So you claim that this capacity depends upon some condition of reciprocity	Yes
(70)	What is the condition of receiprocity	Fulfilment of the biological needs of dependence from birth

Although some of the above has crossed paths with the Dual-Aspect theory of Spinoza[205] and Neutral Monism (Mach, James Russell) and its later exponents (Chalmers: 1996; Heil: 2013; Banks: 2014; Strawson: 2016); a firm dividing line is drawn between ML and those schools of thought. The dimensions and implications of that line are reserved to a later Chapter.

Let us seek to encapsulate the above Q &A table in the following *Experience*.[206]

Experience: Circle

Set up:

- Locate a window or other visual access point to outside;
- Ideally an outside view with lots of length in your field of vision, and natural objects, or a street scene with people.

Look at the scene. Relax your eyes. Just breathe and look. Be thinking and feeling anything at all which comes to you.

Now:

- Think of a geometric circle. Nothing special about it. Just a circle;
- Put it in your minds eye and project that circle into the landscape.

---

[205] (1623-1627)

[206] For readers unfamiliar with the Author's use of Experiences, please refer to *I Want To Love But*, and *Worlds Ends*

Now:

- Return to your field of vision at the start;
- Try to hold that same field of vision and try to see that circle in that same field of vision.

Stop.

Discussion:

(1) You could not hold your ordinary field of vision and the circle, *at the same time;*

(2) The more you tried, you found you were fractionating your point of view. Pulling your focus away from the field of vision to the circle, and back.

Reader: but that's just probably some fact of the matter when it comes to optics. Something to do with focus and objects.

Response: It is understandable you might think that. Let us experiment. Please repeat the Experience, only this time, instead of a circle, think of a *tree, or a dog, or cat.*[207]

(3) Result 2: You are able to project and hold the natural object in the same field of vision with only an initial effort to manufacture the projection. Thereafter, maintining the projection is effortless.

---

[207] Anything occurring in nature, which fits (in size) the landscape.

Reader: OK. So what?

Response: You have just shaken hands with a reality which Matrixial Logic articulates.

Analytic Explanation:

(1) In casual viewing, you were in your default state of thoughts and feelings: whatever they were.

(2) Insofar as your thoughts were engaging with your perceptible domain, you were experiencing E Space: (A) ≠ (na) = E

(3) When you made the claim upon your attention in thought of the circle, you engaged thinking in I Space.

(4) The circle is an artefact of I Space,[208] because it is the expression of a Universal:[209] (A) ≠ (-A) = I

(5) Your thoughts can, and do, co-exist in E Space and I Space

(6) However, when you try to "see"[210] an artefact of I Space in an FoR which is of E Space, you encounter difficulty.

Whatever your epistemological view, you *know* that each aspects of this Experience were completely real.[211] Twisting now in your academic chair, you might wish

[208] Technically, the reference point of a Moment

[209] Refer to ACS (A) at the beginning of this Chapter

[210] To think of

[211] This point is made in a slightly brow-beating way, for deliberate emotional effect

to claim that you do not "know" any such thing. Fair enough, we can defer that till a later Chapter.

What you probably cannot gainsay, is that each element within that Experience felt just as "real" as any other. The circle may feel more "real" than the landscape, for some; and vice versa. But you will admit that you experienced them as co-existent reality.

The Mereological Mistake

An essential feature of Matrixial Logic is the hylomorphic quality of existential relationships: whether those are in Substance in [E] or Moments in Infinity [I].

How we look at any particular idea of something affects whether we treat that thing as Substance, or Moment. This does not render those concepts atomic, separated from any particular form of content. Hylomorphism instead recognises the contingent, yet holistic character of any concept.

This is no more than to say that we discover the nature of any thing, by reference to its relationship with things which are in opposition to, or negation of, that thing.

The discussion of Russell's paradoxes illustrates general principles:

- We can usefully and logically speak of plural parts of a whole, in relation to each other under the form of that whole

- We can't usefully and logically speak of a single part in relation to a whole, under the form of that whole.

The latter inevitably leads to unimorphism: which is merely abstraction in subjectively bounded infinity.

Any unimorphic account of existential relations effects mereological separation of [E] from [I], and indeed from [C].

Such that, where a fake (abstracted) [E] or [I] is pressed into service, such abstraction always ends up having to perform dual functions.

The hidden reality of ACS and ACSY reasoning is dimorphic interaction of [E] and [I] FoR's. Unimorphic abstractions result inevitably in paradoxical forms of forms. Singularities are then sought to rescue such paradoxes.

It does appear that the self-limiting project of First Order Logic has taken such unimorphic extraction to a degree of sophistication which is astonishing. Yet it is not even metaphysical. There are no angels or pin heads in FoL: merely iterations of ideas of them, subsumed under extrusions from a plane of equality.

A plane such that (A) even hardly equals (A). Even that may be a conclusion too far, since it implies the imposition of truth values: which are no longer the province of logic.

It is clear from the Organon what Aristotle would have made of such renunciation of reality as an object of logic. To be fair, we do know what FoL makes of Aristotle.

Paraconsistent Logic

For readers not familiar with this exploding field, [212] here is a standard summary:[213]

> A *paraconsistent* logic is a way to reason about inconsistent information without lapsing into absurdity. In a non-paraconsistent logic, inconsistency *explodes* in the sense that if a contradiction obtains, then everything (everything!) else obtains, too. Someone reasoning with a paraconsistent logic can begin with inconsistent premises—say, a moral dilemma, a Kantian antinomy, or a semantic paradox—and still reach sensible conclusions, without completely exploding into incoherence.
>
> Paraconsistency is a thesis about *logical consequence*: not every contradiction entails arbitrary absurdities. Beyond that minimal claim, views and mechanics of paraconsistent logic come in a broad spectrum, from weak to strong, as follows.
>
> On the very weak end, paraconsistent logics are taken to be safeguards to control for human fallibility. We inevitably revise our theories, have false beliefs, and make mistakes; to prevent falling into incoherence, a paraconsistent logic is required. Such modest and conservative claims say nothing about *truth* per se. Weak paraconsistency is still compatible with the thought that if a contradiction were *true*, then everything would be true, too—because, beliefs and theories notwithstanding, contradictions cannot be true.
>
> On the very strong end of the spectrum, paraconsistent logics underwrite the claim that some contradictions really are true. This thesis – dialetheism - is that sometimes the best theory (of mathematics, or metaphysics, or even the empirical world) is contradictory. Paraconsistency is mandated because the dialetheist still maintains that

[212] A pun a day keeps the logician away

[213] https://iep.utm.edu/para-log/

> not everything is true. In fact, strong paraconsistency maintains that all contradictions are false—even though some contradictions also are true. Thus, at this end of the spectrum, dialetheism is *itself* one of the true contradictions.

Graham Priest and his co-authors also provide an outstanding summary:

> Priest, Graham, Tanaka, Koji and Weber, Zach, "Paraconsistent Logic", *The Stanford Encyclopedia of Philosophy* (Summer 2018 Edition), Edward N. Zalta (ed.), URL = <https://plato.stanford.edu/archives/sum2018/entries/logic-paraconsistent/>.

Having pointed the reader to the field of paraconsistent logic ("PL"), the reader is invited to turn back.

Weak PL seems to be an effort to make predicate and propositional logic softer at the edges, so that paradoxes, and other non-wff[214] fitting phenomena, can be explained.[215]

From the standpoint of matrixial logic, Russell's denotation theory was quite as paraconsistent as the theories of postwar logicians.

Why PL may interest the reader is, that work in this field for the last 80 years or so, demonstrates that there is a coherent body of opinion that classical logic[216] does not, or cannot, do all that logic could.

---

[214] Well formed formul

[215] Or perhaps, explained away

[216] The lineage from Aristotle to Nagel

Matrixial Logic agrees with that fundamental conclusion. Although for very different reasons.

Contradictions Can Be True

Graham Priest's Dialetheism school has become a leading exponent (emphases and hyperlinks in original):

> A *dialetheia* is a sentence, AA, such that both it and its negation, ¬A¬A, are true. If falsity is assumed to be the truth of negation, a dialetheia is a sentence which is both true and false. Such a sentence is, or has, what is called a truth value *glut*, in distinction to a *gap*, a sentence that is neither true nor false. (We shall talk of sentences throughout this entry; but one could run the definition in terms of propositions, statements, or whatever one takes as one's favourite truth-bearer: this would make little difference in the context.)
>
> *Dialetheism* is the view that there are dialetheias. If we define a contradiction as a couple of sentences, one of which is the negation of the other, or as a conjunction of such sentences, then dialetheism amounts to the claim that there *are true contradictions.* As such, dialetheism opposes—contradicts—the *Law of Non-Contradiction* (LNC), sometimes also called the Law of Contradiction.
>
> The Law can be expressed in various ways; fixing the precise formulation is itself a topic of debate (Priest et al 2004, Part II). Thomas Reid put the LNC in the form 'No proposition is both true and false'. A strong (modal) statement of the LNC is: for any AA, it is impossible that both AA and ¬A¬A be true.
>
> In Book of the *Metaphysics*, Aristotle introduced (what was later to be called) the LNC as "the most certain of all principles" (1005b24)—*firmissimum omnium principiorum*, as the Medieval theologians said. Since Aristotle, there have been few sustained attempts to *defend* the law. The LNC has been an (often unstated) assumption, felt to be so fundamental to rationality that some claim it *cannot* be defended, e.g. David Lewis 1999. As a challenge to the LNC, therefore, dialetheism assails what most philosophers take to be unassailable common sense, calling into question the rules

for what can be called into question (cf. Woods 2003, 2005; Dutilh Novaes 2008).

Since the advent of paraconsistent logic in the second half of the twentieth century, dialetheism has been developed as a view in philosophical logic, with precise formal language. Dialetheism has been most famously advanced as a response to logical paradoxes, in tandem with a paraconsistent logic. The view has been gaining, if not acceptance, the respect of other parties in the debate; one critic writes that dialetheists have shown,

> as clearly as anything like this can be shown, that it is coherent to maintain that some sentences can be true and false at the same time. ... [A]nd that perhaps is a radical conclusion, and a major advance in our understanding of the issues (Parsons 1990).

Priest, Graham, Berto, Francesco and Weber, Zach, "Dialetheism", The Stanford Encyclopedia of Philosophy (Fall 2018 Edition), Edward N. Zalta (ed.), URL = <https://plato.stanford.edu/archives/fall2018/entries/dialetheism/>.

Liar, Liar

Logical paradoxes have both haunted and entertained the practice of logic since even before the time of Aristotle.

It is a commonplace that the resolution of such paradoxes, or perhaps more accurately stated, the re-positioning of logic to reduce or obviate their impact, was the initial motivation behind paraconsistent logic and dialetheism.

Priest explains (emphases and hyperlinks in original):

> Probably the master argument used by modern dialetheists invokes the logical paradoxes of self-reference. It is customary to distinguish between two families of such paradoxes: semantic and set-theoretic. The former family typically involves such concepts as *truth*, *denotation*, *definability*, etc; the latter, such notions as *membership*, *cardinality*, etc. After Gödel's and Tarski's well known formal procedures to obtain non-contextual self-reference in formalized languages, it is difficult to draw a sharp line

between the two families, among other things, because of the fact that Tarskian semantics is itself framed in set-theoretic terms. Nevertheless, the distinction is commonly accepted within the relevant literature.

Russell's paradox is the most famous of the set-theoretic paradoxes; it arises when one considers the set of all non self-membered sets, the *Russell set*. Cantor's paradox arises in connection with the universal set. The most famous of the semantic paradoxes is the Liar paradox. Although cases for the existence of dialetheias can be made from almost any paradox of self-reference, we will focus only on the Liar, given that it is the most easily understandable and its exposition requires no particular technicalities.

Beall and his co-authors give a useful account of two exemplar paradoxes:

> The first sentence in this essay is a lie. There is something odd about saying so, as has been known since ancient times. To see why, remember that all lies are untrue. Is the first sentence true? If it is, then it is a lie, and so it is not true. Conversely, suppose that it is not true. We (viz., the authors) have said it, and normally things are said with the intention of being believed. Saying something that way when it is untrue is a lie. But then, given what the sentence says, it is true after all!
>
> That there is some sort of puzzle to be found with sentences like the first one of this essay has been noted frequently throughout the history of philosophy. It was discussed in classical times, notably by the Megarians, but it was also mentioned by Aristotle and by Cicero. As one of the insolubilia, it was the subject of extensive investigation by medieval logicians such as Buridan. More recently, work on this problem has been an integral part of the development of modern mathematical logic, and it has become a subject of extensive research in its own right. The paradox is sometimes called the 'Epimenides paradox' as the tradition attributes a sentence like the first one in this essay to Epimenides of Crete, who is reputed to have said that all Cretans are always liars
>
> ...
>
> 1.3 Liar cycles
>
> Consider a very concise (viz., one-sentence-each) dialog between siblings Max and Agnes.

- **Max**: Agnes' claim is true.
- **Agnes**: Max's claim is not true.

What Max said is true if and only if what Agnes said is true. But what Agnes said (viz., 'Max's claim is not true') is true if and only if what Max said is not true. Hence, what Max said is true if and only if what Max said is not true. But, now, if what Max said is true or not true, then it is both true and not true. And this, as in the FLiar and ULiar cases, is a contradiction, implying, according to many logical theories, absurdity.

Beall, Jc, Glanzberg, Michael and Ripley, David, "Liar Paradox", The Stanford Encyclopedia of Philosophy (Fall 2020 Edition), Edward N. Zalta (ed.), forthcoming URL = <https://plato.stanford.edu/archives/fall2020/entries/liar-paradox/>.

So, let us examine these two forms of paradox:
(1) The Contradictory Self-Declaration ("CSD")
(2) The Contradictory Referent ("CR")

Before running formal logic, the reader is asked to participate in the following Experience.

We are about to prove this to you in one of the most important *Experiences* in this book.

The proposition of non-Singularity
is the keystone of Matrixial Logic.

It has implications which extend back 2,500 years to Aristotle. Implications, for logic, philosophy, psychology, physics and much more.

By way of example, the medieval Scholastics thought that the property of "red" is in a thing. Descartes refuted them. Yet in all that 400 years since, we have made little further progress.

The discussions of qualia have been insightful, brilliant and artfully argued on all sides. Yet they have taken us no further than Descartes.

The following *Experience* is the foundation of why this has been so.

Whether you are a nuclear physicist, or a 7 year old: you will ecounter the next *Experience* in exactly the same way.

You are about to discover that the contents of your Mentation, the long-laboured fruits of your education, and thinking, provide you with no power over the Matrixial Logic architecture of the universe.

Experience: Soloclip

Step 1:

- Close your eyes
- Imagine a blank surface: a sheet of paper

Step 2:

Now: eyes still closed

- Imagine a single paperclip appearing on the blank sheet. Any colour or no colour;

- Put it facing north-south a spictures, just for convenience of explanation;
- [You can run the *Experience* as many times as you like, and alter where you put the clip starting position].

Step 3: Eyes closed

Without touching or sliding the clip, or adjusting the paper

- Move the clip.

Stop.

<u>Discussion</u>

(1) You can't move the paperclip

Notes:

If you feel that you can; then

Try to move the sheet of paper

You can't[217]

Step 4: Eyes closed

Without touching or sliding the clip, or adjusting the paper

- Make the clip 'point' towards the left top corner;
- Make the clip turn 90°.

Stop.

<u>Discussion</u>

(1) You can't do these things

and: same Notes as above

Step 5: Eyes closed

- Make the object disappear.

Stop

<u>Discussion</u>

(1) You can't .

(2) Even if you can "drag" the clip away from the paper, the paper-clip relationship remains "sticky".

(3) It feels stressful to try and maintain that "elastic band" situation

and: same Notes as above

---

[217] There is a class of personality which can move the paperclip, or seem to. We will look at this in *Secret Self*

Step 6: Eyes closed

- Now see **two clips** on the paper. Any different colours.

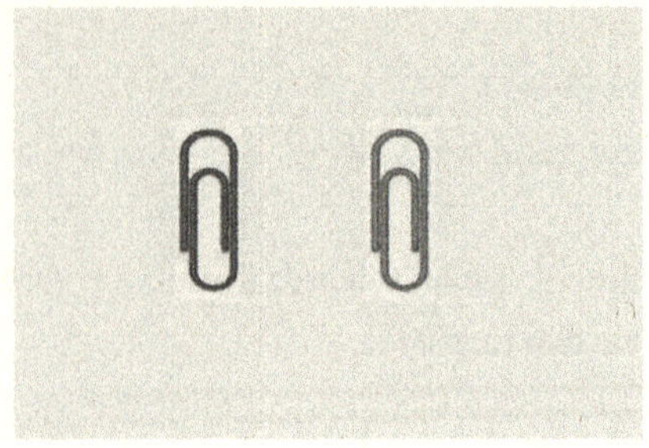

For ease of explanation, we will assume red and blue clips.

- Make the Blue object disappear

Bring it back

- Make the Red object disappear

Bring it back

- Make the Blue object 'point' towards the left top corner, or the right top corner

Stop

- Make the Red object 'point' towards the left top corner, or the right top corner
- Make the Blue object turn 90°

Stop

- Make the Red object turn 90°

Stop

<u>Discussion</u>

(1) Now, it is easy to manipulate either paperclip

This is how we respond to Singularity. The logical force of your mind cannot resolve the paradox which we create by Singular extrusion[218] from your Frame of Reference.

You know perfectly well that the paper and the clip are simply products of your mentative imagination. You imagined them: how can it be possible that you cannot control such mentative figments at will?

Yet it is so: you are unable to manipulate your own mentative products, no matter how hard you try.

This is a classic paradox. The paperclip is a lie about itself:

- It is a mere figment of your will, and has no real existence;
- Yet it defies your efforts of will.

So, it is that which is figmented by you, but you are unable to act towards it as a figment.

What rescues the domain of figmentation is when Singularity becomes Plurality.

- In Singularity, there is no FoR in which you can stand, and "see" you ad the clip in relationship. There is merely an (A). It is a "solitary marble".[219]
- The clip is not acting as Substance
- However, immediately an FoR is created by plurality, such that (A) ≠ (nA) = [E], you can act on that FoR.

[218] There will be much more on "extrusions" in the next Chapter
219 See Chapter 2

You are immediately free to manipulate your mentative figments.

With this insight, we can see that all the Liar paradoxes are merely clever curtains hiding Singularity. They simply posit a Scalar axis[220] of some hypothecated "existence". They then extrude a single "paperclip" equivalent.

They challenge you to reconcile the extrusion with the axis: and of course you can't, because one is conditional upon the other. And because you cannot find an FoR which includes both extrusion and axis, so as to operate a truth-value between them.

How Not to Explode

As Priest states the matter:[221]

> A logical consequence relation is *explosive* if, according to it, a contradiction entails everything:
>
> *ex contradictione quodlibet*:[222]
>
> for all A and B: A,¬A B
>
> *for All A and B: [A, not-A] Entails B*

It will be immediately apparent to the reader that we are here treading on terrain very familiar to Matrixial Logic.

However, we can't translate this ECB formulation to ML. It is simply incoherent, mixing Content elements of an

---

[220] See Chapter 4

[221] op. cit.

[222] "ECQ: *from contradiction all is possible*

FoR with a hypothetical FoR.

Marginal Note:
The symbol for negation and not have seldom been distinguished in formal logic. The same symbol has tended to serve for both, and that symbol has changed in usage over the centuries:

~ meaning 'not' eg in Leibniz
¬ meaning 'not' in modern SL
∄ There does not exist (same as ¬∃)

By contrast, ML uses two completely different symbols, representing two very distinct concepts:
n or NOT meaning "not"
- meaning "negation".

With the greatest of respect to Graham Priest's intent and obvious intellectual ability, this has no more prospect of success, than shuffling deckchairs on the Titanic. When you can simply walk on the water instead.

Priest reports:[223]

2.1 Dialetheism in Western Philosophy

Aristotle takes a number of the Presocratics to endorse dialetheism, and with apparent justification. For example, in Fragment 49a, Heraclitus says: "We step and do not step into the same rivers; we are and we are not" (Robinson, 1987, p. 35). Protagorean relativism may be expressed by the view that man is the measure of all things; but according to Aristotle, since "Many men hold beliefs in which they conflict with one another", it follows that "the same thing must be and not be" (1009a10–12). The Presocratic views triggered Aristotle's attack in *Metaphysics*, Book Γ. Chapter 4 of this Book contains Aristotle's defence of the LNC. As we said above, historically Aristotle was almost completely successful: the LNC has

[223] Op. cit: and entirely accurately, in the Author's view

been orthodoxy in Western Philosophy ever since.

Now we have the lesson of the *Soloclip Experience*, let's solve a Liar paradox.

Let's start with the classic:

This sentence is not true.

So, let's find the solo paperclip:

- In the Soloclip paradox, you were trying to think "This clip is not here: it's existence here is not true, as a function of existence."
- So, the trick is obvious: the phrase is using "is" as an extrusion from existence:

The paradoxial contradiction is simply in saying: that which exists under an assumed frame of reference, does not exist under that assumed frame of reference:

It is not true that this paperclip is not here

≈

It is not true that this paperclip is here

The conjuring trick is:

(1) In assigning the word "true". There is no truth-value in respect of that which has no existence.; and

(2) using the fake word "sentence": because the phrase is not a "sentence", as defined by rules of grammar.

There is another, linguistic, reason why this trick works. It is attempting to argue with the shopping basket cashier in Theta Θ: but we will get to that in the *Languini* Chapter.

The liar paradox:

This sentence is not true.

is thus simply:

There is no truth-function of a curation of words, which curation is predicated upon their non-existence in the same frame of reference as that truth-function.

All you have to do, in other words,[224] to destroy the paradox is: add a second paperclip.

This sentence is not true. But this sentence is.

No dialetheism needed.

[224] Literally

The Greatest Paradox

ποταμοῖσι τοῖσιν αὐτοῖσιν ἐμβαίνουσιν
ἕτερα καὶ ἕτερα ὕδατα ἐπιρρεῖ[225]
potamoisi toisin autoisin embainousin
hetera kai hetera hudata epirrei.

On those stepping into rivers staying the same other and other waters flow.
(Cleanthes from Arius Didymus from Eusebius)
Graham, Daniel W., "Heraclitus", *The Stanford Encyclopedia of Philosophy* (Fall 2019 Edition), Edward N. Zalta (ed.), URL = <https://plato.stanford.edu/archives/fall2019/entries/heraclitus/>.

Waters have flowed the course of two and a half millenia, sibce these words were written. Or perhaps not, according to Zeno of Elea[226] and his paradoxes.

It is notable that, despite this fugive calendared denotation, the paradoxes remain unsolved. The mathematical brilliance of Cantor, and his successors,[227] seems only to displace the paradoxes, not resolve them.

In using Matrixial Logic, it becomes obvious that there was

[225] c.f. Barnes Transl: *On those who enter the same rivers, ever different waters flow.* DK. BK12
http://www.heraclitusfragments.com/files/ge.html
[226] (495–*c.* 430 BC)
[227] See: Huggett, Nick, "Zeno's Paradoxes", *The Stanford Encyclopedia of Philosophy* (Winter 2019 Edition), Edward N. Zalta (ed.), URL = <https://plato.stanford.edu/archives/win2019/entries/paradox-zeno/>.

an essential nature of such paradoxes. They simple created fake Singularities as extrusions from fake infinities: then acted surprised that the ACS (A) (Categorical Universal) would not solve.

It won't solve, because it's not really a categorical universal at all. It decouples the hylomorphic linkage in "man", "all men" and mortal", and creates an illusory abstraction.

In other words, the solutions to such paradoxes are not found in the moves within the game. They are inherent within the artificial structure of the game.

Andrew Gallois wrote, in 2016:

> Irving Copi[228] once defined the problem of identity through time by noting that the following two statements both seem true but, on the assumption that there is change, appear to be inconsistent:
>
> 1. If a changing thing really changes, there can't literally be one and the same thing before and after the change.
> 2. However, if there isn't literally one and the same thing before and after the change, then no thing has really undergone any change.
>
> Gallois, Andre, "Identity Over Time", *The Stanford Encyclopedia of Philosophy* (Winter 2016 Edition), Edward N. Zalta (ed.), URL = <https://plato.stanford.edu/archives/win2016/entries/identity-time/>.

Is the identity problem, identified 2,500 years ago,[229] a matter of mind or matter? Is that a meaningful question? And if it is, then how would we go about solving it?

---

[228] (1917-2002)

[229] And much longer before: as it is the genesis of all religion

David Egan commented in 2019:[230]

> What, then, inclines us – and Heraclitus – to regard this expression of an attitude as a metaphysical fact?
>
> Recall that Heraclitus wants to reform our language-games because he thinks they misrepresent the way things really are.
>
> But consider what you'd need to do in order to assess whether our language-games are more or less adequate to some ultimate reality. You'd need to compare two things: our language-game and the reality that it's meant to represent. In other words, you'd need to compare reality as we represent it to ourselves with reality free of all representation.
>
> But that makes no sense: *how can you represent to yourself how things look free of all representation?*

The shadows on Plato's cave[231] were of us, in our human condition:

> Socrates: All in all, I responded, those who were chained would consider nothing besides the shadows of the artifacts as the unhidden. Glaucon: That would absolutely have to be.

Yet those escaping their shackles would, upon their return to the matrix of the cave, be blinded by the true sunlight of forms, and mocked by the Morlocks.[232]

Matrixial Logic correlates to the position of a cave escapee who can survey unrepresented reality, through the juxtapositions of inequalities, which millenia of paradox solutions have occluded.

---

[230] https://aeon.co/ideas/can-you-step-in-the-same-river-twice-wittgenstein-v-heraclitus Author emphasis added

[231] *Republic* (c375 BC)

[232] *The Time Machine*. HG Wells (1885)

This Chapter, like the others, is intended to be a heuristic narrative. By testing established paradigms of logic, the unique properties of Matrixial Logic, are uncovered.

In the cause of clarity, we have been a little unkind to our illustrious predecessors. That is, intentionally, to aid understanding. With a soupçon of the vanity attributable to novelty.

Chapter 4

FORMS OF INEQUALITY

Atomic Equality

We might metaphorically say that the bivalent logic of Forms provides the "toolkit" with which we can begin to unlock other logics.

Unlike the First / Higher Order logic divide, ML is holistic: *hylomorphically reciprocal.*

In plan English, ML works as the interaction of its parts.

It may be worth addressing, at this point, the phenomenon of *atomism*. Russel's system, like Leibnitz's was explicitly *atomic*: blocks constructed of single elements, and arranged into a model.

In both models, the essential building block is [A=A]: the law of identity. With its congruent laws: non-contradiction and excluded middle.

Essentially, these *atomic models* take:

$$\forall = [A=A]$$

as the atom, and then build from there, through the linear portal of:

$$\textit{Every } \forall \text{ is B.}$$

$$\approx$$

$$\forall\, x{:}\ P\,(\mathrm{x})$$

This gives you Set Theory,[233] and on to *modern logic*, syncopating around Wittgenstein, Godel and Tarski.[234]

In an *atomic model*, the atoms are undifferentiated, save as to linguistic "colour". Each atom has existence independent of the curriculum in which each atom appears. Each atom is unperturbed by its inclusion in a curriculum. There is erected a structure, a lattice, of such atoms.

The atomic lattice is uniplanar. As if each lattice were written on a sheet of paper, and the sheets of paper marshalled upon a table top. So conceived, the derived frames of reference are both arbitrary, and fixed.

- They are arbitrary: since you may shuffle the latices around as you please. That is, after all, why you engaged in the construction of the atoms in the first place;
- But they are fixed by the assumed existential property of each atom.

What the atomic model achieves is a bounded infinity. It is a *shadow* of infinity ∞.

It ought then, perhaps, to be unsurprising that the enterprises in logic derived from the assumed existential atom have proved:[235]

- Incapable of solving the paradoxes of movement, of dynamic

[233] Boole, Cantor, Frege, Peano, Russell, Hilbert

[234] Just the space between some landscape peaks, not the landscape

[235] Repeatedly and since inception in 350 BC

- To create their own paradoxes;[236]
- Incable of representing the world of the real, whether in psychology, biology, physics or quantum mechanics.

It is regrettable, but inevitable, that such logic which ought to have been the Ace of sciences, has becomes mired in the undistributed middle[237] of the pack. Limited and expressly self-limiting. And all because of the existential atom, and its tautological law of identity.

ML works differently. A fundamental principle of Matrixial Logic is that:

nothing Is

Less obscurely, that any Thing which Is, enjoys that property of Existence only in plural juxtaposition to that which is not (nA), or negates (-A).

Existence is no more a property of abstraction into an *atomic* entity, then the necessary plurality of Things (^{NOT}) / (-), which populate existence.

Existence in neither a void, nor a plane, from which extrude (or are intruded) these (Things$^{NOT\ /\ -}$). Existence and (Things$^{NOT\ /\ -}$) are reciprocally inherent features of each other.

[236] The empty set, the set of sets, and so on

[237] A syllogism joke

To posit $\forall = [A=A]$, is to posit a nullity. To operate with that nullity in logic, is to generate a fiction: a shadow of an (A).

Experience: Marked

Step 1:

- Imagine a bare and blank table top (say 1 foot square);
- Imagine taking a one inch square;
- Write A on that square;
- Locate that Asquare anywhere you wish on the table top.

Step 2:

- Create another Asquare: but this time;
- As a perfect replica of the one you marked;
- Place that other Asquare anywhere else on the surface you wish.

Step 3: Round One

We are actually playing a game. Of you against your opponent. The only rule of the game is: whatever move your opponent makes, is within the rules of the game.

So:

- Your opponent takes away this perfect replica Asquare.

End.

<u>Discussion</u>:

(1) Question: How do you feel about your opponent's move?

Answer: Obviously, you do not care. You can just produce more copy Asquares ad infinitum.

Step 4: Round Two

This time:

- Your opponent takes away the Asquare that you marked

End

<u>Discussion</u>:

(1) Question: How do you feel about your opponent's move?

Answer: This time, you do care. It *feels different* to Round One.

But this is odd. The Asquares, the table: they are all figments of your mentation. There is no reality here.

Yet:

(1) We have to conclude that the original Asquare is different to you then the copy Asquare.

(2) You feel differently about these two items.

Suppose you say: "Yes of course, because one is original and one is a copy. So they are not really an example of ∀."

Experience: Unmarked

Step 1:

- Imagine a bare and blank table top (say 1 foot square);
- Imagine a one inch square, with A marked in it;

- The Asquare is already there, on the table.

Step 2:

- Create another Asquare, that is a perfect replica of the Asquare at step 1.

Stop.

Discussion:

(1) You have no doubt here that A=A. There is complete, perfect identity.

(2) But you find that you just can't manufacture the iteration of Asquare.

(3) You try to cheat and:
draw a "line" connecting A ___ A
or
imagine A moving to the position which you want A to occupy.

(4) But the cheats don't work. You begin to have feelings of frustration.

(5) Your A simply will not multiply into a perfect identity.

This is not a trick of the mind, or the mind's eye. That fact is evidenced by changes in your emotional state. You are experiencing reality perfectly clearly.

It is simply that reality does not work as you might have been imagining. It does not work that way for you. It does not work that way for any reader. It does not work that way for anyone, and never has. *Reality does not work in identity.*

Experience: Written

Step 1:

- Take this space below:

- write A on it, anywhere: (example)

A

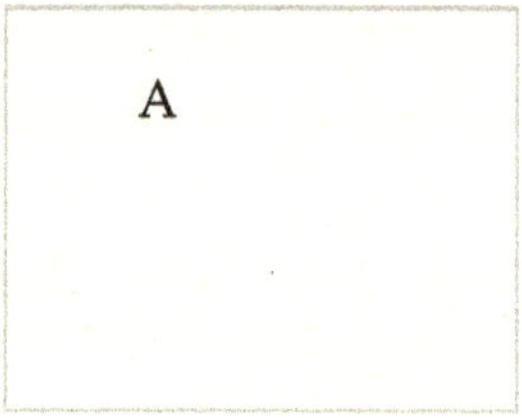

- write another A on it, anywhere: (example)

A

A

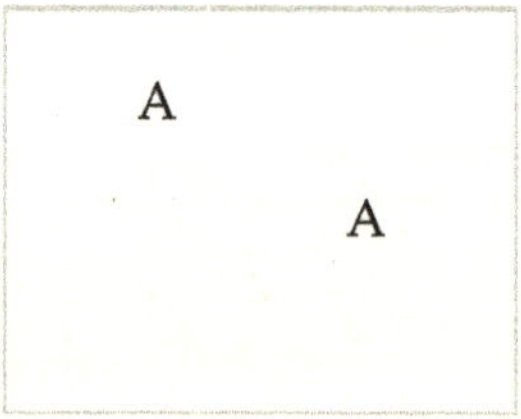

Discussion:

(1) That was easy. Write one A. Write another A. The "same" A.

(2) Yet, from all else read so far in this book, you know from the *Unmarked Experience,* that these graphics are not the same as the A you tried to, but could not, manipulate in your head.

(3) You also know from the *Marked Experience,* that your marked A was not the same as the A you just imagined.

But we do logic with written-down A=A: ∀. We have done that logic for 2,500 years. Indeed, we hold those two A A written down, as above, in our heads, and we manipulate them, in our heads. Then we decant the results of those manipulations into extraordinary feats of logical computation.

Yes, that is how we do logic. That is how we seem[238] to get computers to do logic (by using the law of contradiction, such that $1 \neq 0$). So it works. Except when it doesn't.

All of which leads to these enquiries:

(1) There must be some process in our minds which allows us to think of an idea of "sameness", of identity, such that we can think of *writing* A A. So, how does that thought process arise: *where does the*

238 But only seem

idea of identity come from?

(2) Are we using differerent thought processes when we think of matters, and when we write matters down?[239]

(3) If so, how does the difference arise and how does each such process work?

(4) Is our "in my head" language, a type of thinking like the writing, or like the internal thinking, or something else?

(5) Is our "spoken out loud" language, a type of thinking like the writing, or like the internal thinking, like the "in my head" language, or something else?

(6) Which of these is "real"? Or is reality a composite of all of them?

If any of these are useful questions, what then, has Matrixial Logic to do with any of the answers? The first answer is: that it is Matrixial Logic which compelled the questions to be asked.

This Chapter addresses question (1).

where does the idea of identity come from?

Once we have the answer to this Question, the rest unfolds.

[239] We can already see that this is so, in some way

Notation

We will be building on the notation and equation forms introduced over the last 2 Chapters. No doubt, you can do Matrixial Logic using other notation, and formulating other equations.

Whatever format you decide to invent, you will however have to use ≠, and symbols whiuch differentiate Substance in Opposition in Being, from Moments in Negation in Becoming. These are concepts which allow ML to operate.

What we should be able to show you in this Chapter is how the internal logic of Forms of Inequality can lead analysis using deduction and induction, to results which are interesting, and challenging.

PART (1) NODE FORMATION

Form and Node in [E] Being

Actuality □

We already know that a Substance is extended in spacetime. No Substance exists as a singularity: an 'atom'. Existence is exclusively a quality of an opposing plurality of Substance extension in spacetime.

Substance operates as Actuality □.

$$(A) \neq (nA) = [E]$$

Actuality □ is a Dynamic of things in being

Actuality □ is Dynamic Substance under Form.

Actuality in plurality of Form interacts as Event <∑E>.

There are no rules of Matrixial Logic as to the terms and conditions of such interactions. Thise are matters of inexhaustible contingency in the world.

Event<∑E>

Event <∑E> is not a singular occurrence.

Event <∑E> is an aggregation ∑ of multiple [E] FoR_n:

$$((A^1) \neq (nA^1) = [E^1]) \neq ((A^2) \neq (nA^2) = [E^2]) \; [\sim] \; \sum(FoR_2)$$

for which we can now substitute and simplify:

$$(\square 1) \neq (\square 2) \; [\sim] \; \sum(FoR_2)$$

$$>$$

$$(\square 1) \neq (\square 2) \; [\sim] \; <\sum E>$$

It is somewhat distracting to think of <∑E> as "an" event. This implies a plurality of Event(s). Although we have become used to thinking in pluralities, this is where such mode of thinking takes a temporary respite.

When an Event <∑E> interacts with something else, there forms a Node.

Form of Forms: Node <∑E>

With <∑E> concept, we reach a watershed. Until now, we have been considering Forms:

Substance creates Form as [E].

Moment creates Form as [I].

In Chapter 3, we allowed some interpolations of Forms. This was provisional, so as to aid explanation, by reference to extant problems arisen in logic.

We now tread more carefully, and definitely.

These Forms [E] and [I] are *deduced* FoRs.

- This means that each Form is created by its own Contents; each Form is the aggregated "sum" of its Contents;
- At the same time, each Form is *reciprocally hylomorphic*: Content and Form stand in relation to each other as both contingency and satisfaction.

What we mean by *deduction* is what we mean when we consider a[240] Categorical Syllogism: the major and minor premises together direct the conclusion.

The syllogism is circular reasoning founded in a set of assumptions. The Conclusion is also the expression of that matrix which allows the direction to occur.

240 Barbara form

The following graphic attempts to make that syllogism analogy more clear:

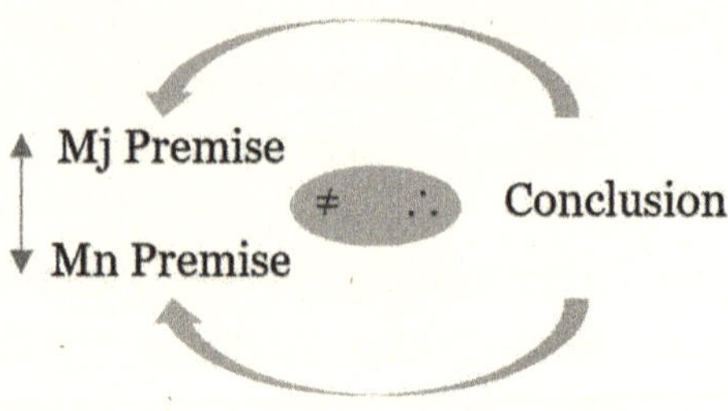

It can only be an approximation. The Premises play a role in relation to the Conclusion which is in a sense like Actuality □.

But what this graphic cannot show is the Forms of Forms upon which the syllogism really relies.

This is a Node: (□1) ≠ (□2) [~] <∑E>

In ML, we do say "form of forms" because it has a nice ring. But it is, strictly, an oxymoron. That is why there is no aequalis = joining the two sides but instead a Nodal [~].[241]

A Node <∑…] is
any convergence of two Forms in the same domain.

[241] For the avoidance of doubt, this is not the ~ meaning non or negative in classical logic.

Thus:

<∑E> is a Node
<∑I> is a Node[242]

but
Substance and Moment cannot together [~] create
Nodality <∑…]

Instead: they together create a completely differenet
phenomenon: as we will see.

These are ML Rules. There are very good reasons for the rules, which we will explore later.

Summation in [E]

Substance (A) ≠ (nA) = [E]
Form [E:] a Frame of Reference FoR

As Actuality □ □
As Event Dynamic <∑E>.

Event <∑E>
is not a singular occurrence.
is an aggregation ∑ of multiple [E] FoR_n

---

242 As a theoretical matter: although we will get used to seeing [I] Nodes only as Interferences.

Thus:

$$(\square 1) \neq (\square 2)\ [\sim]\ \Sigma(\mathrm{FoR}_2)$$

$$>$$

$$(\square 1) \neq (\square 2)\ [\sim]\ <\Sigma E>$$

Finally:

[~] means Nodel; creating a Node

thus:

<∑E> is a Node

Inequality and Nodality: ≠ \ [~]

It may be apparent that a Node is unstable. Compared to what?

The equation (A) ≠ (nA) = [E], describes a completely stable dynamic. Again, it is difficult to analogise with precision. However, consider the stable dynamic of the Hygrogen atom (using the Thompson model just as a thought-guide):

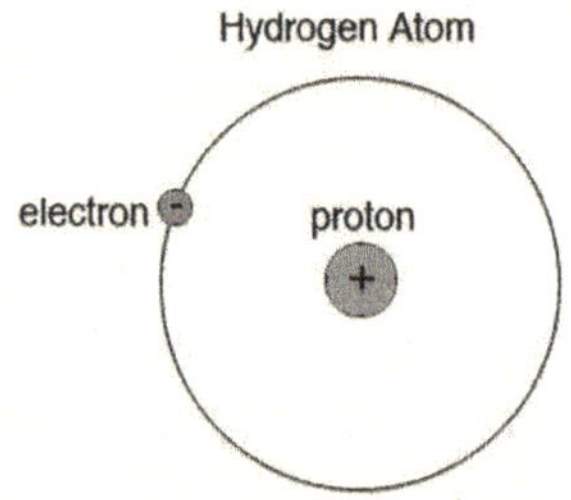

If we conveniently label the Poton as (A) and the Electron as (nA), then we see the dynamic state:[243]

(Proton) ≠(Electron) = H[E]

We use this analogy because every reader will be familiar with basic atomic theory.[244] Unlike the A A written upon a page tablet, the P and A are not static morphs. The picture is in reality a snapshot of a video: the electron is "circulating" the proton.[245]

The point of the analogy is that this is a *stable dynamic of inequality.*

We do not mean by this that an [E] Form is some perpetual perfect Platonic ideal, existing in some nether-domain. Inded, it is the fictions of aequalis (A=A) which permit Platonic idealism: indeed, necessitate it.

The word "organic" (as contrasted with idealised), may help in clarifying what we mean by a *stable dynamic of inequality.*

---

[243] We can state this in the quark-neutrino particles of the quantum Standard Model: the analogy is the same

[244] Most readers will no doubt be very more familiar than the Author

[245] Yes: peturbations in quantum energy fields etc

Form and Node in [E]

Form [E]	**Node <∑E>**
• in Actuality	• of Actuality
• mediated by ≠ with	• interacting [~] with
• Substance (A), (nA)	• Form □ □
• stable	• unstable
• dynamic	• static
• self-actualising	• externally actualised
• deductive	• inductive

Node relates to the Contents of its constituent Forms, only indirectly. When we simplified:

(1.1) $$((A^1) \neq (nA^1) = [E^1]) \neq ((A^2) \neq (nA^2) = [E^2])\ [\sim]\ \sum(FoR_2)$$

$$\approx$$

(2.1) $$(\square 1) \neq (\square 2)\ [\sim]\ \sum(FoR_2)$$

$$>$$

(2.2) $$(\square 1) \neq (\square 2)\ [\sim]\ <\sum E>$$

the underlying cross-dynamics at (1.1):

are present in the Node as the function of
its existence in Subtstance, but present differently:

[E] Form is in Actuality.
<∑E>Node is of Actuality.

This is not saying that we cannot "map" Nodes onto the real world. We can. It is analogous to saying, "That table there is comprised of atoms, which constitute peturbations in quantum energy fields."

We intend by that analogy to convey the idea[246] that you cannot "see" the table, the atom and the quanta at the same time, and in the same way. You choose your perspective. Then you may seek to unify those perspectives.[247]

In like manner, you may in ML choose at what location in reality you decide to begin the deductive process from Substance extended in spacetime, which realises an inductive Node <∑E> .

From what we have already said, the characteristic of [E], that it is self-actualising, should be self-evident. That is merely another way of stating: hylomorphically reciprocal.

Thus, we have done [E]. To the state of decease characteristic of a flogged horse, the reader may feel. A guilty plea is entered to that charge. In mitigation is offered:

- The explanation of [I] is rendered more swift and self-evident;
- When we come to the Vectored Collapse of Infinity Potential, the reader will be glad of fluency gained through slowly layered explanation.

---

246 With which any reader is familiar

247 Whther by an internal or external explanation

With all that said: [E] is the domain of Being. We have analysed the generation of Nodes of Being <ΣE> .

The domain of Becoming [I] works in quite another way. We will consider that, after this.

PART (2) VECTOR AND INFINITY

Vectored Collapse of Infinity Potential

We now turn to examine this phenomenon:

m̄ **Collapse**
of an Infinity ∞ Potential Δ^n
at a Vector /V/ of Node <ΣE>

The reader may recall us stating:

A Node <Σ...] is
any convergence of two Forms in the same domain.

Thus:

<ΣE> is a Node
<ΣI> is a Node
but
Substance and Moment cannot together [~] create
Nodality <Σ...]

When an Event <ΣE> interacts with something else, there forms a Node.

Vector /V/ : Crossing Domains

Actuality □ interacts meets with Potential Δ, only indirectly.

Actuality □ requires a Node <∑E> to interact with Potential Δ. Potential is, of course, the becoming of Moment under a Form [I].

A Vector /V/ is the mode by which
a Node <∑E> interacts with a Potential Δ.

Marginal Note:
Vector
In standard scientific usage, a *Vector* is an object that has both a magnitude and a direction. Geometrically, we can picture a vector as a directed line segment, whose length is the magnitude of the vector and with an arrow indicating the direction. The direction of the vector is from its tail to its head.

The word *Vector* derives from the Latin word carrier, for *vehere* "convey"

In conceptualising Vector /V/, the Latin is helpful. The /V/ is what conveys the Node <∑E> to interaction with a Potential Δ^n.

What then, do we mean by "interaction"? The specification of the properties of interaction are matters for physics, biology, psychology and so on. They are the contingencies of the real world, the world in which we each have our processes of feeling and thinking.

We mean "convey" in like manner.

Which brings us to declare that Vector /V/ interactions and conveyances are matters of induction, not deduction.

- We cannot deduce /V/ from Actuality □ in any Node <ΣE>.
- Nor can we deduce /V/ from Moment Potential Δ.

We can observe and induct Vector /V/ interactions and conveyances. In like manner, the laws of quantum mechanics are not necessarily based upon direct observation of actuality. They are mathematical and statistical derivations arrived at by induction from that which can be observed.

We can illustrate a hypothetical Interaction:

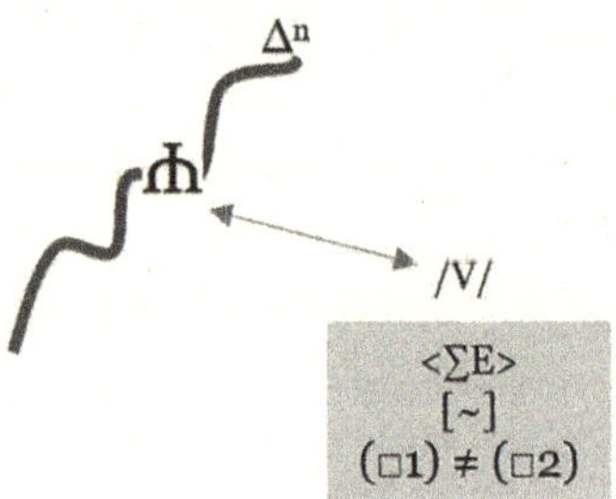

Here, we witness explicitly[248] for the first time:

domain interaction in the matrix

Plurality [E] in Being, aggregated in Nodal forms, meets Plurality [I] in Becoming: which requires no aggregation,

[248] In Chapter 3, we saw how implicit and unrecognised Nodes fundamentally destabilise Aristotelian logic

since that Plurality [I] is coherent as Potential Δ^n in its own process of Infinity ∞.

Such /V/ interaction is VCIP:

Vectored Collapse of
an Infinity ∞ Potential Δ^n

We do not say "effects", or "causes". To do so would be to introduce Scalar founded concepts,[249] when there is not yet any matrixial foundation for them.

The notation for VCIP is self-explanatory:

<∑E> ⇔ ጠ

It can equally be written as:

ጠ⇔<∑E>

To repeat, there is no concept of cause or effect represented in this notation.

The VCIP Perspective

We are now well used to thinking in terms of Frames of reference: FoR.

FoRs apply only in aspect of Forms: [E], [I]. FoRs are deductive. That is a functional attribute of their hylopmorphic reciprocity.

---

249 See later in this Chapter

As we have seen, FoRs in plurality, and in relation to each other, constitute a Node. A Node is not an FoR: it is a state of affairs of correspondent FoRs. Perhaps FoR_1 and FoR_2 are coherent. Perhaps they are in conflict. It depends upon the matters involved.

We can be sure that an FoR in [E] and FoR in [I] do conflict. They do not cohere.

These different FoR domains meet only in Vector interaction.

But:

$$[<\sum E> \Leftrightarrow 帀] \approx [帀 \Leftrightarrow <\sum E>]$$

This has attributes of the Chaotic.[250] It is Perspectival. Our Perspective is inducted from ⇔ and ≈. But nothing specifies in advance the rules or principles of such induction.

Is this something to do with subjectivity? It is certainly something to do with what we have chosen to debate in and as the domain of subjectivity.

The notations ⇔ and ≈ tell us nothing, because there is nothing here that we can in any "objective" sense talk about.

[250] Please return to this, when you have read *Chaotic Order*

We may think whatever we may each wish. We may then seek to communicate those thoughts, within ourselves and between ourselves.[251]

The mode of communication of such unmediated VCIP Perspective is not obvious. By what means does a Perspective, unmediated by anything like what we could call a rule, have effect?

Perhaps that question does not matter. This is a world, a reality, without humans: without human minds, or brains, or bodies. It is simply the world in Forms of Inequality.

There is no present Perspective, because there is nobody who needs one. These interactions do not require an observer.

PART (3) SCALAR SOLUTIONS

The Human Scale

What, then, if there be a population of persons? A collection of mediations through rules is a requisite for the transformation of un-Ruled Perspective into: humanity.

We must turn Vector /V/ relations into Scalar /S/ solutions.

[251] If there is a difference: which there may or may not be. We shall see later

Marginal Note:
In standard scientific usage, *Scalar* is a physical quantity that is completely described by its magnitude. Examples of scalars are volume, density, speed, energy, mass, and time. Other quantities, such as force and velocity, have both magnitude and direction and are called vectors.

The word *Scalar* derives from the Latin word *scalaris*, an adjectival form of *scala* (Latin for "ladder"), from which the English word *scale* also comes.

In ML, Scalar /S/ is an *inducted* Perspective.

VCIP

Let us return to Vectored Collapse of Infinity Potential.

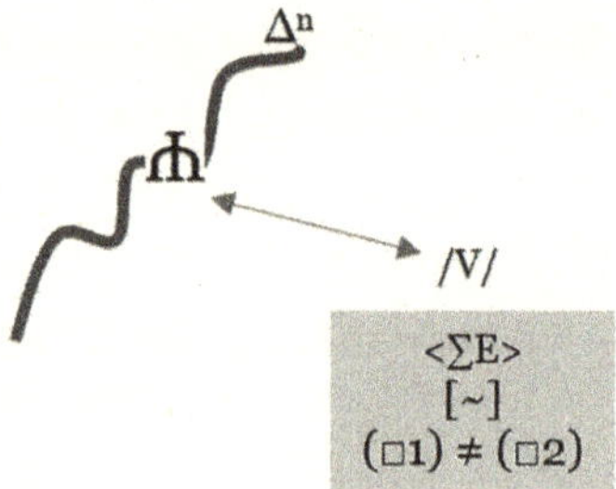

Can the reader please:

- Pause for some moments;
- Consider the ML gnosticism "nothing exists";
- Consider the ML fundamental proposition that existence is exclusively an attribute of plurality.

Experience: River

- Please do not turn the page, yet.

- Please invest a few minutes just looking at this:

- See VCIP Point m̄ in Δ^n

and

- Look in the graphic for new plurality

Stop.

<u>Discussion</u>:

(1) Note that the plurality cannot be the m̄ and the "river".

(2) That would involve creating an abstract quasi-form, which functions in (m̄) ≠ (river) = <?> [?]

(3) Indeed, this very error is at the core of the Aristotelian syllogism.

Counting in the Caves

Hopefully, in your thinking over this *Experience,* there has emerged the *Scalar Solution*:

(-m̄) ≠ (+m̄) = /S/

With the Scalar Solution, we have a novel form of equation in ML: one that uses quasi-arithmetic operators:

Arithmetic - Arithmetic +

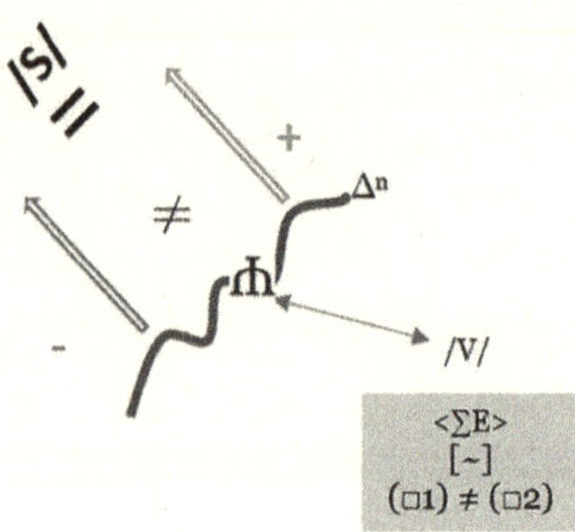

After all the reader has read, this may well, at first sight, appear bewildering. How, it may be asked, can quasi-arithmetic operators suddenly appear? What, the questioner asks, if their foundation?

Let us consider the matter under reasoning by exclusion:

(a) Infinity Potential Δ^n has collapsed under interaction with a Vector /V/.

(b) That Collapse "freezes" Δ^n at point ᵯ.

(c) Suppose that we denote (-ᵯ) as [nᵯ]. That cannot work, as that which is not [nᵯ] is just ᵯ.

(d) Suppose that we denote (-ᵯ) as [-ᵯ], where its antigone is □. That cannot work either.

What we are looking for, then, is a set of operators which:

- Refer to point ᵯ
- Are antigones
- Allow us to "escape" from point ᵯ

- Are self-cancelling, such that point m̄ is unaffected by the antigones.

We know already that one antigone cannot be *identical* to the other. It takes no ML reasoning to appreciate that the perspective of one side of an incident (a point m̄), whether in time or space, or both; is not identical to the other.

The antigones must accordingly relate under a *form of inequality* ≠.

We are already well used to this idea:

$$(A) \quad \begin{bmatrix} \neq \\ = \\ [E] \end{bmatrix} \quad (nA)$$

When we seek a correlation between Substance extended in spacetime, we have nothing to work with except (A), (nA). Nothing except their extensions in spacetime, which give us co-ordinate reference such as to be able to create a form from their inequality.

When we seek a correlation between Moment (A), (-A) in infinity ∞, we have infinity itself to work with. The potential Δ^n of Moments gives us the form of those Moments, in hylomorphic reciprocity.

To find Form in respect of that plurality which is not point m̄, we need only look to: that plurality which is not point m̄.

In assigning the quasi-arithmetic operators, we are not specifying any particular other point in Collapsed Δ.

The positions indicated in the graphic are arbitrary:

(-m̄) ≠ (+m̄) = /S/

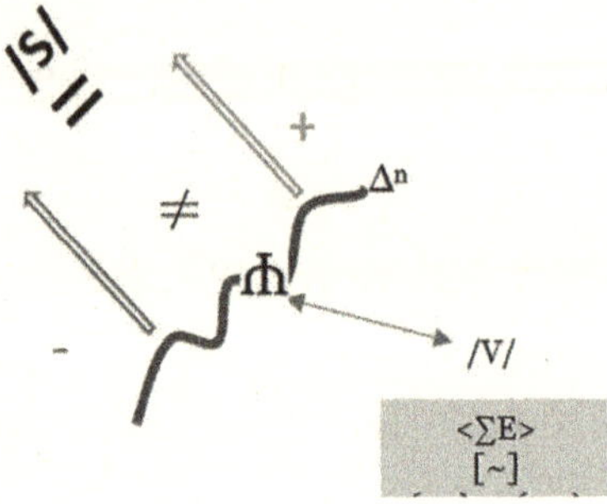

Each quasi-arithmetic operator [-], [+] merely denotes an undefined infinity of process.

If one were to imagine an arrow extending southwest in this imaginary topography, that arrow would be of indefinite, infinite, length. Likewise if one were to imagine an arrow extending northeast in this imaginary topography: indefinite and infinite.

The *order* of the operators [-], [+] is completely arbitrary:

(-m̄) ≠ (+m̄) = /S/

≈

(+m̄)≠(-m̄)= /S/

In the domain of Infinity Δ^n, divided by Collapse ₼, the *placement* of the operators [-], [+] is completely arbitrary. You can arbitrarily assign the location of [-], [+] "anywhere" on either side of point ₼ in the ∞ continuum. Indeed there is no "where", save by reference to being ${}^{\text{NOT}}$₼.

What we also see is that ≠ translates ₼ from static, to active. You cannot "do" anything with ₼.

You cannot even predicate "existence" of ₼. The domain of Existence [E] is way back, beyond the Vector /V/ interaction, and yet back beyond the node <∑E>.[252]

The equalisation:

(-₼) ≠ (+₼)

=

allows you to do anything, and everything. From making a bone hand tool, to constructing a space station.[253]

Because you have the Scalar Solution, as a form of inequality.

[252] Thus: farewell Heraclitus. You had a good 2.5 millennia run

[253] *2001: A Space Odyssey*. Kubrick. Clarke (1968). We are jumping ahead slightly

The Scalar Plane

Opportunity Is Freedom

Perhaps the Scalar Plane is unnecessary. Let us discover for ourselves.

Experience: Point

- See m̄ intersecting Δ:[254]

- Now, move m̄

Stop.

<u>Discussion</u>

(1) As with the paperclip,[255] you were unable to move m̄

(2) Sometimes, as you tried to move m̄, you ended up also moving Δ: as if they were stuck together.

(3) You tried creating shadow m̄,[256] but you knew that it was just a shadow. The real m̄ had not moved.

---

[254] Ensure that your visual field is limited to the "squiggle" and m̄

[255] See Chapter 3

[256] Like the Asquare on the table: earlier in this Chapter

We can conclude that the Scalar Plane is necessary. Let us evidence that necessity and opportunity:

Experience: Scale

- See ₥ intersecting Δ:

- Now, move ₥

Stop

<u>Discussion</u>

(1) It is now child's play to move ₥

(2) You can move ₥ "up" and "down" Δ

(3) You can even move ₥ off Δ altogether and allow ₥ freely to traverse the space bounded by the (-₥) ≠ (+₥) arrows.

We can conclude that the Scalar Plane:

- Is necessary;
- Provides opportunity.

Let us continue with the experiments, so as to demonstrate more features of the Scalar Plane.

Experience: Equals

- See ⫙ intersecting Δ

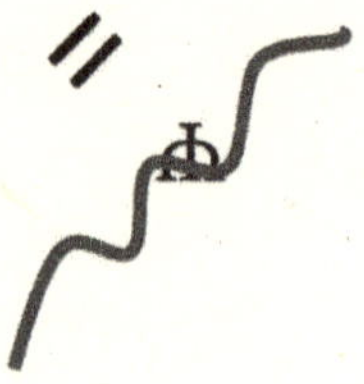

- Now, move ⫙

Stop

Discussion

(1) It is again easy to move ⫙

(2) You can move ⫙ with, in parallel, to =

(3) You can move ⫙ at angle to =

(4) You can move ⫙ off the path of Δ entirely

We can conclude that equality [=] in the Scalar Plane:

- Activates ⫙
- Allows free manipulation of ⫙

Equality Determines Freedom

We have established:

(-⫙) ≠ (+⫙) = /S/

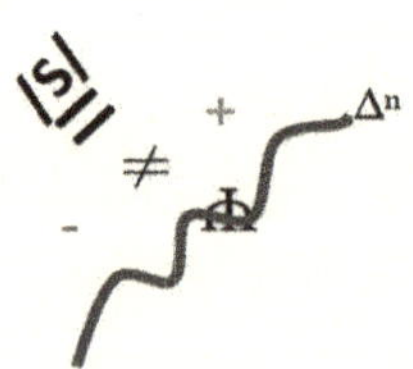

We now have =/S/. Let us examine what we can say of and do with /S/.

Firstly, remind ourselves:

A Vector /V/ is the mode by which
a Node <∑E> interacts with a Potential Δ.

This is:
domain interaction in the matrix

Let us contrast /V/ and /S/:

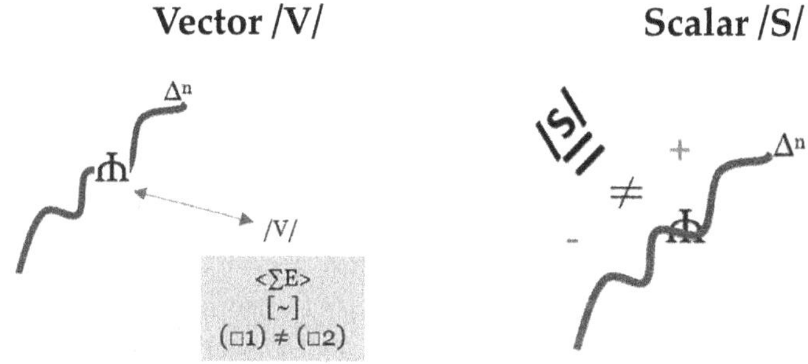

It is apparent that:

- /V/ has a direct and unilinear relationship to Δ^n. It is isomorphic.
- /S/ has an indirect. mediated, relationship to Δ^n
- Indeed /S/ only has relationship to a Collapsed Δ^n
- That relationship is polymorphic.

Now, would we say that the Form/Content relationship in /S/ is *hylomorphic*? Are either of [(-ሐ) ≠ (+ሐ)] anything like Substance or Moment?

Manifestly, no Substance or Moment is pre-determined. To ascribe prior determination to:

- An atom, or a quark, or any even "smaller" entity [(S) ≠ (nS)] appears to misuse the ascription.
- A moment [(M) ≠ (-M)] in any process give rise to difficulties of infinite regress, and its correlative infinite succession.

Moreover, the concept of cause and effect still has not yet entered ML. We have used only *descriptive correlation.*

Let us again be careful to bear this latter point in mind. In no way do we declare that /V/ "causes" $\bar{m}$, nor that $\bar{m}$ is the effect of /V/. No more than we say that the leaf is caused by the branch, or the effect of the branch.

Of course, without the tree, there would be merely a void in the landscape. Yet, to know that there is a tree, is simultaneously[257] to know, and be able to describe the interconnections between, the branch and the leaf.

It is well-established that correlation is not causation. Neither is connection: which is nothing other than a specific case of correlation.[258]

Thus: no Substance or Moment is pre-determined.

---

[257] The reader might find thoughts turning to "*spooky action at a distance*": Einsten to Born (1947)

[258] See preceeding Fn

But /S/ most assuredly is pre-determined in its limits:

that which can be /S/ is fixed by that which is already m̄.

By that, we do not mean fixed "in place", or "location". Point m̄ could, after all, occur anywhere in the domain of [I] ∞ in process Δ^n. That is as "large" and indeterminate a domain as it gets.

Indeed, rather than visualising ∞ in Δ^n as a "river": [259]

it may be more helpful to view ∞ in Δ^n as an expanse:

or in this view:

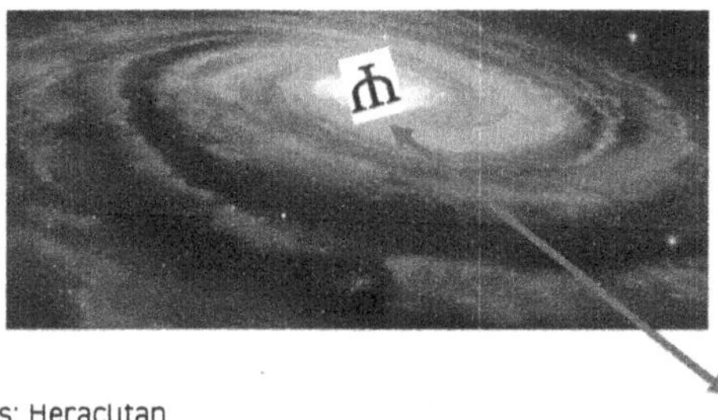

259 Yes: Heraclitan

or in this:

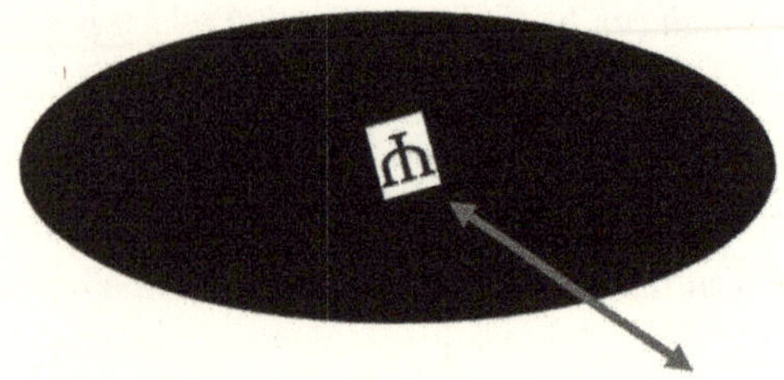

or in this:

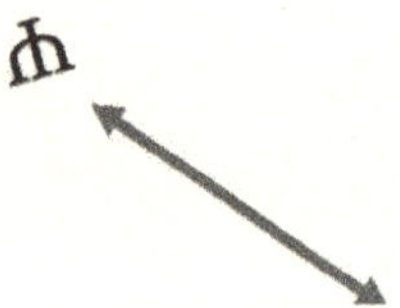

If the reader feels that the Author is perhaps twitching the *Wizard of Oz* curtain: that is so. These are deliberate choices of ∞ in Δ^n as expanse.

They each relate to our most profound, seemingly unbidden thoughts and feelings about our world and universe. These unbiddens repeat in every generation in history, are cross-cultural and have remained the unsolved[260] mysteries at the core of human civilsation.

260 Until the advent of Matrixial Logic?

Experience: Expanse

- See ጠ intersecting Δ, in each of the above representations[261]

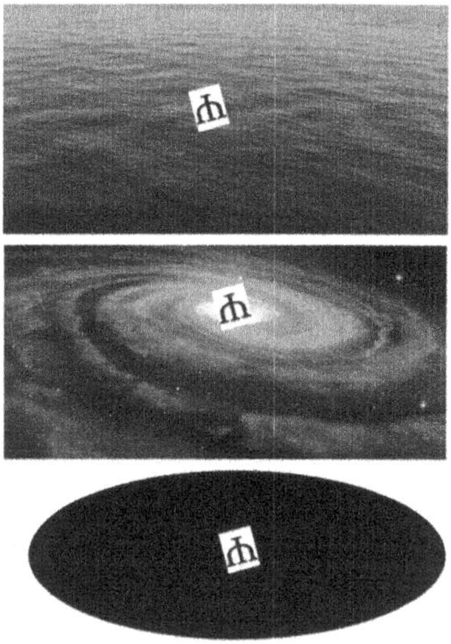

- Now, move ጠ, in each separate expanse.

Stop.

<u>Discussion</u>

(1) As with the paperclip,[262] and the Δ^n river, you were unable to move ጠ

(2) Because of the additional visual points in the starfield, moving ጠ was easier

(3) Moving ጠ in the whitespace was utterly impossible.

---

[261] Ensure that your visual field is limited to each separate "expanse" and ጠ

[262] See Chapter 3

You stare:[263]

- Into empty space. You feel like a dot. A point in infinity;
- At the ocean. You feel like a speck on the horizon. A point in infinity;
- Into the infinite darkness of soul, or death. You feel yourself to be a mere point in the darkness;
- Into the light, of afterlife or truth. You feel yourself to be a mere point in illuminated infinity.

This *Experience* $/S/^n$ is clearly telling us something, possibly something profound and significant about these terrains, landscapes, planes, surfaces, expanses. But what?

That is reserved for later Chapters.

Form and Node in [I] Becoming

Potential Δ

The domain of [I], the domain of Becoming, is Infinity ∞.

Moments Δ^n operate in Infinity ∞.

$$(A) \neq (-A) = [I]$$

Moments are in infinite succession and regression, simultaneously. To analogise imprecisely again, Moments can be thought of as if they were a *fabric of Time*.

Moment operates as Potential Δ^n in Infinity ∞

$$(A) \neq (-A) = [I]$$

[263] Maybe this reader doesn't: but it is commonplace that billions of human kind do, and have

Potential Δ^n is a Dynamic of Moments in Becoming.

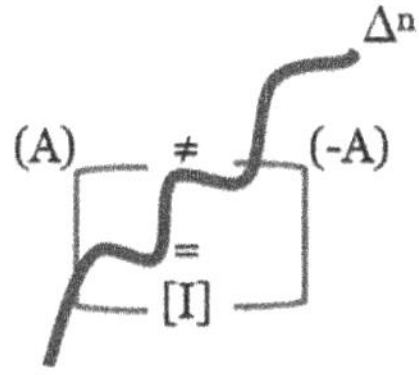

Interference <∑I>

An Interference <∑I> is not a singular occurrence.

Interference <∑I> is an aggregation ∑ of multiple [I] FoR_n

$$((A^1) \neq (-A^1) = [I^1]) \neq ((A^2) \neq (-A^2) = [I^2]) \; [\sim] \; \sum(FoR_2)$$

for which we can now substitute and simplify:

$$(\Delta 1) \neq (\Delta 2) \; [\sim] \; \sum(FoR_2)$$

$$>$$

$$(\Delta 1) \neq (\Delta 2) \; [\sim] \; <\sum I>$$

Like Event, Interference is perhaps self-evidently, only effected in plurality.

There are no rules of Matrixial Logic as to the terms and conditions of such *interferences*. Those are matters of inexhaustible possibility in the world.

Form of Forms: Node <∑I>

As stated earlier in this Chapter:

A Node <∑…] is
any convergence of two Forms in the same domain.

Hopefully, it is self-evident from all that has been written before, that:

an Event <∑E> and an Interference <∑I>,
cannot possibly be the same sort of phenomenon:
whether in the world, or in perception.[264]

Tabulating the essential differences between [I] and <∑I>

Form and Node in [I]

Form [I]	**Node <∑I>**
• in Potential	• of Potential
• mediated by ≠ with	• interacting [~] with
• Moment (A), (-A)	• Form Δ Δ
• stable	• unstable
• dynamic	• static
• self-actualising	• externally actualised
• deductive	• inductive

Node relates to the Contents of its constituent Forms, only indirectly.

When we simplified:

(1.1) $$([(A^1) \neq (-A^1) = [I^1]]) \neq ([(A^2) \neq (-A^2) = [I^2]]) \; [\sim] \; \Sigma(FoR_2)$$

$$\approx$$

(2.1) $$(\Delta 1) \neq (\Delta 2) \; [\sim] \; \Sigma(FoR_2)$$

$$>$$

(2.2) $$(\Delta 1) \neq (\Delta 2) \; [\sim] \; \langle \Sigma I \rangle$$

[264] Here intended to be a vacuous catch-all word for whatever one might term as "subjective", or occuring inside human beings, their minds, their feelings

the underlying cross-dynamics at (1.1)

are present in the Node as the function of its Potential in Infinity, but present differently:

[I] Form is	in	Potential.
<∑I>Node is	of	Potential.

Again, this is not saying that we cannot "map" <∑I> Nodes onto the real world. We can.

Interference, Infinity and Potential

Moment (A), (-A) is a Potential Δ^n in Infinity ∞.

It is tempting to give examples from the "real world". But any example we give at this time will serve only to detract from the explanation.

Interference Φ

We have seen that Interference between Potentials Δ Δ, can create <∑I> Nodes.

Potentials may *interfere* with each other, producing a simalacrum of an Event <∑E>.

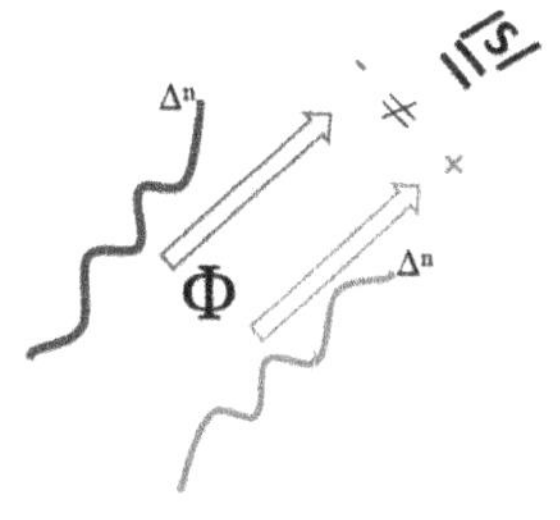

Interference does not prompt or effect Collapse ₼ of any Potential.

To bring forward an analogy which may already have occurred to the reader, imagine 2 waveforms. The underlying content of the waveforms could be anything: tidal currents, electricity, sound, and so on,

We are very familiar with the interference phenomenon, as applied to such waves. Interference may:

a. Sum

b. Subtract

c. Cancel

It is easy to see that in cases (a) and (b), the wave Potentials:

- Interfere with each other
- Do not Collapse

It is less easy to see, but no less the case that in (c) there is not a Collapse ₼, but an annihilation: a cancelling. You do not imagine the waveform suddenly freezing in place: standing still.

By contrast:
Just such "freezing" interruption of dynamic, of becoming, is what we mean by Collapse ₼.

Scalar Operations

We called /S/ a Perspective, and contrasted that with [E]

and [I] Frame of Reference.

We said that /S/:

- Is not hylomorphic;
- Is point-determined (or determinate).

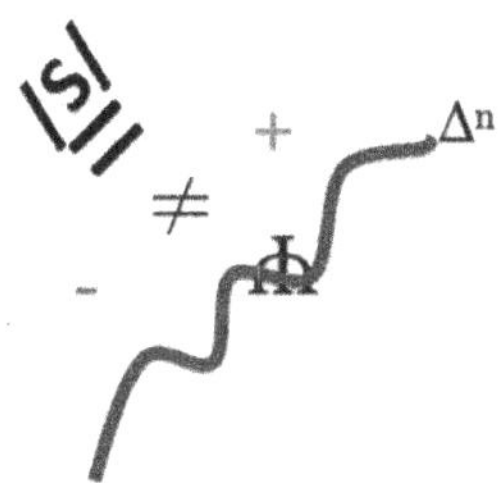

We deny hylomorphism to =/S/ because no part of:

(-m̄) ≠ (+m̄) =

- Is anywhere in Being;
- Is in any process of Becoming.

We can be content, at this point, to recognise that /S/ is an inducted fiction. We may not be so happy to admit that later.

By all means, we can say that /S/ is bivalent. The valencies are there in [+] [-]. And we have of course acknowledged that those bivalencies are arbitrary.

As previously stated, we only require of any bivalent set of operators that they:

- Have m̄ point reference
- Are antigones
- Allow us to "escape" from point m̄

- Are self-cancelling, such that point ⋔ is unaffected by the antigones.

So, we could just as easily assign as bivelent nominals "T" and "F". Or (p V q). Which makes /S/ a truth gate.

And, as we find more $/S/^n$, we can populate our truth tables. So, how do we go about finding more $/S/^n$?

Well that is obviously easy.

- We simply extract a plane of /S/

- and assign "drop down" points to the ⋔ point Collapsed Continum Δ^n

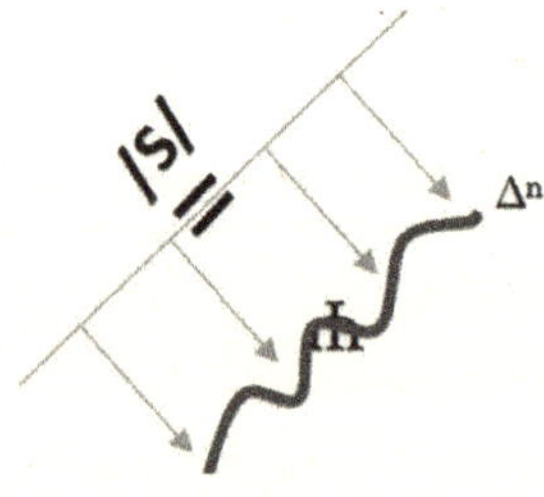

Scalar Extrusions

Ex Nihilo

Please assume that the orange lines, as drawn, are not accidental in their placement, or orientation.

The Author congratulates any reader who does not need to read this section. You have arrived at the conclusion by a process of necessary deduction.

For those who do not see the conclusion, [265] or who baulk at it, let us proceed by careful steps in Matrixial Logic.

Scalar Plane

We now arbitrarily orient the Scalar Plane /S/±

orthogonally lateral to the margin.

Experience: Rook

Step 1:

- Please ascertain that you can, in your mind, move the ± plane line horizontally, with complete freedom.

Step 2:

- You can also extend the ± line freely, in either direction,

[265] Which is entirely the Author's fault in failure of explanation thus far

with no effort of will, merely small concentration.

Step 3:

- Now, try to move the ± line so that is pointing up and down the page;
- Try to move the ± line as a diagonal.

<u>Discussion</u>

(1) As with the paperclip,[266] and the Δ^n river, and the ⫪ expanses, you were unable to move ⫪

We can conclude that the Scalar Plane ± has some, although very limited utility. We are still stuck in the plane of [=]. We can move sideways, and up and down.

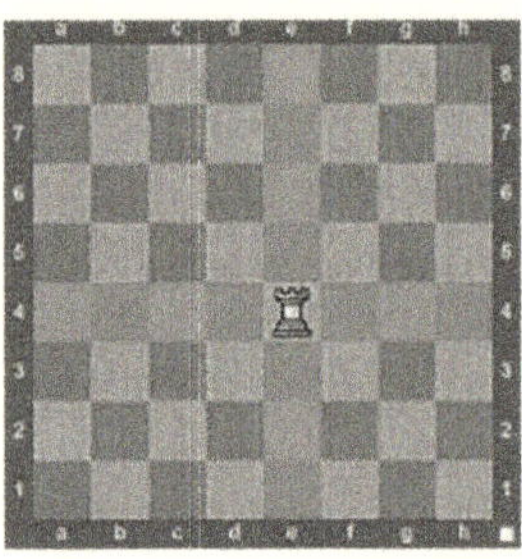

We are rather like a rook or castle in chess. We do not have two dimensional freedom.

Scalar Extrusions

Can the reader now please imagine dropping these ⊤

[266] See Chapter 3

arrowed lines orthagonally to ±

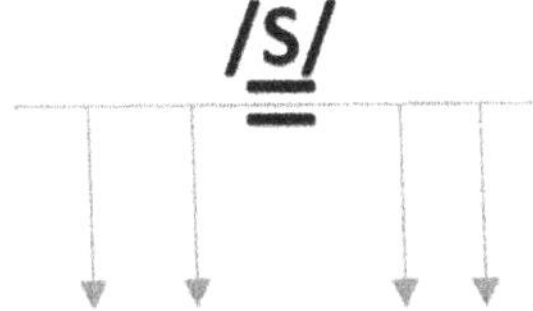

Please cover over the ⊤ arrowed lines, or use this as your focal point:

The reader will have seen that it is child's play to effect the arrowed orthagonal ⊤ extrusions.

Marginal Note
You will recall that the Scalar Plane line ± has a length. It is not an infinite length. However ± extends, that extension can never be infinite: because $\pm^n$ is an iteration, not a process Δ^n.

However, it is interesting that, no matter how long you make the arrowed lines, and no matter how many you make, their total length will always equal ± in length.

In this paradox, you will find the solution to the *completeness problems*,[267] wrestled with by Hilbert and Gödel, and finally abandoned by Turing.

This paradox applies to both Gödel Theorems.[268]

[267] https://en.wikipedia.org/wiki/G%C3%B6del%27s_incompleteness_theorems
[268] This is a large claim. For those mathematical logicians who are reading, please try it

Experience: Queen

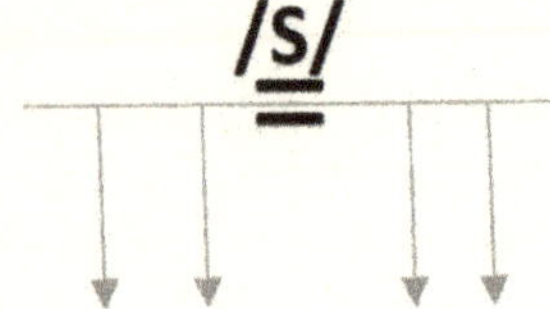

Step 1:

- Please ascertain that you can, in your mind, move the ± plane line, diaganolly, between the two dots:

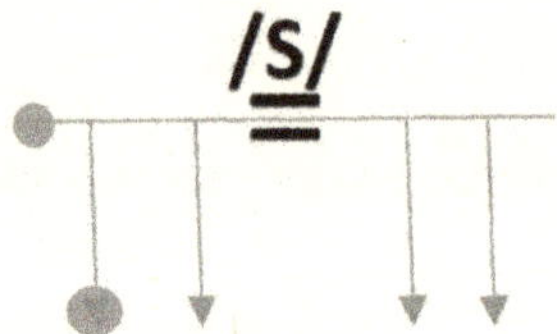

Step 2:

- Now, experiment on playing with the ± plane line.

We can conclude that the Scalar Plane ±, with the addition of has $[\top^{n}]$ has much greater utility. We are still stuck in the plane of [=].

But now we can move like the Queen in chess.

We have acquired two dimensional freedom.

Extrusions Arithmetic

We have these elements:

/S/   ±   [⊤^{n}]

What purposes may we make them serve?

We can add numerals, as follows.

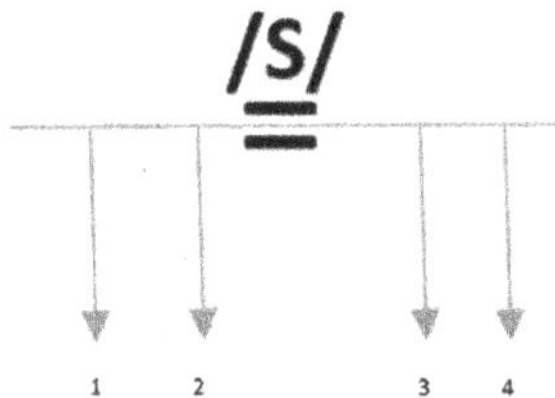

These numerals are just marks. They do not mean anything. We could easily mark:

- \ \ \ \

or

- \ \\ \\\ \\\\

They all merely answer to the question: *how many* ± planar extrusions ⊤ exist?

Exist where? Perhaps we should put an arbitrary marker (x) on the scalar plane line ± :

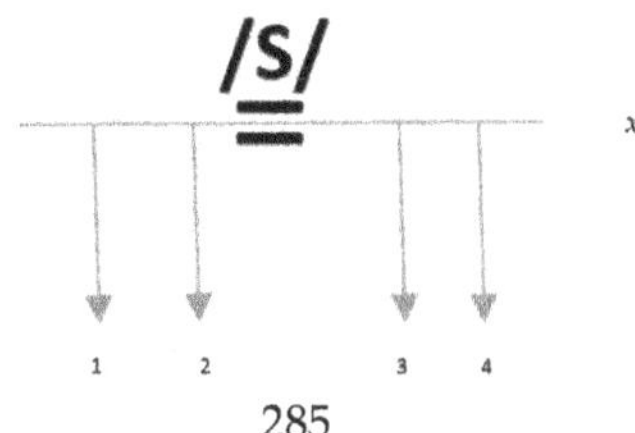

Perhaps now we have something we can really cook with:

[\x] ≠ [\\x] ≠ [\\\x] ≠ [\\\\x]

That does not tell us much.

But, look: suppose we just shunt marker [\\x] in the direction of [\x]. We can easily imagine doing that: it is just a lateral Castle move.

Then ≠ [\\x] disappears. So now:

State Rabbit in ±(x): [\x] ≠ ~~[\\x] ≠~~ [\\\x] ≠ [\\\\x]

>

State Mouse in ±(x): [\x] ≠ [\\\x] ≠ [\\\\x]

The reader's three year old child can answer this question: *what is the difference between State Rabbit and State Mouse?*

And the answer is:[269]

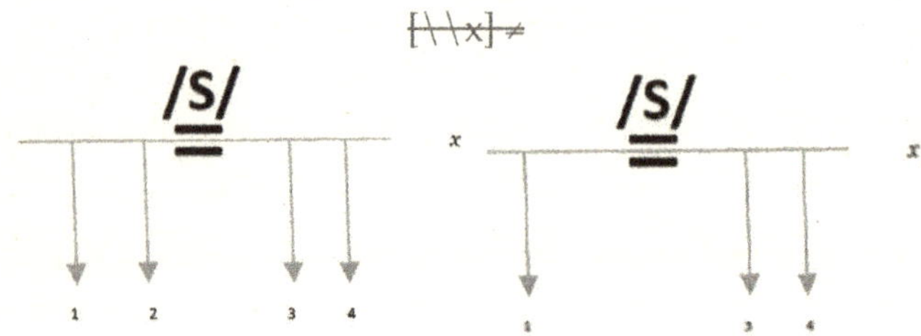

Now, each ±(x) is \ line.

So, how many ±(x) lines do we have at State Rabbit?

[269] We are using numerals instead of sticks, just for ease of illustration. Just think sticks

= (\ \ \ \)(x)

So, how many ±(x) lines do we have at State Mouse?

= (\ \ \)(x)

So: what is another way of saying the difference between State Rabbit and State Mouse?

= (\ \ \ \)(x)

≠

= (\ \ \)(x)

>

= (\)(x)

Great. That is elementary deductive reasoning. If we assign *truth values*, we have the kernel of an Aristotelian syllogism. Or a Fregeian or Godelian calculus.

There is a lot of clutter in those lines. Why don't we simplify?

- We can dispense with the first two = symbols: don't they really just mean "here there is on the ±(x) scalar line". Well, we know that. We can leave the last = symbol in, just to be sure that we know what plane or dimension we are in: just as we clssically do in ML. So:

= (\ \ \ \)(x)		(\ \ \ \)(x)
≠		≠
= (\ \ \)(x)	≈	(\ \ \)(x)
>		>
= (\)(x)		= (\)(x)

- We don't need that > symbol.
- So now, it all fits neatly on one line:

(\ \ \ \)(x) ≠ (\ \ \)(x) = (\)(x)

We can swap the line around:

(\)(x) = (\ \ \)(x) ≠(\ \ \ \)(x)

And say it out loud:

	=	≠	
	is	the difference between	
(\)(x)	(\ \ \) (x)		(\ \ \ \)(x)
	and		and

Let's simplify some more:

Instead of saying (again and again) for symbol [≠] "the difference between", let's shorten that.

How about instead of:

(\)(x) is the difference between (\ \ \)(x) and (\ \ \ \)(x)

- we get rid of all those repeated (x) (x) (x)

and

- we just say:

(\ \ \ \) *minus* (\ \ \) *is* (\)

and

- instead of saying "is", why don't we just put the = symbol and say "aequalis", or "equals".

Now, we have made *Scalar Extrusions Arithmetic* simple, and repeatable:

Three ⊤ minus Two ⊤ equals One ⊤

We don't need to repeat ⊤ ⊤ ⊤, do we?

So now let's have it simple:

Three *minus* Two *equals* One

≈

$3 - 2 = 1$

But this is awesome.

From this, we can deduce the whole of arithmetic.

Review

Let's just remind ourselves of the journey to this point:

(1) A Node <∑E> interacts with Infinity ∞ Potential Δ^n

(2) The /V/ point ᗰ in ∞ alters $\infty\Delta^n$

(3) We see Vectored /V/ Collapse ᗰ of $\infty\ \Delta^n$

(4) A continuum of Moments Δ^n in ∞ is now subject to ᗰ point interference

(5) The "river" is frozen, into parts [-ᗰ] ≠ [+ᗰ]

(6) A Form of Inequality can subsist: = /S/

(7) Scalar /S/ is, in itself, useless

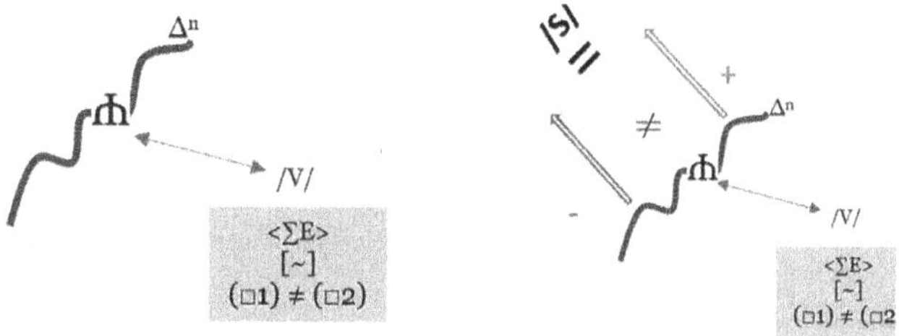

But then:

(8) we can extend /S/ as a plane ±

(9) /S/ plane ± can only move like a chess Castle

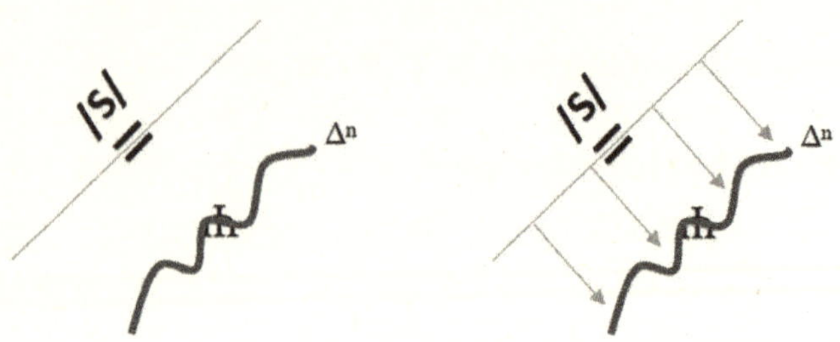

And then:

(10) we can extrude from /S/ plane ± to create □^{n} extrusions

(11) Now, we can move like a chess Queen: we have 2-dimensional freedom

(12) We can (and must) nominate the $\top^n$ extrusions as being of /S/ ±: and we can abbreviate /S/ ± to (x)

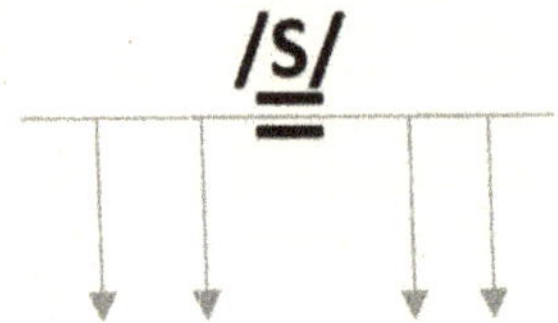

Which allows that:

(13) We can add arbitrary denominations to those various extrusions $\top^n$ (x): such as sticks \ \\ \\\ \\\\

(14) We can perform experiments with the sticks, as extrusions of (x)

(15) We quickly see that the sticks relate to each other under Rules

(16) We can call these Rules in (x), or just "Rules"

(17) So long as we remember that we are in $\top^n$ (x), we can create denominations for the sticks

(18) The denominations allow us to simplify

(19) We arrive at the abstract rules of arithmetic

(20) All of which work because we are manipulating extrusions from the Scalar plane.

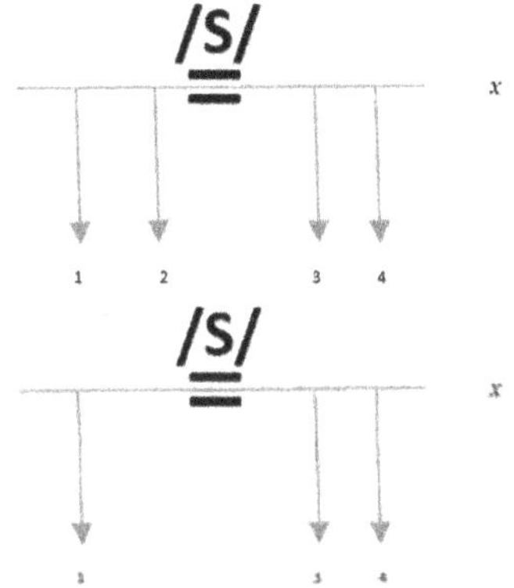

(\ \ \ \) *minus* (\ \ \) *is* (\)

Three $\top$ minus Two $\top$ equals One $\top$

$3 - 2 = 1$

This transition from a Nodal point, to the calculus of arithmetic make justly feel magical, transcendent. But it is not a mystery.

Extrusions Geometry

Let us begin again with:

/S/

We know we can imagine:

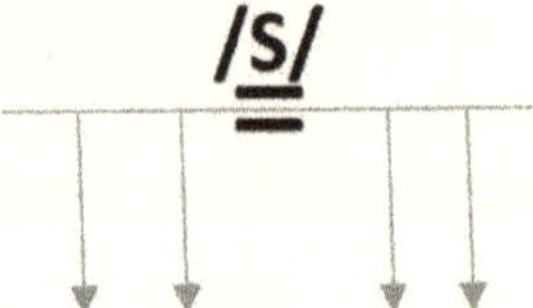

Try imagining a move

from: to

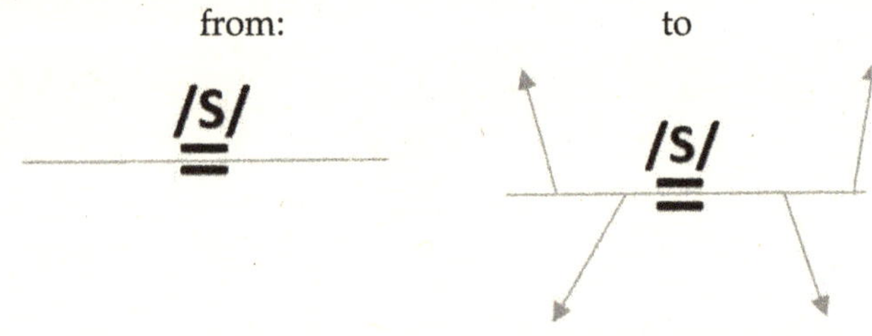

Again, the Scalar plane ± anchor allows these extrusions.

And, we didn't trouble about it before. But when we dropped the stick lines, we extruded from, rather than along the Scalar Plane (x). So let's call *extrusions from* (y).

So, we can do this:

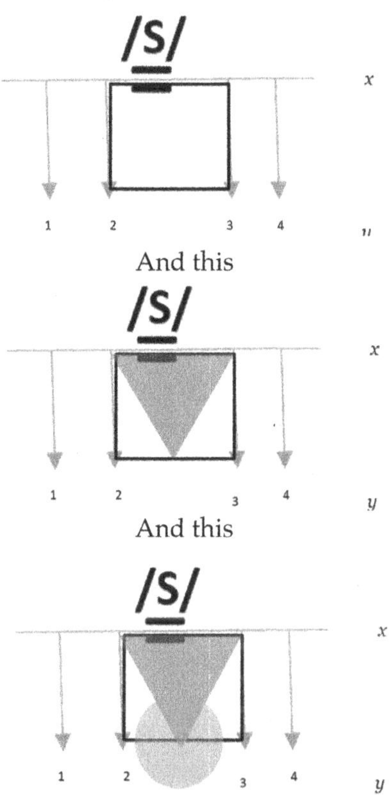

We can even plan the pyramids of Giza:

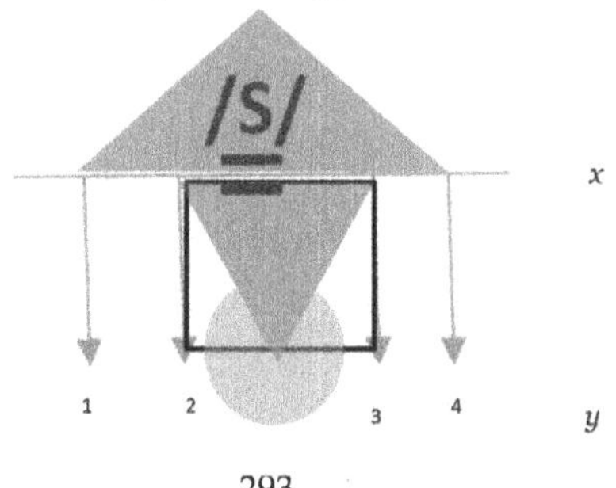

Scalar Extrusions give us (x) and (y): they give us the Queen's moves in chess. They give us Geometry.

We can use (x) and (y), in Arithmetic, Geometry and Algebra to do:

- Weights and measures
- Accountancy
- Calculus
- Maps
- Astronomy

and much more.

We can do:

- Euclid
- Aristotle
- Leibnitz
- Lambert
- Newton
- Boole
- Russell
- Godel

and much more.

All from extrusion at an angle from the Scalar plane ± .

PART (4) THE TRAP

The Burdens of Identity

It can then be seen that there is an upside, and a downside to manipulating of extrusions in (x) and from (y) the Scalar plane.

The upside is, for starters, all the technological attributes of pre-electric and atomic civilsation. That's essentially, the year 12,000 BC,[270] to 1820 AD.

The downside is:

The Law of = Identity

Recall our primitive arithmetic. Which we were doing, with sticks, in Bordeaux caves in over 10,000 years ago.

The reader's three year old child
can answer this question:
what is the difference between State Rabbit and State Mouse?

(\ \ \ \ ) *minus* (\ \ \ ) *is* (\ )

We say:

[*** | ****] IS (*)

We equal, we equate, we *identify*.

There is one rule, just one, which is the fundamental, the

[270] Or longer, per Graham Hancock, Michael Cremo et al

alpha and omega of manipulating of [⊤^{n} extrusions] in and from the Scalar /S/ plane ± :

[⊤] = [⊤]

Because one Extrusion IS another Extrusion.

Every Extrusion is a manifestation of /S/ plane ±.

There IS nothing else.

We can arbitrarily assign rabbit and mice, and (x) and (y) and (\)^n nominators for any of the Extrusions, or their parts, or their relationships.

We can form the most elaborate geometric shapes, run the most internally complex calculus. We can add Truth values to Euclid's geometric arithmetic, and get Aristotle. We can recombine them to engage in the construction of Boolean binaries, and Frege's calculus.

We can spend volumes of *Principia Mathematica,* trying to resolve empty sets, and sets which are members of themselves. We can talk Tarski to Godel.

And all together, all the genius, the brilliance, the two and a half thousand years of it.

And much more. Such as the claim that they really built

the pyramids and wrote the Book of the Dead, without syllogistic logic, and its foundations in geometric arithmetic?[271]

That archeo-technical claim cannot any longer be taken seriously. Not when we have just proved that all of it, every element, derives from the extension in a single plane of one line from a point created by a single cross-dimensional peturbation of natural infinity.

It is apparent to the Author that the builders of the ancient megalithic structures were far closer to understanding of the origins of "scientific" thought, than were the Ionians and Athenians.

Human civisation. All of it.
It all seems to depend upon one fundamental:
The Law of = Identity

The /S/ planar foundation in *Identity* of [⊤] = [⊤], is exactly why:

- We cannot step in the same river twice;
- /S/ cannot be the set of itself;
- /S/ cannot be an empty set;
- The self-referring Liar paradoxes work;
- Numbers are not infinite;

---

[271] We know from the *Rhind Papyrus* that the Egyptians did have this mathematics technology, a millenia before the Greeks; and the Sumerians has it around 3,000 BC: a millenia before that

- No information can travel faster than C.[272]

But is that supposed dependence real?

Matrixial Logic comes, not to destroy the law of identity.

But to liberate equality from those chains, forged from fourteen millenia celebrated in the amnesia of success.

Or, more soberly, because nobody thought of it, till now.

[272] Irresistably leaping ahead somewhat

CHAPTER 5

LANGUINI

Everything Is Meaning

We would need to spend all of this book, and an entire library of books, to traverse the landscape of linguistics.

However, may we quickly pour some snow over one Tarskian mountain peak. Encode words as numbers then manipulate the numbers: but what about the spaces between words?

"I feel fine" is not the same in meaning as "I feel… fine". That pause may actually negate the meaning of "fine". But you can't put numbers on gaps. Nor can you put numbers on cadence. Yet only robots speak without cadence. In many languages and linguistic uses, cadence (or lilt), is essential to meaning.

As we widen the field of focus from written words, we appreciate that setting provides its own semantic framework. The same words mean very different things in different settings. We have cultural and micro-cultural rules for what these meanings are.

Language is grammatical rules about communicating context.By using context, we can bend and play with the rules, so as to create inverted and new meanings. If language was reducible to the grammatical rules of language, it's difficult to think how language could ever change at all. Essentially, whatever the entirety of our individually experienced changing worlds, from moment to moment, is what any use of language means to each of us, and to each to whom we communicate.

Which brings us to the halting realisation that there are no determinable Chompskyan "universals". No Platonic library of semantics, syntax or social meaning out there beyond the cave entrance.

That doesn't seem to stop people trying. For example, this 1981 analysis by Robert Port:[273]

> It seems to be widely agreed that there must be some universal inventory of possible speech sounds and that these sound elements are inherently segmented despite evidence of many nonsegmental phonetic structures.
>
> It is argued here, however, that if a language is viewed as fundamentally a communication device for a community and phonetics is viewed as the signalling space for the language, then certain general constraints for speech sounds follow. For example, speech sounds must be invariant between gesture and sound and across members of the community.
>
> By employing the notion of symmetry, it is possible to define the notion of phonetic space in a theoretically useful way without having to specify any particular elements in it.

In 2005, Port and Leary published their challenge to half a century of what may be termed "behavioural-cognitive" linguistics: the attempt to create universal rules out of

[273] Lingua Volume 55, Issues 2-3, October–November 1981, Pages 181-219

language usage:[274]

> Chomsky and Halle (1968) and many formal linguists rely on the notion of a universally available phonetic space defined in discrete time. This assumption plays a central role in phonological theory. Discreteness at the phonetic level guarantees the discreteness of all other levels of language.
>
> But decades of phonetics research demonstrate that there exists no universal inventory of phonetic objects.
>
> We discuss three kinds of evidence: first, phonologies differ incommensurably. Second, some phonetic characteristics of languages depend on intrinsically temporal patterns, and, third, some linguistic sound categories within a language are different from each other despite a high degree of overlap that precludes distinctness.
>
> Linguistics has mistakenly presumed that speech can always be spelled with letter-like tokens. A variety of implications of these conclusions for research in phonology are discussed.

Which may leave the reader considering the rational prospect of this Chapter ending right here.

As Tim Rayner writes:[275]

> 'That whereof we cannot speak, thereof we must remain silent', Wittgenstein intoned in the closing passages of the *Tractatus.*
>
> To become a philosopher, one must learn to hold one's tongue. Logical positivism was a powerful movement that defined the shape of analytic philosophy well into the 1960s. However, it was undercut by the work of the same man who was its founder. By the 1930s, Wittgenstein had decided that the picture theory language was quite wrong. He devoted the rest of his life to explaining why. 'Resting on

---

[274] Port, Robert F and Adam P Leary. "Against Formal Phonology." *Language*, vol. 81 no. 4, 2005, p. 927-964. *Project MUSE*, doi:10.1353/lan.2005.0195.

[275] https://philosophyforchange.wordpress.com/2014/03/11/meaning-is-use-wittgenstein-on-the-limits-of-language/#more-11377 (2014)

> your laurels is as dangerous as resting when you are walking in the snow', he commented. 'You doze off and die in your sleep'.
>
> Wittgenstein's shift in thinking, between the *Tractatus* and the *Investigations*, maps the general shift in 20th century philosophy from logical positivism to behaviourism and pragmatism. It is a shift from seeing language as a fixed structure imposed upon the world to seeing it as a fluid structure that is intimately bound up with our everyday practices and forms of life. For later Wittgenstein, creating meaningful statements is not a matter of mapping the logical form of the world. It is a matter of using conventionally-defined terms within 'language games' that we play out in the course of everyday life. 'In most cases, the meaning of a word is its use', Wittgenstein claimed, in perhaps the most famous passage in the *Investigations*. It ain't what you say, it's the way that you say it, and the context in which you say it. Words are how you use them.
>
> Communication, on this model, involves using conventional terms in a way that is recognised by a linguistic community. It involves playing a conventionally accepted language game.

So, if there is anything useful to be said about the human use of language, can Matrixial Logic say it? And what does it mean?

Vectored Collapse of Infinity Potential

Framework

We draw the reader back to Chapter 4 *Forms Of Inequality*.

The reader may recall the journey from atoms, rocks, bones and space-shuttles bumping around E-Space to the creation of the Scalar Plane Perspective ("ScPP", or "Sc").

For ease of exposition, we took certain matters for granted.

We did not trouble to ask where the words came from:

One | equals | Three | minus | Two

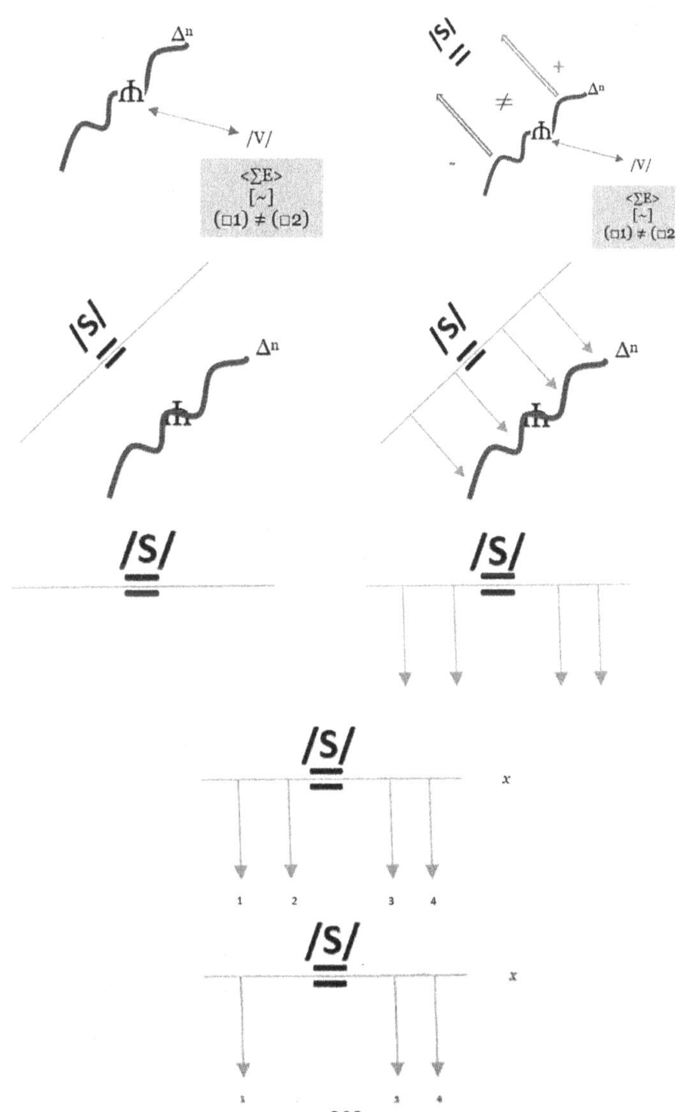

(\ \ \ \) *minus* (\ \ \) *is* (\ )

Three ⊤ minus Two ⊤ equals One ⊤

$$3 - 2 = 1$$

So, let us now ask.

I Am Not A Number, I am A Free Man[276]

Number One wanted *information.*[277]

> late 14c., *informacion*, "act of informing, communication of news," from Old French *informacion*, *enformacio* "advice, instruction,"
>
> from Latin *informationem* (nominative *informatio*) "outline, concept, idea," noun of action from past participle stem of *informare* «to train, instruct, educate; shape, give form to». The restored Latin spelling is from 16c.
>
> Meaning "knowledge communicated concerning a particular topic" is from mid-15c. The word was used in reference to television broadcast signals from 1937; to punch-card operating systems from 1944; to DNA from 1953. ***Information theory*** is from 1950; ***information technology*** is from 1958 (coined in "Harvard Business Review"); ***information revolution***, to be brought about by advances in computing, is from 1966. ***Information overload*** is by 1967.

Suppose the Prisoner, narrowly avoiding a balloon on Portmeirion beach, were to provide information. Perhaps the safe code back at ministry headquarters in London.

He could drop sticks on the beach, and order them. Or he could use a finger and draw SC equations in the sand. That is clearly the communication of information.

---

[276] *The Prisoner*. Patrick McGoohan (1967-1970)

[277] https://www.etymonline.com/word/information

Is that numero-graphic communication of information a language? Is it linguistical?

Suppose the Prisoner whistles: perhaps in morse code (replacing numbers for words). Is that auto-graphic communication of information a language? Is it linguistic?

Is it all merely a matter of definitions? Is it that any communication between people, which consists of information transmission, is always a language?

The author is going to come immediately off the fence and state that it is of no matter to Matrixial Logic how anyone chososes to answer these questions. You may call modes of human information transmission ("HIT") whatever you like; and categorise them however you please.

Indeed, going back to Tarski, and beyond, it seems that every attempt to provide a language-based explanation of HIT founders on rocks constructed from the assumptions of what the definition of "language" consists of.

Known Unknowns

We know that HIT involves some kind of iterative framework. But it is not merely repetitive. It is heuristic. By doing, we learn.

HIT is also hylomorphic: meaning derives from forms both constrained by and expressing contents.

Another obvious attribute of HIT is that it is holistic. Not merely in the life experience of the individual as its is expressed in a present moment of communication. But also socially holistic, and indeed environmentally holistic.

- Heuristic;
- Hylomorphic;
- Holistic.

The 3H's: this looks to be an obvious part of the Matrixial Logic landscape.

We know that newborn babies effect HIT. There is a vast literature on the development path from crying, movement, and eye contact, through to first words, and beyond.

How, exactly we subsume the stages of that development path under a model of language, and indeed whether we can do so, have remained open issues.

In a very useful analysis, *Early Language Acquisition: Cracking The Speech Code,*[278] Patricia K. Kuhl explains:

> Abstract | Infants learn language with remarkable speed, but how they do it remains a mystery.
>
> New data show that infants use computational strategies to detect the statistical and prosodic patterns in language input, and that this leads to the discovery of phonemes and words.

[278] Nature Reviews | Neuroscience Volume 5 | November 2004 | 83

Social interaction with another human being affects speech learning in a way that resembles communicative learning in songbirds.

The brain's commitment to the statistical and prosodic patterns that are experienced early in life might help to explain the long-standing puzzle of why infants are better language learners than adults.

Successful learning by infants, as well as constraints on that learning, are changing theories of language acquisition.

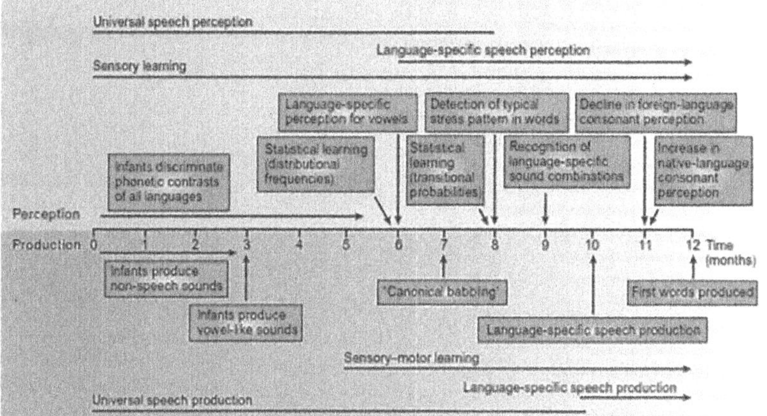

Figure 1 | The universal language timeline of speech-perception and speech-production development. This figure shows the changes that occur in speech perception and production in typically developing human infants during their first year of life.

...The encryption problem: sorting out the sounds.

The world's languages contain many basic elements - around 600 consonants and 200 vowels.

However, each language uses a unique set of only about 40 distinct elements, called Phonemes, which change the meaning of a word (for example, from 'bat' to 'pat').

These phonemes are actually groups of non-identical sounds, called Phonetic Units, that are functionally equivalent in the language.

The infant's task is to make some progress in figuring out the composition of the 40 or so phonemic categories before trying to acquire words on which these elementary units depend.

So we know that:

- There is a developmental environment from birth to full language acquisition;
- There is a (perhaps surprisingly) limited set of phonetic "building blocks" in any language;
- Newborns need to effect HIT and do so;
- Newborns are able to discriminate between authors of HIT[279]

Let us know see whether we can capture the 3H's operating as HIT.

In Outline

We can imagine a motion picture of thought translating into language:

- See dog barking;
- Say "dog is barking", or "dog barks".

So simple. But how on earth do we get there? Like all good language theorists, let's break the complex whole down into parts. But not parts of an assumed model linguistic whole. Parts of an operation.

Let's go on a journey of discovery: how the mind

[279] Whether for linguistic or neuro-biological reasons is obviously difficult to disover

processes information about a dog, and its barking. We will discover much about language, and acquire greater understanding of how ML explains us, and our world.

Experiencing

See some thing (A). Like anything in [E], [280] the (A) can only be seen to exist in opposition to (nA) under the Form of their extension.

Let that thing be a noun: a named thing. We will concern ourselves with the naming conventions later.

Let the thing be *Dog*.

We locate *Dog* in [E]. Dog is, for our present HIT purposes, a Thing in Being, not a Process in Becoming.

Baby first words might include *Dog*. Pensioner[281] words might include *Dog*. Baby is satisfied just with *Dog*. Pensioner probably wants to say a bit more than a mere nominative.

But how? How do we say more? How do we move on from the domain of [E]?

---

[280] See Chapter 2 *Framework*

[281] Retiree in American idiom

Experience: Outline

Step 1:

- Look at this outline:

- Put the outline into your head, behind closed eyes;
- Turn your head;
- Get up and walk around a bit (don't look back at the graphic).

Stop.

<u>Discussion</u>:

(1) Inside your head you could feel a "connection" to the outline dog?

(2) As if there were "wires" connecting the outline dog and the inside of your head?

(3) You could still feel those connections when you got up and walked around, even with your eyes open.

Step 2:

- Look at the outline agin;
- Close your eyes; pause. Open eyes;

- Go look out of the window;[282]
- Put the outline in that landscape.

Stop.

Discussion:

(1) Just like the circle and the landscape,[283] you couldn't "see" both at the same time.

Step 3:

- Look at the outline again;
- Close your eyes; pause. Open eyes;
- Look at your personal space: up to around 2 arms lengths from your body;
- Put the outline on the floor, in that personal space;
- Move the outline around;
- Bring the outline to life (without adding features to it).

Stop.

Discussion:

(1) That was easy. It felt fun.

(2) Somehow, there is a difference in our ability to project an outline image (which is artificial),[284] depending upon whether we try to locate it in personal space, or beyond.

(3) In our personal space, we have control of the projection, which we don't in wider space.

---

282 Your environment can be urban or rurual

283 See Chapter 3

284 Yes: clue

Experience: Real

Step 1:

- Look at this image:

- Put the image into your head, behind closed eyes;
- Open your eyes and look again at the image;
- Stop. Put the image into your head, behind closed eyes.

Stop.

Discussion:

(1) It is a completely different experience to the outline dog.

(2) Once you close your eyes, the "connection" is gone[285.]

Step 2:

- Look at the image;
- Imagine, with your eyes open, "filling in" some real-world background landscape for the image[286;]

---

[285] Female readers will find this less so. There are neuro-biological reasons fo that gender difference

[286] A dog bed, sofa, room park, street: whatever

- Put the whole enhanced image into your head, behind closed eyes;
- Relax.

Stop

Discussion:

(1) It is a different experience again;

(2) With the dog image located in some environment, even just a dog-bed, the "reality image" sits in your head, just like anything else;

(3) You don't need to concentrate. You can recall that "reality image" anytime. It's there in[287] your short term memory.

Step 3:

- Look back at the image;
- Create a background;
- Pull the package "into" your head.

Now

- Let the dog run around in the landscape;
- Change the landscape.

Stop.

Discussion:

(1) This is like Step 2, extended.

(2) As you let the dog play, and extend landscapes.

(3) You can also swap landscapes.

---

[287] Another discussion

Step 4:

- Look back at the image and create a background;
- Pull the package "into" your head;
- Now let the dog run around in the landscape;
- Change the landscape.

Now

- Separate the dog image from any landscape.

Stop.

<u>Discussion</u>:

(1) It all goes OK until the final Moment.

(2) As we try to separate dog from landscape, it becomes difficult.

(3) We experience fractionation between the two[288].

Just to make sure this isn't something specifically canine:

Experience: Pull

Step 1:

- Look around you and identify any "thing"[289] in its background.

Now:

- Float the Thing free of the background;
- Pull the Thing "into" your head.

Stop.

---

[288] As in other previous experiences

[289] Chair, curtain, book: whatever

Discussion:

(1) We just can't separate the Object from the background.

(2) The Object seems "stuck" to the background.

(3) Yet, when it's a cut-out dog, we can't "pull" that cut-out into our head.

We can conclude from these Experiences that:

- Our "inside head" perceptions work differently for artificial outlines, than for real images;
- We feel "connected" to artificial outlines, in a way that we don't for real images;
- We can't easily "see" an artificial outline out there in the real world environment;
- But we can easily"see", and even bring to life an artificial outline within our personal space;
- An image that communicates a thing (like a dog) is inaccessible to us, in our head, until we give the thing image a context: a visual environment;
- Separating in our head an image which we have extracted from its "out there" landscape is possible, but is not easy;
- Separating "out there" something from its landscape, and pulling it into our head, is impossible.[290]

And, from all this, it can be seen that, when we said:

> See some thing (A). Like anything in [E], [291] the (A) can only be seen to exist in opposition to (nA) under the Form of their extension.

[290] Although not: as we will see later

[291] See Chapter 2 *Framework*

and you said "Oh no it isn't. Why so?"

Well, thanks to our canine friend, we just proved it, didn't we. You could not think of the dog,[292] the dog as thing (A) in the world, without conjuring up some ≠ (nA) to go with it.

Only then did you have (Dog) ≠ (nDog), which allows you to create in your perception =[E] DogThing.

Being is Sticky

Let's return to the *Pull Experience.*

Please undertake a few Experiences of this Step. Try it with various types of Thing: animal, vegetable, mineral. It must be something you can actually see: like in the "I Spy" game.

Don't use Things that are Tools: cutlery, phone, cup. We will catch up with these later.

Then run the history of your head, and you will find that:

(1) When you were successful in "cutting out" the Thing with a pair of mental scissors, there was actually a landscape somewhere else in your mind, which you were using, to allow the cut out to exist.

(2) You were using a Wizard of Oz curtain, to hide

[292] Not "the idea of the dog"

> that mental landscape from the "space" you were trying to exhibit the Thing in.

We arrive at a large point: the [E] domain is "sticky".

We need a (^{NOT}A) to make the domain of Being work. But such a (^{NOT}A) is difficult to shake off. We find it difficult to abstract the (A) from its (nA)=[E] equation.

And this is not merely an equation written down. *The [E] equation actually describes what is really going on in our headspace relationship with the real world out there.*

So, unlike classical modern linguistic theory this is not modeling words, or numbers standing for words, or "phonetic gaps". It is not conducting exercises in extrusions of equality from the Scalar Plane.[293]

Through these Experiences, we realise that relationship is not at all simple. We can, or can't extract Things from the environment we see, depending on what those Things are.

---

[293] See Chapter 4

Experience: Naming

Please click the link, if it helps to see a bigger version of the picture: https://www.ikea.com images/3b/7c/3b7c8714486b447f542403634e58e697.jpg

- Look around you and identify any "thing"[294] in its background;
- Or, use this simple picture. Lots of ordinary things in here, from a perfectly ordinary part of the real world.

Now

- Look at your room, or the picture;
- Allow the name of the Thing to come to you;
- Make the name disappear (it will keep coming back, it's sticky).[295]

[294] Including a tool

[295] Yes: clue

Now

- Bring the Thing plus background "into" your head;
- Hold the composite image there. All easy so far.

Now: in your head

- Name something in the room;
- Name something else;
- Do this another 3 times.

Stop.

<u>Discussion</u>:

(1) Review the history of your head.

(2) As you brought the name of each Thing into your mind, all the other Things faded away a little; or the Thing you were naming came into sharper focus, "popping" in an extra dimension out of the photo in your head.

(3) Either way, the "space" between the naming Thing and the rest, seemed to increase, in some weird way.

You find that these processes are like keeping plates spinning on sticks, with 2 hands. You work on one, then you have to work the other. Sometimes you're working on 2 sticks with both hands.

Now, whatever is going on here: something is going on.

Experience: Namegame

Please click the link, if it helps to see a bigger version of the picture: https://hips.hearstapps.com/hmg-prod.s3.amazonaws.com/images/blue-velvet-sofa-living-room-1534263798.png

Step 1:

- Look at this simple picture. Lots of ordinary things in here, from a perfectly ordinary part of the real world;
- Just look at the picture, with your eyes open.

Now

- Name something in the room;
- Name something else;
- Do this another 3 times.

Stop.

Discussion:

(1) That was easy. You've been doing that since kindergarden.

Step 2:

- Just look at the picture again, with your eyes open;
- Pick a Thing and name it.

Now

- Put that name word into the picture;
- Not in a speech bubble. Not as if it were typed on the picture.
- Just put the name in your head "into" the picture.

Stop

<u>Discussion</u>:

(1) That was impossible.

(2) You found yourself trying to "pull" the Thing into your head.

(3) Because inside your head, the Thing and the Name "fit" in the same mental landscape.

Again, whatever is going on here: something is going on. Your world has become a little bit stranger. You might find yourself practicing these *Experiences*.

That's great. But please, seriously, do not do this exercise with anyone under the age of 7. They could find these *Experiences* upsetting.

We have proved something very important in these *Experiences*:

Our word association relationship, both to the

> world out there, and the world in our heads *is not isomorphic.*[296]
>
> We do not: see dog | think dog | say "dog".

We have just proved that. In so proving, we have discarded linguistic theory from Tarski to Chomsky, and beyond. In the discipline of linguistics, that is a very big bang indeed.

The name-word association is not *hylopmorphic,* either. The sofa, and the word ˇsofaˇ, are not directly wrapped up in each other as form and content.

That is obviously so, because the word for sofa in French is not the English word. And the way you *say* that word is completely different to mine. And furthermore, nobody ever says the same word exactly the same way twice.[297] This is a truism of forensic phonetics.

What you have experienced is the *Paramorphic* nature of our name-word association with the world.

- There is a Thing in the [E] domain of Being;
- We can attach <name> to that Thing;
- But the <name> operates in a dimension which is not the [E] domain of Being.

---

[296] 1-1 correlation; connection

[297] http://expert-evidence.forensic-voice-comparison.net/doc/Rose%20(2003)%20The%20technical%20comparison%20of%20forensic%20voice%20samples%20(Expert%20Evidence%2099).pdf

This last point is profoundly important. Names are not simply "out there". Yet we tend to believe in common place thinking that names are exactly like that: just like old-fashioned price labels stuck to objects in a shop.

Nor are names just "in here", in our head. Names and the act of naming rely upon some relationship between us and the world. Now let's try to discover what that might be.

VCIP

Recall Chapter 4, where we discussed:

Vector /V/ : Crossing Domains

Actuality □ does not meet with Potential Δ, in direct interaction.

Actuality □ requires Node <∑E> to interact with Potential Δ^n. Potential is the becoming of Moment under a Form [I].

A Vector /V/ is the mode by which
a Node <∑E> interacts with a Potential Δ^n.

And we explained:

> We can observe and induct Vector /V/ interactions and conveyances.

We can illustrate a hypothetical Interaction:

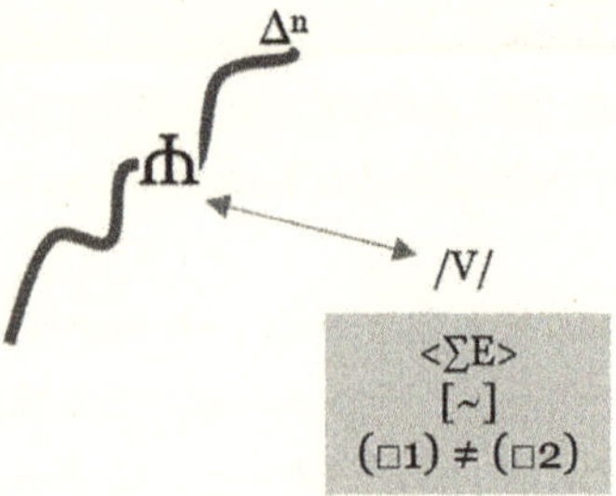

Here, we witness for the first time:
domain interaction in the matrix

Plurality [E] in Being, aggregated in Nodal forms, meets Plurality [I] in Becoming: which requires no aggregation, since that Plurality [I] is coherent as Potential Δ^n in its own process of Infinity ∞.

Such /V/ interaction is VCIP:

Vectored Collapse of
an Infinity ∞ Potential Δ^n

We do not say "effects", or "causes". To do so would be to introduce Scalar founded concepts,[298] when there is not yet any matrixial foundation for them.

The notation for VCIP is self-explanatory:

<∑E> ⇔ ⋔

[298] See later in this Chapter

It can equally be written as:

$$\bar{m} \Leftrightarrow \langle \sum E \rangle$$

To repeat, there is no concept of cause or effect to be represented in this notation.

Let's just quickly run that cause and effect issue by ourselves.

- When you look at the dog, do you think you are "causing" the dog? No;
- Do you think the dog is causing your seeing of it? No;
- Do you think the dog is causing your thinking of the dog? If you think yes, then that's a good trick. This pup is climbing into your head and turning all sorts of mental wheels and gears. Clever boy!

That's why seeing the difference between Isomorphic, Hylomorphic and Paramorphic, is both useful and necessary.

These are different ways we interact with the real world, which are real. And it is reality at both ends. While some functions in reality are about cause and effect, some simply are not.

Logic is: understanding reality. It is lack of understanding of reality which has poor logic, invoking "cause and effect", as reductionist Scalar[299] methodology, to avoid understanding reality.

---

[299] And therefore wrong before they start

Node Hunting

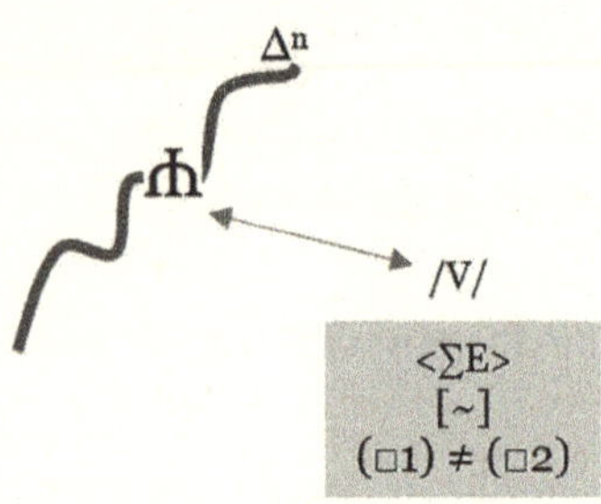

"Dog barks"

Noun plus verb. Not a well-formed sentence, in the logic of linguistics. But pretty simple.

We know from Chapter 4 that:

A Node <∑…] is
any convergence of two Forms in the same domain.

And further to remind ourselves:

Form and Node in [E]

Form [E]	**Node <∑E>**
• in Actuality	• of Actuality
• mediated by ≠ with	• interacting [~] with
• Substance (A), (nA)	• Form □ □
• stable	• unstable
• dynamic	• static
• self-actualising	• externally actualised
• deductive	• inductive

Node relates to the Contents of its constituent Forms, only indirectly. When we simplified:

(1.1) $$((A^1) \neq (nA^1) = [E^1]) \neq ((A^2) \neq (nA^2) = [E^2])\ [\sim]\ \sum(FoR_2)$$

$$\approx$$

(2.1) $$(\square 1) \neq (\square 2)\ [\sim]\ \sum(FoR_2)$$

$$>$$

(2.2) $$(\square 1) \neq (\square 2)\ [\sim]\ <\sum E>$$

the underlying cross-dynamics at (1.1) are:

present in the Node as the function of
its existence in Subtstance, but present differently:

[E] Form is in Actuality.
$<\sum E>$Node is of Actuality.

This is not saying that we cannot "map" Nodes onto the real world. We can.

Let's take <Dog> first.
We are Node hunting:

$$((A^1) \neq (nA^1) = [E^1]) \neq ((A^2) \neq (nA^2) = [E^2])\ [\sim]\ \sum(FoR_2)$$

We found clues in the Experiences: *Real; Name and Namegame*. We found that:

when you run the history of your head:

(1) When you were successful in "cutting out" the Thing with a pair of mental scissors,

there was actually a landscape somewhere else in your mind, which you were using, to allow the cut out to exist.

(2) You were using a Wizard of Oz curtain, to hide that mental landscape from the "space" you were trying to exhibit the Thing in.

We arrive at a large point: the [E] domain is "sticky".

We need a (nA) to make the domain of Being work. But such a (nA) is difficult to shake off. We find it difficult to abstract the (A) from its equation.

And this is not merely an equation written down. The equation actually describes what is really going on in our headspace relationship with the real world out there.

Our <Dog> Node begins to take shape:

$$((\text{Dog}) \neq (\text{nDog}) = \text{Dog Idea } [E^1])$$

$$\neq$$

$$((\text{Curtain}) \neq (\text{nCurtain}) = \text{Landscape } [E^2])$$

Let's paws.[300] By :

$$= \text{Dog Idea } [E^1])$$

we definitely don't mean some Platonic idea "dog" like

[300] Sorry

the trace shape in the Experience: *Outline Dog*. We mean the idea of the actual real dog: the photographic dog: and its landscape.[301] The hylopmorphic <dog> .

What is it that is ≠ (nDog)? We saw in the photo isolate dog, that we need a background, landscape, environment to picture the dog in. So that it can be a Dog for us: a Dog we can operate with in our heads.

Now, we know from the *Experience: Namegame,* that we are operating with that Dog in our heads. We are not putting a word "into" the real world landscape.

So, we are taking the Dog and Landscape, in the real world, and importing them, as a composite, into our heads.

It is that composite ((Dog) ≠ (nDog) = the Dog Idea, in our heads.

Hang on, though. How is a [= Dog Idea] "in our heads" part of [E]? How is that *Substance extended in spacetime*?

Well, unless you're a very serious dualist, and idealist, where else is that [= Dog Idea]? Just floating around in some miasma of "consciousness" separate from the brain, the body, the world?

---

[301] Whether that be just a dog basket, or bone, or anything else which is nDog

You don't have to believe in the fact of Self-Consciousness (or the Self).[302] Indeed, people who stringently commit themselves to the Dennett / Churchland line of thought, and deny one or both of these, are stuck with the same admission: the [=Dog Idea] "in our heads" is indeed part of [E].

Let's be quick and crude about this, reserving nuance for later. There's a [=Dog Idea] "in our heads", and our heads and brains are definitely *Substance extended in spacetime.*

And, if the [=Dog Idea] "in our heads" is not *extended in spacetime,* then how is it that we "see" a dog and its landscape out there, at all? How is it that we are not merely locked in a head-space bubble outside of spacetime?[303]

Now, what about:

((Curtain) ≠ (nCurtain) = Landscape [E^2])

Recall again that *Pull Experience*:

(1) When you were successful in "cutting out" the Thing with a pair of mental scissors, there was actually a landscape somewhere else in your mind, which you were using, to allow the cut out to exist.

---

[302] Although you don't have to "believe": it is a fact of the world, just like that Dog. See later Chapters

[303] Yes, this is 400 years of epistemological argument territory. We will revisit it later

(2) You were using a Wizard of Oz curtain, to hide that mental landscape from the "space" you were trying to exhibit the Thing in.

Experience: Landscape

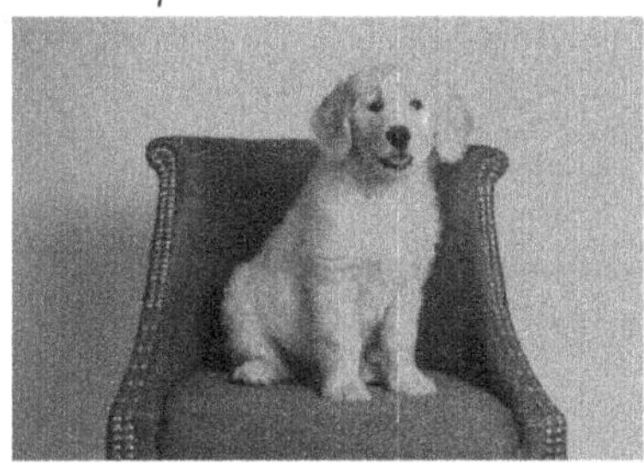

Step 1:

- See Dog and Landscape (chair and room) as a Composite;
- Keep looking, eyes open on the picture.

Now

- Separate the Dog from the Landscape, and simultaneously substutute any other landscape you like.

Stop.

<u>Discussion</u>:

(1) This is very easy.

(2) So long as you switch landscapes, you can get the Dog to go anywhere that your imagination chooses.

Step 2:

- Again: see Dog and Landscape (chair and room) as a Composite;
- Keep looking, eyes open on the picture.

Now.

- Separate the Dog from the Landscape;

and

- Just have the Dog hanging in "no space";
- Don't close your eyes.

Stop

Discussion:

(1) You find this frustratingly difficult.

(2) Even when you get a successful "cut out", you can't get rid of the "sticky" landscape

By the way:

1. Are you still harbouring any doubts about:

   ((Dog) ≠ (nDog) = Dog Idea [E^1])

2. Does any part of you think that

   [≠] "really" just means [=]

   If so, please go write on the blackboard 100 times:

   equality does not equal reality

3. At this point, does any intellectual or emotional part of you believe that:

   (a) We have a noun "dog" (true)

   (b) That noun is unique: no other noun is dog (true)

   and

   (c) ∴ when I see a dog I am thinking exclusively of the noun "dog" and nothing else.

4. Do you think that we simply:

   see dog | think dog | say "dog"

Step 3:

- See Dog and Landscape (chair and room) as a Composite;
- Pull the Dog and Landscape Composite "into" your head;

Now

- Separate the Dog from the Landscape, and simultaneously substutute any other landscape you like.

Stop.

Discussion:

(1) This is very easy. Yes, you have to concentrate a little bit harder than at Step 1. But still easy when you get used to it, after a few iterations.

(2) So long as you switch landscapes, you can get the Dog to go anywhere that your imagination chooses.

Step 4:

- See Dog and Landscape (chair and room) as a Composite;
- Pull the Dog and Landscape Composite "into" your head.

Now.

- Separate the Dog from the Landscape

and

- Just have the Dog hanging in "no space".

Stop.

Discussion:

Then run the history of your head, and you will find that:

(1) You just can't do it.

(2) When you were successful in "cutting out" the Dog with a pair of mental scissors, there was actually a landscape somewhere else in your mind, which you were using, to allow the cut out to exist.

(3) You were using a Wizard of Oz curtain, to hide that mental landscape from the "space" you were trying to exhibit the Dog in.

So, we have established:

$$((\text{Curtain}) \neq (\text{nCurtain}) = \text{Landscape } [E^2])$$

What's more, we have established that matter of Being, both:

- "out there" in what you see in the world, and
- "in there" in what you see in your own head.

Knowing the Node

Refreshing the reader again, because this is uniquely unfamiliar stuff:[304]

(1.1) $((A^1) \neq (nA^1) = [E^1]) \neq ((A^2) \neq (nA^2) = [E^2])\ [\sim]\ \Sigma(FoR_2)$

$$\approx$$

(2.1) $(\square 1) \neq (\square 2)\ [\sim]\ \Sigma(FoR_2)$

$$>$$

(2.2) $(\square 1) \neq (\square 2)\ [\sim]\ <\Sigma E>$

---

304 Whether or not you're a professional logician; and perhaps even more so, in that case

the underlying cross-dynamics at (1.1) are:

present in the Node as the function of
its existence in Subtstance, but present differently:

[E] Form is in Actuality.
<∑E>Node is of Actuality.

We have established:

((Dog) ≠ (nDog) = Dog Idea [E^1])
≠
((Curtain) ≠ (nCurtain) = Landscape [E^2])
[~]
<∑Cut-Out Dog>

Let's abbreviate:

((D) ≠ (nD) = D[E^1]) ≠ (C) ≠ (nC) = L[E^2])] [~] <∑D*>
>
(□D1) ≠ (□L2) [~] <∑D*>

And recall, again:
the underlying cross-dynamics are:

present in the Node as the function of
its existence in Subtstance, but present differently:

[E] Form is in Actuality.
<∑E>Node is of Actuality.

We have found our <∑E>Node for:

(□D1) ≠ (□L2) [~] <∑D*>

Being Rich

This, you may think, looks like an awful lot of nodal bother just for one dog. Do we really go through all this just to begin getting the <dog> word?

To put this exercise in perspective, we are establishing the "grammar", as it were, of language use. We are going in very slow motion.

What we have learned is that we take Things from the world into our heads *in context*. We learned that when we had the dog photo isolate and tried to put that into our heads. With all that we had to conjure up, by way of environment, to allow us take the dog in at all comfortably.

We have learned that our acquisition in our heads of what we see out there is *Paramorphic*.

Which is no way provides any basis to doubt that what is

out there, which we see, is a reality out there, independent of any us actually seeing it. Your foot is there, undoubtedly, in pretty much anyone's philosophy, even when you're not looking at it.[305]

Just on that Paramorphic idea.

- There's a ball, just lying there. Or nestling a tin marked "Tennis Balls". But it's just a bouncy round object. At 4.55pm on a grass court, it's flying through the air at 143 miles and hour towards you and your racket. You are calculating trajectory angles, court placement, racket string tension, grip[306] in a split second.;
- *That is not just a ball. It is a Vector of a vast amount of Paramorphic information in you;*
- Tennis, the ball and you are all real.[307]

We can therefore, at this early stage, say that the analysis is going much better than the de-contextualised systems of linguistic logic. Those systems which take written down words, as if they were things in the world. Or as if individual written down things directly represent things in the world.

It is understood that the Nodal analysis feels chunky and over-complicated, at the moment.

[305] Yes: foot is part of an integrated body including your apparatus of perception. And the world is, or isn't

[306] And countless other things

[307] Thanks to Naomi for this one

The reader is welcome to try and dispense with Nodes here: to try and digest the isolate photo dog, and tunnel that through one's head to what comes next.

The author is only trying to do Matrixial Logic. Smarter people will be able to do it much better.

Unstable

We know that a Node is unstable.

That is why the equation for a Node ends in [~], not [=]. A Node is not a Frame of Reference. A Node is an unstable interface of two FoR's.

Review the history of your head in the dog extracting – landscape using Experiences. You can feel that instability. The act of will which you need to apply, to sustain it.

There is something we do to remedy that instability.

Verbal Binding

We know from Chapter 4 what happens when VCIP ሐ arises:

<ΣE>
[~]
(□1) ≠ (□2)

<ΣE>
[~]
(□1) ≠ (□2

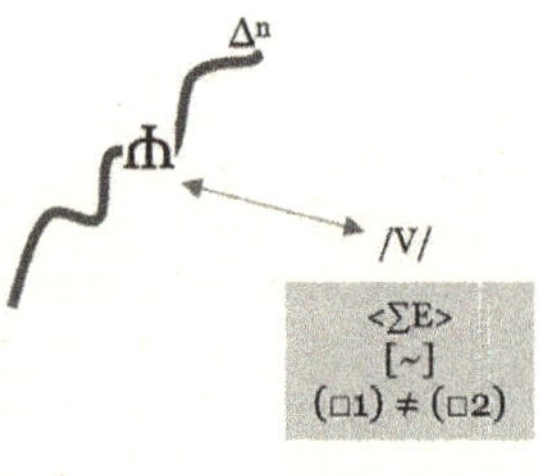

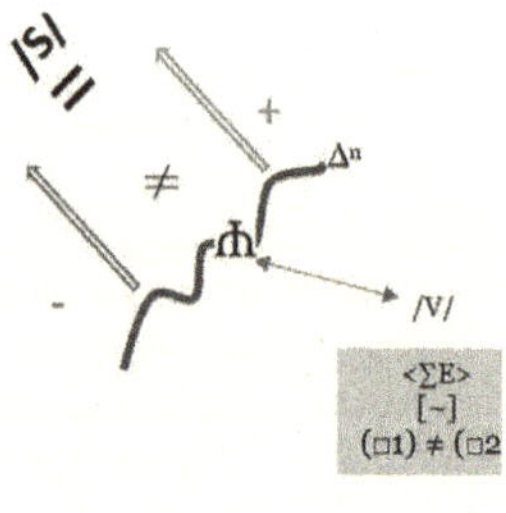

The Scalar Plane manifests.

And we know how the Scalar Plane operates:

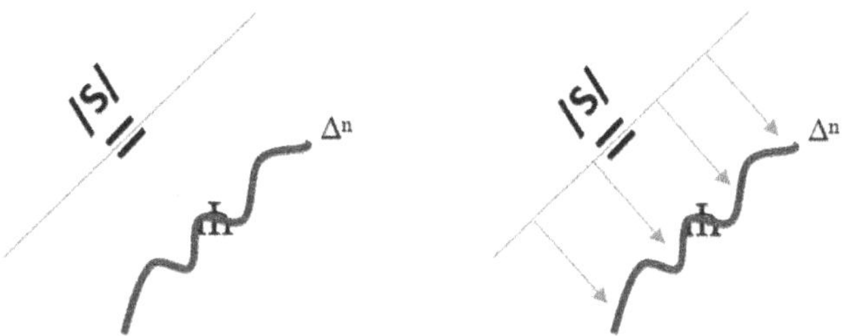

But this time, it's not co-ordinate numbers.

This time, the Scalar Plane is referencing a Node which has fixed a verbal infinity.

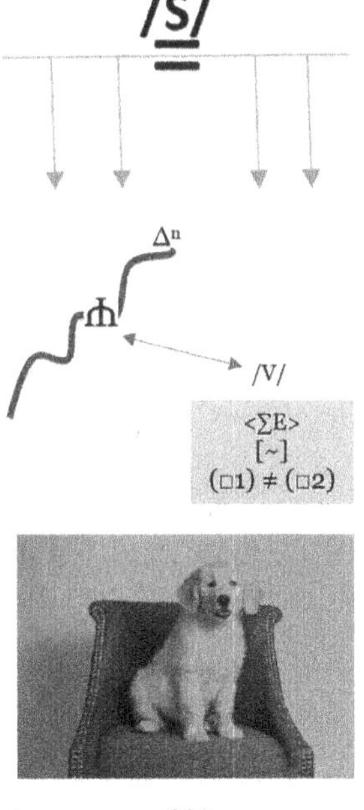

Let's just be clear about something, before we proceed:

1. There is definitely a dog on a chair, out there in the world.[308]
2. That is a "fact", a "reality", an "objective phenomenon": however you like to phrase such things.
3. You are seeing it directly. No doubt about that. [309]
4. Whether you pull the dog composite into your head, and close your eyes, or just keep your eyes open and have that physical interaction with it: makes no difference to your language use.[310]
5. Let's say that last one more precisely: *the operation of your language system is indifferent as between the internal and external domains (or frames, or modes) of reference.*
6. We have shown that in our Experiences. In fact, we have been using those Experiences to train you to see and feel how the bits of the jigsaw fit together, and how odd it feels when we stop you doing what comes automatically.
7. There is not some hyper, or meta "dog" that is undergoing some complex metaphysical set of operations in your head.
8. We are doing *slow motion syntax*. So it does inevitably feel a bit hyper, or meta: that's the result

[308] Well, there was once, as this is a picture. But that could just as well be a real live barking dog a chair in front of you

[309] You could be hearing it directly, and stroking and smelling it directly as well. The word "directly" wil bear some re-examination in Chapter 7

[310] Well, it does, but that is a matter of nuance arising from differently operating feelings

of doing all this in slow motion.

9. What we are describing is the jigsaw puzzle of: how that world out there interacts with your world of thinking, and then saying (or writing) "dog barks".

In the real world, all of this slow motion syntax happens in (literally) milliseconds. Because nothing ever happens on its own: we being alive, an organic being, our very atoms changing moment by moment.

None of our thoughts or feelings ever exist as a singularity, but only as moments in an infinite process Δ^n, interfering with other moments in infinite process Δ^n. [311]

Let's return to canine appearance.

Experience: Binding

- Look at $<\sum D_1*>$
- Allow yourself to think of the word of what $<\sum D_2*>$ is (you actually did that automatically)

[311] More poetic than analytic: as we will see in Chapter 8

Now

- Take <∑D_1*> into your head.

Now

- Say the name word of <∑D_1*> to yourself, over and over again.

Discussion:

(1) Saying the word keeps the image refreshed in your head

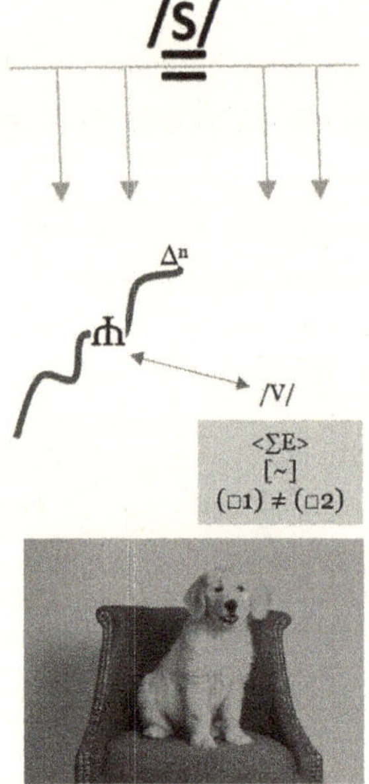

It is because your verb-ing of bark-ing Binds the image in a Scalar Plane by reference to the ⫚ collapse of the verb Potential: Barking [I] Δ^n.

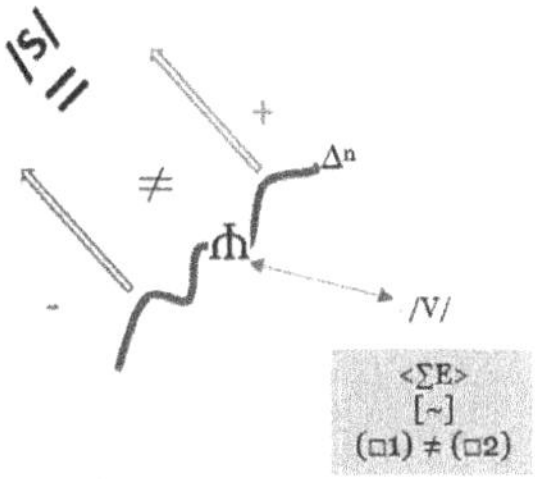

<∑E>
[~]
(□1) ≠ (□2)

The repetition is increasingly Binding the image.

In fact, you can now try this:

Experience: UpDown

Step 1

- Look at <∑D*>
- Take <∑D*> into your head
- Let it rest there
- Allow the <∑D*> word to come into your head
- Allow the <∑D*> word to repeat in your head

Stop.

Step 2

- Look at <∑D*>
- Take <∑D*> into your head
- Let it rest there
- Allow the <∑D*> word to come into your head
- Allow the <∑D*> word to repeat in your head

Now

- Fade the image away as you repeat the <∑D*> word (but don't try to disappear the image).

Stop

Step 3

- Look at <∑D*>
- Take <∑D*> into your head
- Let it rest there
- Allow the <∑D*> word to come into your head
- Allow the <∑D*> word to repeat in your head

Now

- Turn the image sharpness / brightness / size Up as you repeat the <∑D*> word

Stop.

Discussion:

(1) You find that, once you have Binding of the <∑D*> word you can adjust the <∑D*> image, up and down.

(2) It's like you take manipulative control over the image. The <∑D*> word is the tool which lets you play around with the <∑D*> image.

Scaling Basics

Here we are again, with our Scalar Plane /S/±

Let's recall, and repeat:

Experience /S/±

Step 1:

- Please ascertain that you can, in your mind, move the ± plane line horizontally, with complete freedom.

Step 2:

- You can also extend the ± line freely, in either direction, with no effort of will, merely small concentration.

Step 3:

- Now, try to move the ± line so that is pointing up and down the page.
- Try to move the ± line as a diagonal.

Discussion

(2) As with the paperclip,[312] and the Δ^n river, and the m̄ expanses, you were unable to move m̄

We can conclude that the Scalar Plane ± has some, although very limited utility. We are still stuck in the plane of [=]. We can move sideways, and up and down.

We are rather like a rook or castle in chess. We do not have two dimensional freedom.

Scalar Extrusions

Can the reader now please imagine dropping these ⊤ arrowed lines orthagonally to ±

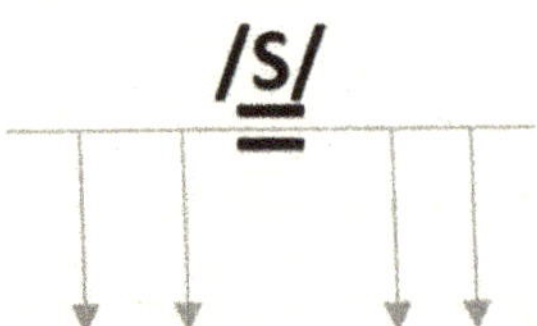

[312] See Chapter 3

Please cover over the ⊤ arrowed lines, or use this as your focal point:

The reader will have seen that it is child's play to effect the arrowed orthagonal ⊤ extrusions.

Experience /S/ ± [$\top^n$]

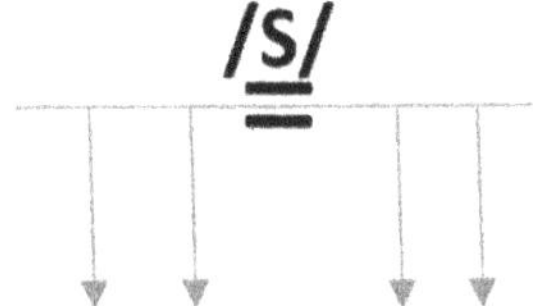

Step 1:

- Please ascertain that you can, in your mind, move the ± plane line, diagonally, between the two dots:

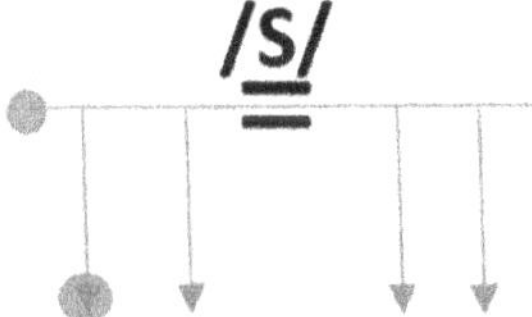

Step 2:

- Now, experiment on playing with the ± plane line.

We can conclude that the Scalar Plane ±, with the addition of has [$\top^n$] has much greater utility. We are still stuck in the plane of [=].

But now we can move like the Queen in chess.

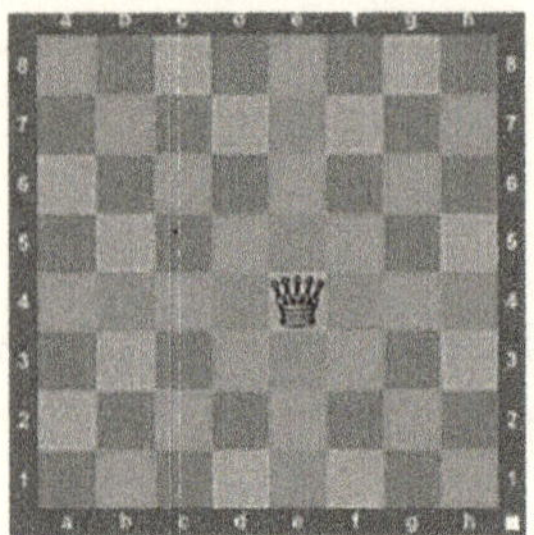

We have acquired two dimensional freedom.

We know this is a Scalar Plane. It only goes ⇔ in a lateral direction. It is a plane of =: +/-; is/is not; exist/not exist.

The Scalar Plane (x) axis is unidimensional. We were stuck with Castle moves in chess. We could extrude from $(x)^n$. But it was only useful for basic arithmetic.

(\\\\) *minus* (\\\) *is* (\)

Three ⊤ minus Two ⊤ equals One ⊤

3 – 2 = 1

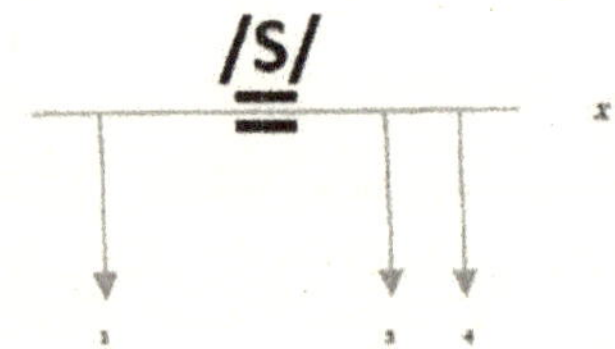

To do geometry, we had to add a (y) axis. We had to make the Scalar Plane bi-demensional.

Then we could build the pyramids.

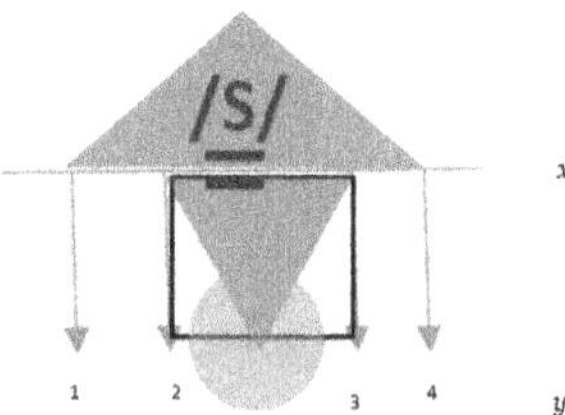

Scaling Equality

Again, let's recall from Chapter 4:

There is one rule, just one, which is the fundamental, the alpha and omega of manipulating of [$\top^n$ extrusions] in and from the Scalar /S/ plane ± :

$$[\top] = [\top]$$

Because one Extrusion IS another Extrusion.

Every Extrusion is a manifestation of /S/ plane ±.
There IS nothing else.

The Scalar /S/ plane ± is the plane of *is*, of *equality*: of *existence*.

Everything that is, exists: in the Scalar /S/ plane ±. There is no half-way house of existence: is or is not; sien oder nicht sein; to be or not to be.

We can add *Cogito ergo sum*: I am thinking, therefore I am existing.[313] Descarte's famous tautolgy, posing as a proof.

Let's clarify what "not" means here.

(1) We have spent much of the last Chapters pounding home the rubric: $(A) \neq (^{NOT}A) = [E]$.

(2) In [E], the $[^{NOT}]$ / [n] means "all that which is not (A) in the domain of [E]." The n is hylomrphic: a necessary element of the opposed content, which is (in extension) the Form.

(3) This is not what /not/ signifies in the /S/ plane.

(4) /not/ is simply a function of existence.

So, here we are with the existence thought:

Above the /S/ Plane :	┘	= Does Exist
Below the /S/ Plane:	┐	= Not Exist

Please note, the symbol ┐ ┘ positionings are arbitrary. You can swap them. It makes no difference to the concepts.

That is all the /S/ Plane function that we need for now.

Let's prove that.

---

[313] *Principles of Philosophy.* Part 1, At 7 (1644) Published originally in French *Discourse on Method* (1637)

Experience: DogGone

Step 1:

- Look at <∑D_1*>
- Think Dog.

Now

- Think Dog /exists/ ┘

Stop.

Discussion:

(1) Very easy to think Dog.

(2) When you were made to add-think /exists/ ┘, how did it make you feel?

(3) You felt it was pointless. Almost as if the idea / exists/ ┘ didn't really belong to the dog.

Step 2:

- Look at <∑D_1*>

Bring it into your head, then

- Think Dog.

Now, in your head

- Think Dog /exists/ ┘

Stop

<u>Discussion</u>:

(1) Very easy to think Dog.

(2) When you were made to add-think /exists/ ┘, how did it make you feel?

(3) It felt a little less strange.

Step 3:

 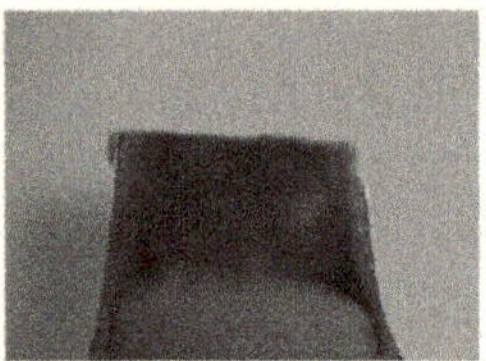

- Look at $<\Sigma D_1{}^*>$ and Empty Chair
- Think Dog.

Now

- Think Dog /exists/ ┘

Stop

<u>Discussion</u>:

(1) This time it's very easy.

(2) Thinking /exists/ ┘ makes sense.

Step 3:

- Look at $<\sum D_1{}^*>$ and Black
- Think Dog.

Now

- Think Dog /exists/ ┘

Stop

Discussion:

(1) This time it's a little strange.

(2) Thinking /exists/ ┘ makes sense.

(3) But the Black is also making you think ¬/exists/

Step 4:

- Look at $<\sum D_1{}^*>$ and Blank
- Think Dog.

Now

- Think Dog /exists/ ┘

Stop.

Discussion:

(1) This time it's a little strange.

(2) Thinking /exists/ ┘makes sense.

(3) But the Blank is also helping you think ¬/exists/

The difference between these two images is, of course, something which psychologists have known about for years.

This is the Word Association test.[314] The test originally began with words, but then moved into contrasts such as these, and shapes, and so on.

We are finding *interference patterns* in our verbalisation.

The Missing Curtain

This last Experience may strike you as extraordinary. By all means, try it again. Most readers will get the same result.[315]

---

[314] starting with *The Association Method*. Jung (1910). First published in *American Journal of Psychology*, *31*, 219-269.

[315] It is an established psuchological phenomenon, and common sense, that repetition starts to clutter and destablilse our thinking. Try repeating "dolphin" ten times and see what happens in your head

In case you were wondering: try the curtain with these different colours:

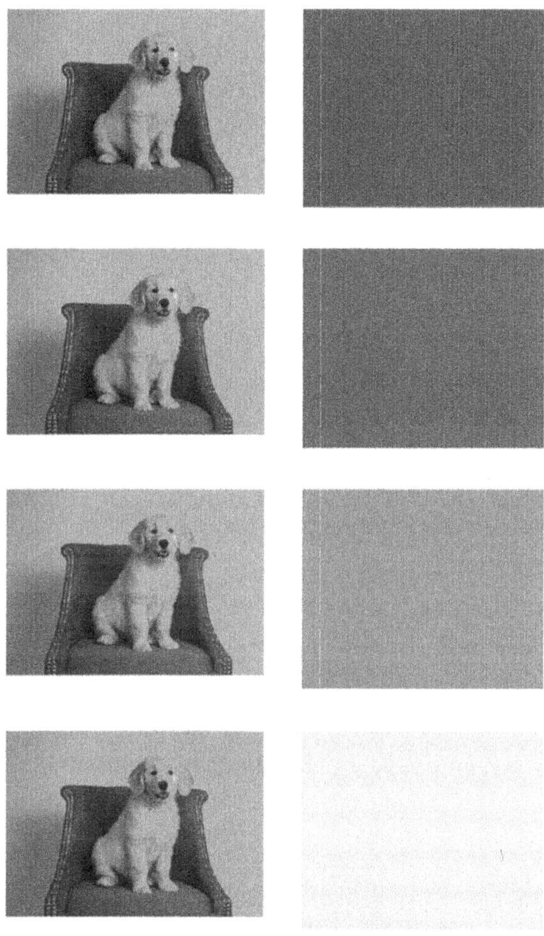

Then try it again with black:

And try it again with white.

The <∑D*> /exists/ ┘ function is indifferent as to blackspace, or primary colourspace. But the function is affected by whitespace.

Yet, if you read Russell, or Tarski, or Chomski: this is where the real world story ends:

dog | barking | is

They then manipulate, by class or numerical coding, these 3 elements "of language".

They add in:

werewolf | barking | is [not]

And this is supposed to provide a "solution" to the problems of language operation, and "truth".

<∑D*> ጠ /S/ /exists/ ┘

Please look at this gram. It's clearly not an equation. There is no stability about it.

We stated, towards the start of this Chapter:

> *Being Rich*
>
> This, you may think, looks like an awful lot of nodal bother just for one dog. Do we really go through all this just to begin getting the <dog> word?
>
> To put this exercise in perspective, we are establishing the "grammar", as it were, of language use. We are going in very slow motion.
>
> What we have learned is that we take Things from the world into our heads *in context*. We learned that when we had the dog photo isolate and tried to put that into our heads. With all that we had to conjure up, by way of environment, to allow us take the dog in at all comfortably.
>
> We can therefore, at this early stage, say that the analysis is going much better than the de-contextualised systems of linguistic logic. Those systems which take written down words, as if they were things in the world. Or as if individual written down things directly represent things in the world.

A Picture Is

Our slow-motion syntax is perhaps jangling with your senses. Let's do a reset.

Experience: There

- Look around your room
- or through a window at a scene, any scene
- Just look. Relax

Stop.

<u>Discussion</u>:

(1) Look inside the history of your head.

(2) When you were simply relaxed looking, *with nothing going on,* no words came into your head.

(3) You might have been playing music in your head, or just enjoying not actively thinking.

We conclude that words do not come to us from our landscape "out there" unless we need them to.

Unless we pay attention to thing (by which we move it in our attention), or something out there changes in a way we notice.

Our default setting with the world is: wordless.

As we said, a few section back:

- The <name> operates in a dimension which is not the [E] domain of Being.

  This last point is profoundly important. Names are not simply "out there". Yet we tend to believe in common place thinking that names are exactly like that: just like old-fashioned price labels stuck to objects in a shop.

  Nor are names just "in here", in our head. Names and the act of naming rely upon some relationship between us and the world.

From the World to a Word

For now, we have this:

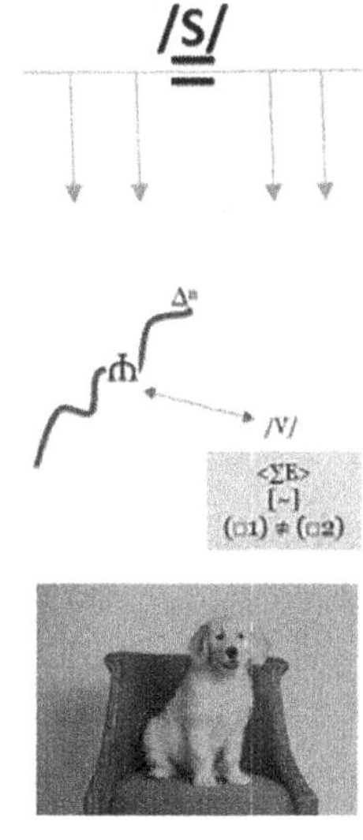

<∑D*> m̄ /S/ /exists/ ┘

We can represent it like this:

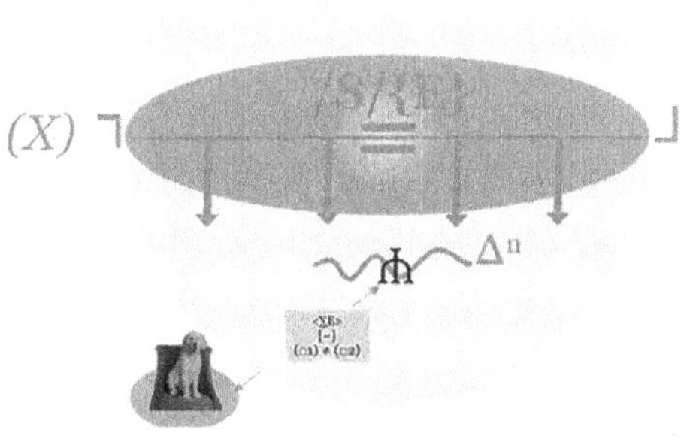

You recall that a Node is unstable.

That is why the equation for a Node ends in [~], not [=]. A Node is not a Frame of Reference. A Node is an unstable interface of two FoR's.

Yet, we seem to have found a way to stabalise a Node <∑D*>.

A stabilisation which seems to rely on something about *Identity*. Let's return to this, once we have found our next Node.

The Curious Incident[316]
"Dog barks."

Now let's get some of that barking action.

Experience: Barking

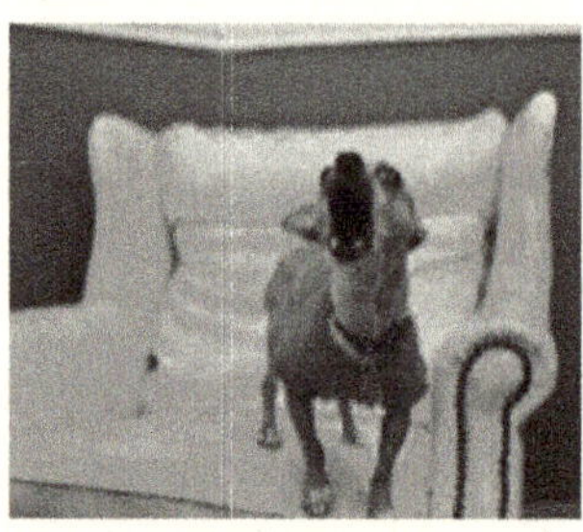

[316] ‹The *dog* did nothing in the *night-time*.'That was *the curious incident*,' remarked *Sherlock Holmes*." *The Adventure of Silver Blaze*. Conan-Doyle (1892)

Step 1:

- See Dog and Landscape (chair and room) as a Composite: [~] <∑D*>
- Keep looking, eyes open on the picture.

Now

- Force yourself not to think of the word for what <∑D*> is doing.

Stop

<u>Discussion</u>:

(1) This is not easy.

(2) That word keeps trying to pop into your head

Step 2:

- See Dog and Landscape (chair and room) as a Composite: [~] <∑D*>
- Keep looking, eyes open on the picture.

Now

- Allow yourself freely to think of the word for what <∑D*> is doing;
- Repeat the doing word to yourself.

Stop.

<u>Discussion</u>:

(1) This is really easy.

(2) The picture feels "whole" now.

Experience: HeadBarking

Step 1:

- See Dog and Landscape (chair and room) as a Composite: [~] <∑D*>

Now

- Take the composite <∑D*> into your head and close your eyes;
- Force yourself not to think of the word for what <∑D*> is doing.

Stop.

Discussion:

(1) This is virtually impossible.

(2) The composite image <∑D*> basically becomes the word you are trying not to say.

Step 2:

- See Dog and Landscape (chair and room) as a Composite: [~] <∑D*>

Now

- Take the composite <∑D*> into your head and close your eyes

Now

- Allow yourself freely to think of the word for what <∑D*> is doing.
- Allow yourself to repeat the word.
- Notice what happens to the <∑D*> image as you repeat the word.

Stop

<u>Discussion</u>:

(1) This is really easy.
(2) The picture feels "whole" now.
(3) The experience feels "natural"

Now, review the history of your head.
When you allowed yourself to repeat the word, the <∑D*> image became sharper and sharper.

Experience: TwoDogs

Step 1

- Look at <∑D_1*>
- Take <∑D_1*> into your head.
- Let it rest there.
- Allow whatever thoughts come into your head, to come, and go.

Stop.

Discussion:

(1) We know now that this is really easy.

(2) The picture feels "whole".

(3) You feel calm.

(4) In fact, it's quite hard to stop the picture just fading away. You have to keep putting in effort to bring it back.

Step 2

- Look at <$\sum D_1$*>
- Take <$\sum D_1$*> into your head.
- Let it rest there.
- Allow whatever thoughts come into your head, to come, and go.

Now

- Watch for when the image starts to fade, and notice yourself bringing it back.

Stop

Step 3

- Look at <∑D2*>
- Take <∑D2*> into your head.
- Let it rest there.
- Allow the doing thought to come into your head.
- Allow the doing thought to repeat in your head.

Stop

<u>Discussion</u>:

(1) The doing thought changes how the picture sits in your head, compared to <∑D_1*> .

(2) The <∑D2*> image stays sharp.

(3) The more you repeat the doing word, the sharper the image becomes (up to a limit).

Step 4

- Look at <∑D_1*> and look at <∑D_2*>
- Allow yourself to think of the word of what <∑D_2*> is doing (you actually did that automatically).

Now

- Keep that doing word in your head.
- Focus on <∑D_1*> and "fade out" <∑D_2*>

Now

- Focus on <∑D_2*> and "fade out" <∑D_1*>

Stop.

Discussion:

(1) Fading out $<\sum D_1*>$ is a bit easier than fading out $<\sum D_2*>$

Step 5

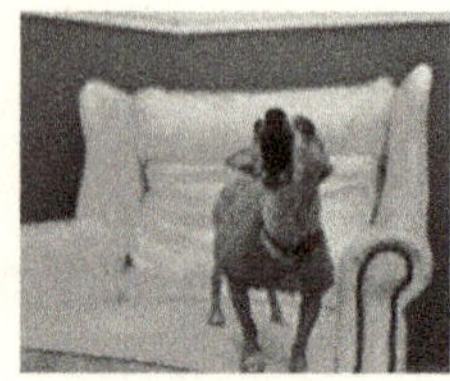

- Look at $<\sum D_1*>$ and look at $<\sum D_2*>$
- Allow yourself to think of the word of what $<\sum D_2*>$ is doing (you actually did that automatically)

Now

- Take them both into your head.

Now

- Focus on $<\sum D_1*>$ and "fade out" $<\sum D_2*>$

Now

- Focus on $<\sum D_2*>$ and "fade out" $<\sum D_1*>$

Stop.

Discussion:

(1) Fading out $<\sum D_1*>$ is now much easier.

(2) Fading out $<\sum D_2*>$ is virtually impossible. The doing word keeps bringing $<\sum D_2*>$ back.

Woof

Barking: a verb. Where shall we find a verb in the world out there?

A doing is not any thing. A thing is a noun: a being word. A verb is not a noun. Surely, a verb is descriptive of things going on, things happening, in the world.

The verb is not a description of how a thing is: big, small, coloured, light, heavy. These are adjectives, and adverbs, and they are all relational to each other.

So, what sort of thought is a verb?

It is easy to say, although once we have even thought it, our world-view sets off down a ski-slope of inevitability: towards a by now familiar destination.

The verb describes a Process Δ^n in Becoming.

Doings are Becomings: not in the Genesis sense of growth. But in the sense of motion. Of Heraclitean change.[317]

We know by now that Becomings happen in the Domain of Infinity. These are moments (A) ≠ (-A) in becoming. They are Potential Δ^n.

317 Yes: clue

Moment operates as Potential Δ in Infinity ∞

$$(A) \neq (-A) = [I]$$

Potential Δ is a Dynamic of
Moments in Becoming

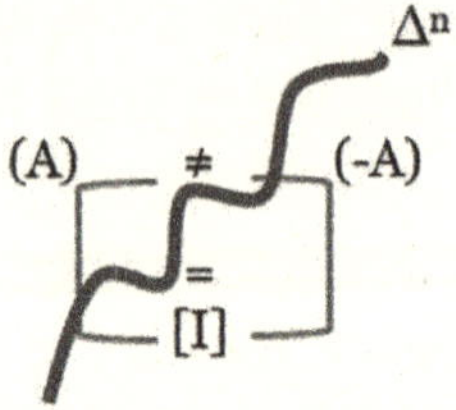

There is no doing without a something. Barking does not happen all on its own. Even a bumping in the night requires a ghost to be doing it.

There is barking going on, because there is a Thing doing it: <∑D*> . There is a sound association.

Now, in this case, we are imagining the barking. We are activating a system in which sense input is attached to words.

We do not have to go Node hunting for a Verb. Interference between Infinity potentials produces its own <∑I> Nodes.

A reminder from Chapter 4:

Interference Φ

> We have seen that Interference between Potentials Δ Δ, can create <∑I> Nodes.

Potentials may *interfere* with each other, producing a simulacrum of an Event <∑E>.

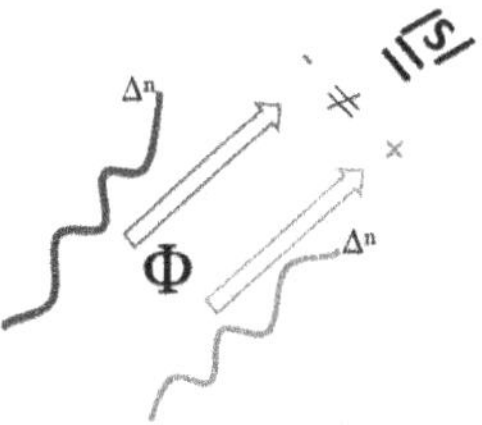

Interference does not prompt or effect Collapse □ of any Potential.

To bring forward an analogy which may already have occurred to the reader, imagine 2 waveforms. The underlying content of the waveforms could be anything: tidal currents, electricity, sound, and so on.

We are very familiar with the interference phenomenon, as applied to such waves. Interference may:

(a) Sum;

(b) Subtract;

(c) Cancel.

It is easy to see that in cases (a) and (b), the wave Potentials:

- Interfere with each other;
- Do not Collapse.

It is less easy to see, but no less the case that in (c)

there is not a Collapse ñ, but an annihilation: a cancelling. You do not imagine the waveform suddenly freezing in place: standing still.

The substrate, already within us, is represented by the blue line (A) ≠ (-A) = [I]. These are Moments Δ in infinity∞. This is all familiar stuff by now.

The sound of barking (or its recapitulation in imagination) needs no Node to lift itself out of its landscape and arrive in our ears. As any dog owner will testify, that <barking> wave arrives in our ears unbidden.

It needs no ML to say that the <barking> is sound wave. Being a wave of this kind, it needs no Node conversion for us to acquire it, for mental use.

Let's just deal with this aside. We undoubtedly see matter though light waves, or rather perhaps, *in* lightwaves. So what, it may reasonably be asked, is the difference with sound waves?

Let's ask it in this very pointed way: what is the distinction in our relationship with the material world between:

- Light waves, being the carriers of visual information; and
- sound waves, being the carriers of sonic information.

Perfectly reasonable question. Here's the experiential answer:

Experience: Woofle
Step 1:

- Look at <∑D2*>
- Take <∑D2*> into your head, with the whole chair and room

Now:

- Give doggy a Name (anything: but a short one, for convenience);
- Turn that name around in your head, so you can see it clearly.

Now:

- Put that doggy name word into the picture;
- Not in a speech bubble. Not as if it were typed on the picture;
- Just put the name in your head "into" the picture.

Stop

<u>Discussion</u>:

(1) Can't be done, can it?

Step 2:

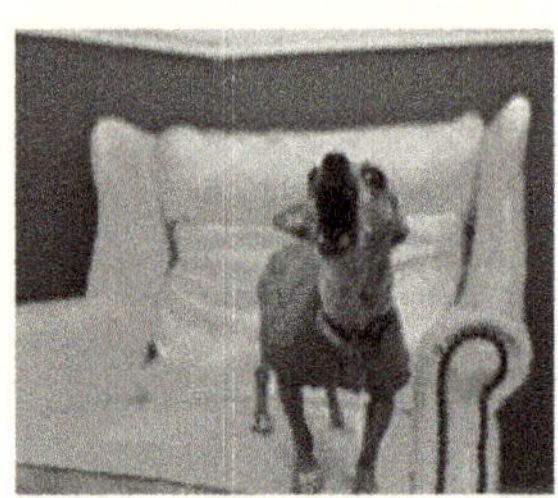

- Look at <∑D2*>
- Imagine the sound. Really imagine that sound.
- Take <∑D2*> into your head

Now:

- Put that sound in your head "into" the picture.

Stop

<u>Discussion</u>:

(1) It's so easy.

(2) In fact, you can't really stop doing it.

(3) It's like there is *wave of sound* connecting you to the picture.

So, the distinction in our relationship with the material world between:

- Light waves, being the carriers of visual information; and
- sound waves, being the carriers of sonic information.

is:

- We form complex composite images from light waves. We decompose them into a Node, and bring that Node into our head, so that we can conduct mental operations on the VCIP resulting from that Node;
- We do not form complex composite images from sound waves.[318] We conduct mental operations on the Φ point (s)[319] of wave Interferences.

Still don't believe this? Then, close your eyes and just listen for a few moments. Think about your experience.

Let's now deconstruct this graphic:

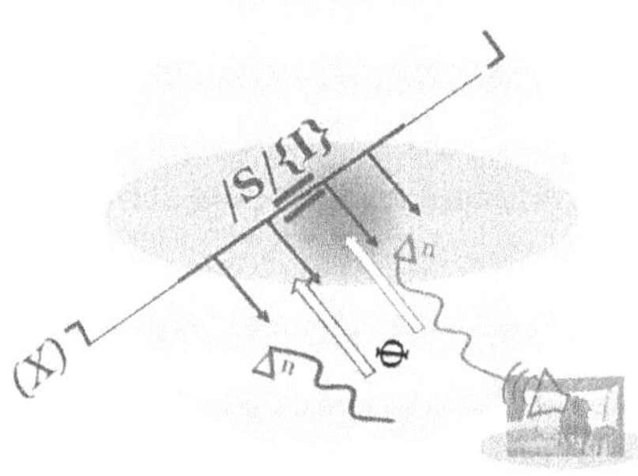

[318] Yes, of course there are harmonics and rhythms in nature and music. But those are equally unbidden and un-formed by us. If it were otherwise, music would never have had the emotional effect which it has
[319] See preceding note

Reminder again of some basic principles:

Interference Φ

> We have seen that Interference between Potentials Δ Δ, can create <∑I> Nodes.
>
> Potentials may *interfere* with each other, producing a simulacrum of an Event <∑E>.

We can all presumably agree that[320] we have sonic systems ready to receive sonic input. The usual arrangement is one on the side of each head.[321]

- The blue "river" accordingly represents that standing wave [Standing] Δ^n.
- The amber line is simply the sonic waveform produced by (or as) the barking.

As we have demonstrated, the sonic wave [barking]Δ^n needs no Node conversion to be accessed by us.

Input of the [barking]ΔB^n wave interferes with our standing wave [Standing] ΔS^n. Exactly how that interaction manifests (as augmentation, denuding, or cancellation) is a matter of fact in every instance.

It is elementary ML that:

$$[(\Delta B^n) \neq (-\Delta B^n)] \neq [(\Delta S^n) \neq (-\Delta SB^n)] = \Phi^n$$

[320] "Somehow", if you persist in mystifying elementary facts

[321] And of course with all the wire-ware through to the brain, and inside the brain

What is interesting, is that Φ^n (a Phi Point) is a simulacrum, a virtual version of a VCIP Node point ጠ^n.

It may be that, as a developmental matter, Φ^n comes first, and then the visual matrix creates a meta-copy.[322]

In either event, both Δ^n interface points needs to function in the same way, otherwise Scalar /S/ operations would not be possible.

Well, we can conceive lots of elaborate ways in which that would be possible. But what is the point in nature making more complicated path which needs to be experienced and learned by a newborn?

Just to clarify "standing wave". We don't mean a waveform which is permanently hanging around, never changing in aspect or dimension. Of course, we are producing these standing waves on a moment by moment basis.

Which is exactly why we *never hear exactly the same sound twice*[323] (no matter how that sound is produced).

We process a Φ^n (Phi Point) in a Scalar /S/ plane, of course. This is /S/{I}, which contrasts with the Scalar /S/ {E} plane for a VCIP <∑E*> Node Point ጠ^n.

---

322 We will come to this

323 Think about it.

Once in the Scalar Plane /S/{I}, the familiar operations in Identity can be conducted: x = y = z{I}. As demonstrated in Chapter 4, with arithmetic in Scalar /S/{E} plane.

Now, there is a clear cognitive gap between your hearing a sound, and thinking about it in language.

Experience: SoundScope

Step 1:

- Look at <∑D2*>
- Imagine hearing the sound <∑D2*> is making;
- Feel that sound in your head;
- Feel your connection with that sound, almost as if you can see the sound waves connecting you to <∑D2*>

<u>Discussion</u>:

(1) Easy and natural feeling.

Step 2:

Now:

(a) Practice saying the words: *There's a good doggy.*

(b) Say this a few times.

Next:

while you are feeling and focusing on that connection

- Say the words out loud: *There's a good doggy*

Discussion:

(1) Can't be done, can it?

(2) You just can't speak, at the same time you are "connecting" with the sound.

Step 3:

Now:

(a) Practice saying the words: *There's a good doggy;*

(b) Say this a few times;

(c) So, you can definitely think these words.

Next:

while you are feeling and focusing on that connection

- Think the words: *There's a good doggy.*

Discussion:

(1) Can't be done, can it?

(2) You can't even think language, at the same time you are "connecting" with the sound.

Once again, your view of reality may have just changed,[324] as a result of this *Experience.*

---

[324] A little, or a lot

You know perfectly well, from your common life experience, that doggy barks and you say, almost instinctively, "There's a good doggy", or "quiet there doggy".

Yet, we have just proven that you can't be speaking, or even thinking language, at the same time as you are actively listening.

You do already know this, in a way. We are all used to the phenomenon that we "tune out" *what a person is saying,* so that we can focus on *what they have just said,* in order to be able to formulate language thoughts in reply to their past utterance.

What we are seeing, again and again, with these experiences, is that there is a *cognitive gap,* a disconnection, between:

- Our experience of seeing, and hearing;
- and
- our linguistic processing of our thoughts, in preparation for speech.

Somehow, what we thought were *reactions* to what we see and hear: aren't.

The things we see and hear, are clearly providing a stimulus. But they are not the train setting off from the station, on rail lines into our head. The visual Node, and the sound wave are being carried on physical tracks in the real world.

Then they come to a terminus. At and in us. Then other processes happen. Yet we feel, outside these *Experiences,* that it is all happening, from doggy sitting and doggy barking, to us, in a *unified* way.

Let's go explain how that all works.

Pawfect

Now we need to put these jigsaw pieces together:

<∑Barking> Φ /S/{I} ┘

?

Something, just something, is missing. We don't have an equation. We have obviously important bits of the jigsaw puzzle.

Now, that missing piece. It must be simple. If we have to go off writing vast volumes about bald kings of france who don't exist, or number-word encodes that would crash a computer, then that's probably not something Patricia Krulls' 6 month old could work out.

Cogito ergo sum: why has the idea lasted for 400 years?

Was Descartes onto something *non-dualist* after all? Who was it thinking that "I" exist?

Please run the original *Experience: NameGame*

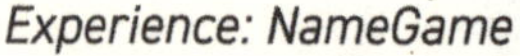

Experience: NameGame

Please click the link, if it helps to see a bigger version of the picture: https://hips.hearstapps.com/hmg-prod.s3.amazonaws.com/images/blue-velvet-sofa-living-room-1534263798.png

Step 1:

- Look at this simple picture. Lots of ordinary things in here, from a perfectly ordinary part of the real world;
- Just look at the picture, with your eyes open.

Now

- Name something in the room;
- Name something else;
- Do this another 3 times.

Stop

<u>Discussion</u>:

(1) That was easy. You've been doing that since kindergarden.

Step 2:

- Just look at the picture again, with your eyes open;
- Pick a Thing and name it.

Now

- Put that name word into the picture;
- Not in a speech bubble. Not as if it were typed on the picture;
- Just put the name in your head "into" the picture.

Stop

<u>Discussion</u>:

(1) That was impossible;

(2) You found yourself trying to "pull" the Thing into your head;

(3) Because inside your head, the Thing and the Name "fit" in the same mental landscape.

Now, whatever is going on here: something is going on.

Now, try this version:

Experience: YouTwo

Preparation:

This is really important

You need to focus and really try here

Imagine a second you. A perfect replica copy of you. This is secondYou: You_2

For this *Experience*:

- Imagine you're on a tennis court;
- Umpire the room at the other end;
- And You_2 where the umpire would be;
- On your right-hand side.[325]

Step 1:

- Just look at the picture again, with your eyes open;
- Pick a Thing and name it.

---

[325] Your left-hand side, if you're left-handed

Now:
imagine You_2 looking at You seeing the room
You_2 can see You
You_2 can see the space between You and the room
You_2 can see You looking at the room

And **with You_2 watching you**:

- You_1 put that name word into the picture;
- You^1 do another one;
- If it's easier, let You_2 do it;
- And again.

Stop.

Discussion:
(1) That was easy. Child's play.

Step 2:
With You_2 watching you:

- You project that name word into the picture;
- You do another one;
- And again.

Discussion:
(1) That was easy again.

Notice how you didn't need to keep thinking of $You_{2.}$ You just needed to perform the projection once: and You_2 stayed there. Occasionally you popped the You_2 reference thought into your head. But *so long as you knew You_2 was there,* you could name away: putting names freely into that space.

It's like the room became "yours": completely under your control. That relationship, inside your head, with You$_2$ "other you" changed everything.

Step 3:

- Just look at the picture again, with your eyes open;
- Pick a Thing and name it;
- Now: bring it into your head.

You can refresh the image by opening your eyes to look

Now:

imagine You$_2$ looking at You seeing the room

And **with You$_2$ watching you**:

- Put that name word into the picture;
- Do another one;
- And again.

Stop.

<u>Discussion</u>:

(1) It was like You_2 wasn't even there.

(2) You really had to try hard to notice You_2 "in the corner of your mind".

(3) It felt at some moments like You_2 "came over" and "rejoined" You_1

(4) Then the whole image stability faded. You had to start again.

In *I Want To Love But, Realising The Power Of You,*[326] the author describes a perfect system, which is born into everyone who ever lived. This is both an emotional and mental balancing and control system.

You have the same language comprehension, creation, and control system. You are not quite born with it. You are born with the developmental pathways to it. Just as Patricia Krull so brilliantly observed it.[327]

A Perspective of Matter

Back to Chapter 4, to remind us:

> ***The VCIP Perspective***
>
> We are now well used to thinking in terms of Frames of reference: FoR.
>
> FoRs apply only as aspect of Forms: [E], [I]. FoRs are deductive. That is the functional attribute of their hylopmorphic reciprocity.
>
> ...
>
> In ML, Scalar /S/ is an *inducted* Perspective.

---

[326] (2019)

[327] op. cit

When you read that, back then, you may have thought: well the author states that, and OK: but so what?

The so what is:

Frames of Reference do not harmonise
Perspectives do

It was clashing, non-harmonic FoR's, which were causing you so much trouble:

Shaking off outline dog:

Now try it with You$_2$

Bringing cut out dog into your head

Now try it with You$_2$

The object name stickiness

Now try it with You$_2$

Putting the name into the picture

Now you've tried it with You$_2$

Separating dog from the background

Now try it with You$_2$

Thinking <∑D*> and /exists ¬ all at the same time

Now try it with You$_2$

Why your black curtain worked

Now try it with You_2

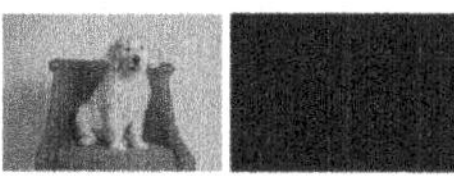

Why your coloured curtain worked

Now try it with You_2

That is also why you had no trouble at all putting the "bark" into this:

Let's get spooky. How can the Author possibly put together these Experiences in this book? Discounting telepathy, or mass hypnosis with millions of readers, [328] it may seem mysterious that the Experiences relate to what actually happens inside readers' heads.

The answer is that the Author is using inductive reasoning from *coherence experienced between the Subjective and the Objective Perspective.*

Now, what follows next is going to irritate, or distress, or otherwise frustrate many in academic philosophy.[329] Although, when we come to discuss *Matrixial Time,* they may feel different.

[328] Hope springs eternal: *An Essay on Man.* Alexander Pope. (1734)
[329] Sorry: not sorry

What we have been describing is these processes:

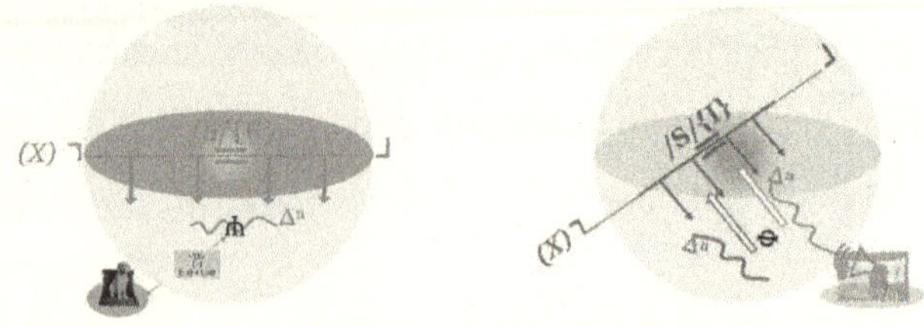

This perception of matter all happens in *Subjective Space*.

Subjective Space

What we mean by this, is that information processing, which happens with reference only to one's own internal attributes.

The Scalar mapping of the <∑D*> Node happens by reference to a Phi point ф in the "river" of an infinity ∞ process Δ^n :

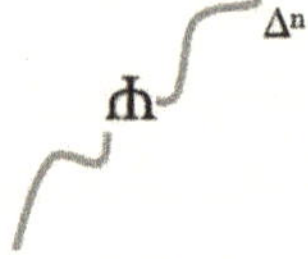

But, where and what is that "river"?

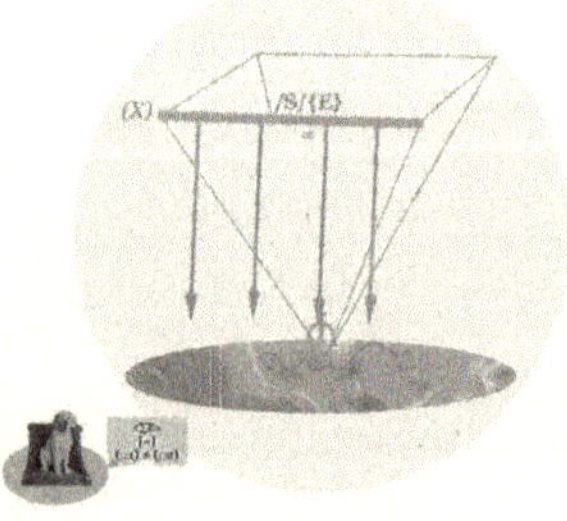

This graphic is a place-holder. There is much more to be explored, particularly in later chapters.

For now, the essential point is that <seeing the dog> is a process which is happening within the whole of you. You are seeing: using the human faculty of sight. The functions of your eyes and optic nerves. These are all well-established in neuroscience literature.

Everyone's eyes are different. Everyone's brain (as a purely physical matter) is different. Everybody's emotional state, from moment to moment, is different: and different from every other person alive or previously alive.

Which is exactly why we *never see exactly the same image twice*[330] (no matter how that image is produced).

The seeing-dog-photo experience is: *unique* to you, during the time that you do it.[331] We don't have to be doing Matrixial Logic to know that, of course.

Bringing back some Chapter 4 material:

> Thus: no Substance or Moment is pre-determined.
> But /S/ most assuredly is pre-determined:
> *that which can be /S/ is fixed by that which is already* ሐ.

By that,

[330] Think about it.
[331] Yes: clue

we do not mean fixed "in place", or "location". Point ₼ could, after all, occur anywhere in the domain of I ∞ in process Δ. That is as "large" and indeterminate a domain as it gets.

Indeed, rather than visualising ∞ in Δ^n as a "river": [332]

it may be more helpful to view ∞ in Δ^n as an expanse:

We can see that:

/S/ is fixed by that which is already ₼.

That also helps to explain verbal binding.

Experience: Ocean

Step 1:

[332] Yes...

- You are hovering (no gravity) over this vast ocean
- This VCIP point m̄ has arisen. The point is here represented by a piece of paper, there only the ocean.
- There are no waves.

Now: in your mind

- Move the paper.

Stop

Discussion:

(1) The m̄ paper won't move.[333]

Step 2:

Now:

- You are the m̄ paper;
- Project yourself into it.

Now: in your mind

- Move you.

Stop.

Discussion:

(1) This is even harder. The You m̄ paper won't move.[334]

[333] For some readers it will: because you are using You$_2$

[334] For some readers it will: because you are using You$_2$

Step 3:

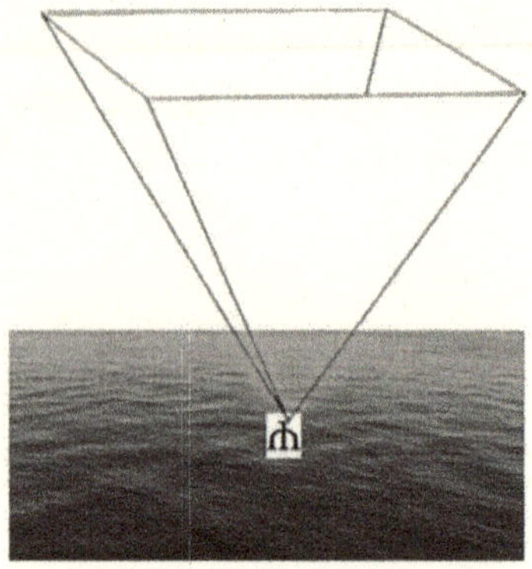

Now:

- You can be the ₥ paper, or be hovering, as at Step 1.

Now: in your mind

- Move you.

Stop

<u>Discussion</u>:

(1) This is instantaneous.

(2) You move up to the top of the pyramid, or onto its surface.

Your perspective of matter changes fundamentally, when the brain-ware (represented by the pyramid) is there to "lift you out" of the ocean.[335]

---

[335] Yes: clue

It is in the ocean that there are "rivers", the currents, produced by you as a living organic being:

The [dog-landscape] Node <∑D*> produces a perturbation in that ocean current, which has sufficient effect to collapse the infinity potential: you have VCIP: ф̄

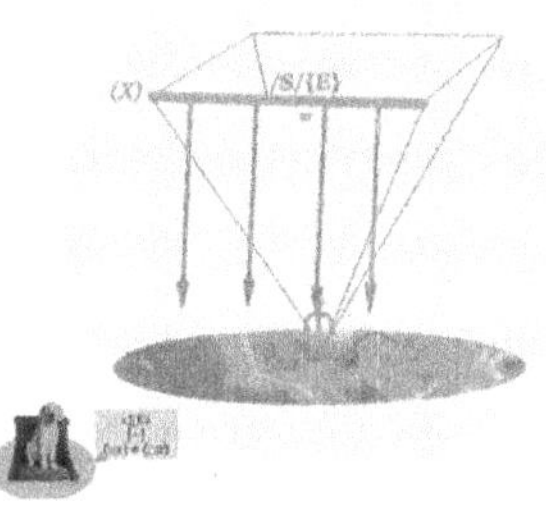

This allows Scalar interaction for Noun type words.

And here is the Verb version:

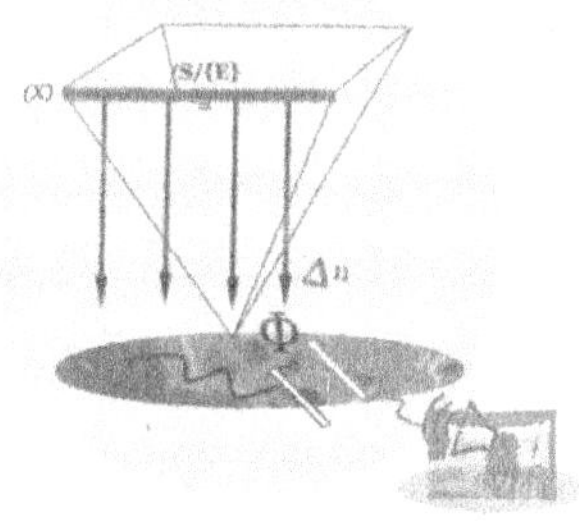

We are still in subjective space.

To give a rough analogy: this [dog-landscape] Node <∑D*> is acting like a taste of something: you're not sure of it's sweet or salty. We can look later at the brain process for that experience. But we can agree for now that we don't have to trouble your Scalar /S/ verbal plane with that matter.

Of course, that sweet or salty tasting is completely *subjective*. Whatever any philosophers wish to argue about epistemology and qualia, that can be common ground: sweet or salty taste, is *subjective*.[336]

That analogy is even clearer in relation to hearing (or imagining hearing) the <barking>.

Now, if that were all there is to it, many readers could be excused for exiting here, and asking for half their money back. Or all of it, if they are Cromwellian uber-materialists.

For we have, on that view, ended up after these interesting Experiences, with a dead end: a cul-de-sac of subjectivity. A domain which logic is helpless to penetrate or regulate.

---

[336] Even if that subjectivity is reducible to, or explicable by, objective criteria

Is that really our final destination? Or perhaps there is something rather more radical coming round the corner.

The Missing Person

What is it? Where is it? How is it?

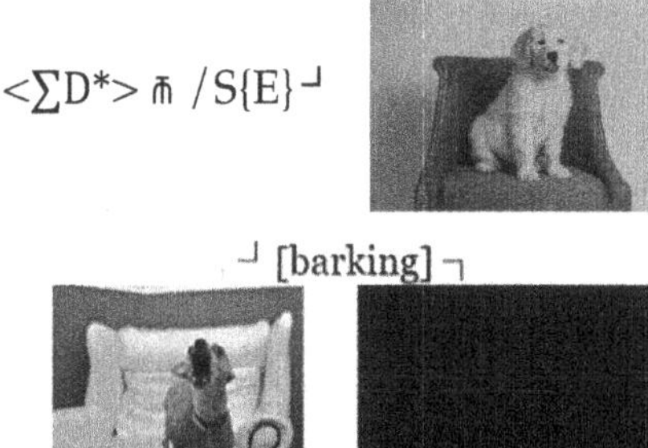

<∑Barking> Φ /S/{I} ┘

?

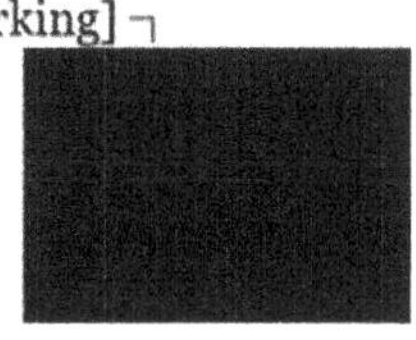

As we said at the start of this Chapter:

Known Unknowns

We know that HIT involves some kind of iterative framework. But it is not merely repetitive. It is heuristic. By doing, we learn.

HIT is also hylomorphic: meaning derives from forms both constrained by and expressing contents.

Another obvious attribute of HIT is that it is holistic. Not merely in the life experience of the individual as its is expressed in a present moment of communication. But also socially holistic, and indeed environmentally holistic.

So what we seek needs to be:

- heuristic
- hylomorphic
- holistic

And, it must be something so simple that a newborn develops it within 6 months.

To arrive at the answer, just ask one question:

Who was the other person
in the Experience: Name Game Round 2?

In all the *Experiences* we were doing, we were actually taking you through the developmental path which you, as newborn to 6 months, took. Which everyone ever born has taken.

We had to misdirect you here and there. Because your:

- heuristic
- hylomorphic
- holistic

language comprehension system is so powerful.

Now, we can fill in the missing piece of the jigsaw: and it was so obvious, right fom the very beginning. That's why you missed it, right up until the final *Experience.*

The missing piece is: You$_2$.

Considering You$_2$

We had arrived at this:

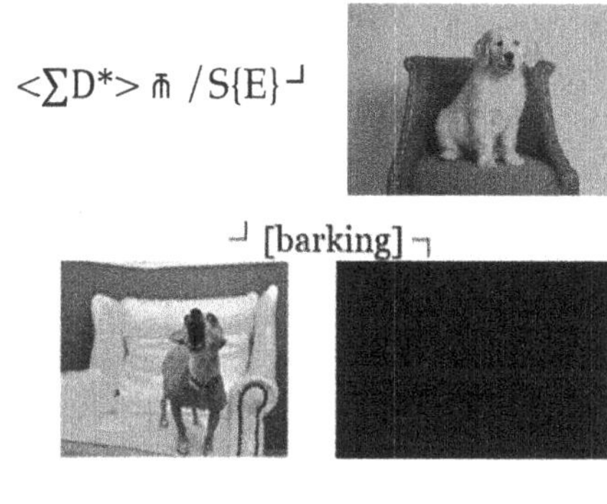

<∑Barking> Φ /S/{I} ┘

?

incomplete elements of an equation.

Let's put the pieces together. While we are doing it, keep thinking about You$_2$. About how it feels when you notice the activity of You$_2$.

Consider, using your own reasoning:

- Is You$_2$ real

or

- Just some mentalist illusion?

But, if You$_2$ is just some illusion, then how can You$_2$ perform real functions? After all, you can have no doubt that before you accessed You$_2$, you just couldn't put names into the landscape, or pull objects out of the landscape.

Yet, as soon as you did access You_2, all this became not only possible, but easy.

Where is this You_2? Obviously, it's somewhere inside your head. Yet it also has real effects on your mental relationship with the world outsie your head.

We were looking at pictures. But you can run these *Experiences* looking out at the actual world.[337] The same results occur.

Please also consider: why is it that You_2 works as a perfect copy, duplicate, replica of You? Not an adjustment of You, but an iteration of You?

What is it about seeing the world with You in it, as part of the world, which immediately changes your relationship with the world?

It's You_2, But How

We have reached the half-way turning point in Matrixial Logic.[338]

We have been preparing you, with ideas and *Experiences,* to reach this point. We had to. If we had started here,

[337] And please do try

[338] You may be glad, or sad, to learn

you may have understood it, but you wouldn't have experienced it.

Matrixial Logic is not external to You. It is a process in thought of unwrapping what is You: your dynamic of interaction with yourself and the world.

It is about understanding reality: the only reality that you have.

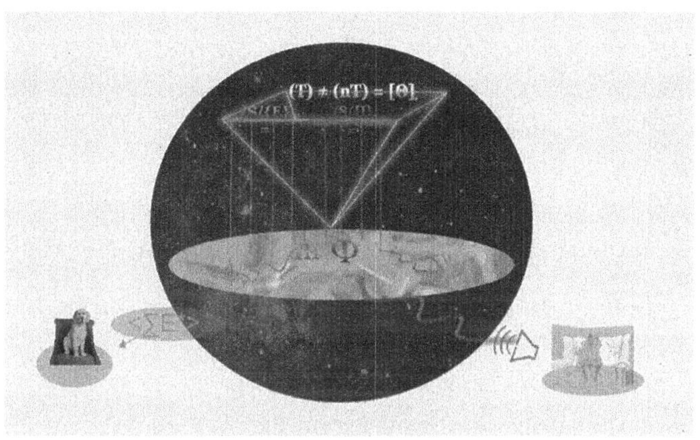

This is the domain of your Mentation
Your interacting axes fields of thought

{E}

We can see the /S/{E} Scalar Plane axis. The realm of the "ᕼ Node Point of Visual [E]" perspective. The one that:

- Can't extract backgrounds
- Can't attach names to things in their own landscape
- Can only:
  - (a) Identify.
  - (b) Attribute.

{I}

We can see the /S/{I} Scalar Plane axis. The realm of the Φ Phi Point perspective. The one that:

- Is directly connected to the outside world;
- Can attach sounds to things in their own landscape;
- Can't think or speak while engaged in doing that
- Can:
  - (a) Identify ;
  - (b) Attribute.

Then we have, making its first appearance in Matrixial Logic:[339]

{Θ} Theta Axis

$$(T) \neq (nT) = \Theta$$

This may look very familiar, as an equation form.

It is where we started, back in Chapter 2:

[339] although it has been hovering like Banquo's ghost all the way through

$$(A) \neq (nA) = E$$

This is, of course, the equation for Substance extended in spacetime [E]. That is the equation for the world of Substance, "out there".

The {Θ} Theta Axis provides the equation for the world of Substance, "in here".

How is it possible to state that there is a world of Substance, "in here": unless one is some kind of reductionist uber-materialist? That looks like a sensible objection. So it deserves an answer.[340] But that will have to wait to a later Chapter.

The {Θ} Theta Axis is what connects you to the world, in the matter of your speech behaviour in the world.[341]

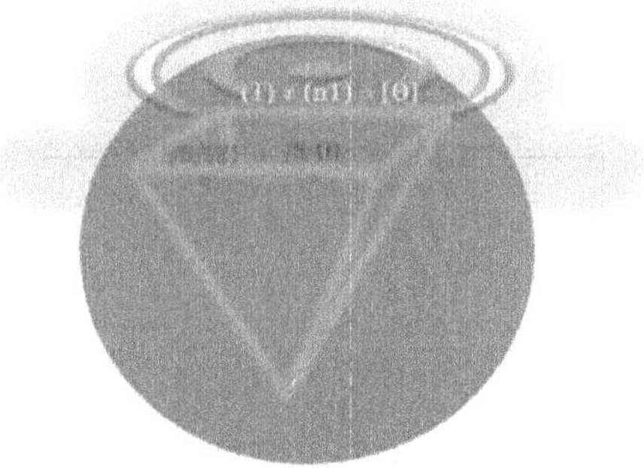

[340] Sensible is never guaranteed

[341] Not including "owww" or other emotion-controlled outbursts

<∑D*> m̄ m̄ /S/{E} (⍰)

≠Ω (T) ≠ (nT) = Θ

<∑Barking> Φ /S/{I} (⍰)

Note: we will explain this notation (ᴄ) ≠(ɔ) = Ω at a later point. Suffice to say for now, that it represents operations in the Emotional domain.

We have already proved to you that the {Θ} Theta Axis perspective exists.

This is You$_2$.

The You$_2$ that:

- Can extract backgrounds (because it is World Connected);
- Can attach names to things in their own landscape (because it is World Connected);
- Can do all the amazing things that the You$_2$ perspective achieves;
- Can put Noun and Verbs together: because they are all part of the world, and the You$_2$ Theta Θ Perspective is World Connected.

How does the You$_2$ Theta Θ axis get to be "world connected"? A later Chapter for that one please.[342]

As you might expect for a 3H system, just like all those

[342] In fact, four later Chapters

3H systems which run your body, there is perspective harmony, in dynamic, which learns through dynamic harmony. It is a syncopated system, relying on two connected processing capacities.

Bi-dimensional processing systems: eyes; ears; nostrils. Hands, feet, arms, legs, lungs. The /S/{E} and /S/{I} perspectives axes. This is not logical metaphysics, spooky disembodied "consciousness". It's systems analysis.

<ΣD*> ⫚ ⫚ /S/{E} (⍰)
≠Ω (T) ≠ (nT) = Θ
<ΣBarking> Φ /S/{I} (⍰)

The equations now look a bit better:

$$(\mathrm{T}) \neq (\mathrm{nT}) = \Theta$$

= L/T: ˆDog Is Barkingˇ

≈

"Dog Barks"

L/T simply means language in thought.

Contrasted with:

- L/S (language in speech)
- L/W (language in writing)
- L/R (language in reading)
- L/M (language in motor: touch, signs, et. al).

There are obvious differences in how we process these different modes of L/. We will get to those later, once we have finished the basics of L/T.

Now, let's explore the {Θ} Theta Axis.

The Existential Domain

We have just been introduced to:

{Θ} Theta Axis Field

$(T) \neq (nT) = \Theta$

This is the realm of the Objective. Or, as we more functionally and glamorously say, the Theta Axis is:

Universal Television (1985-1989)

The Θ Theta Axis takes Subjective states, which are Scalar /S/ modified from the original disparate subjective sea inputs, and *equalises* them.

We can't of course show you this function going on, in real time, in the brain: in its chemistry or anatomy. But then, we can't do any of these things with most of what we over-generously refer to as scientific inspection of reality.[343]

However, we are aware, by induction from the observed facts, that this *equalising* function occurs. Otherwise human kind would not still be here. For all human civilisation is due to it.

We can actually prove the existence and operation of the *equalising* function of the Θ Theta Axis. In fact, we have already done so repeatedly in the *Experiences*. The ones which required You_2 to solve them. We'll say more about this, shortly.

The *equalising* function of the Θ Theta Axis needs to be childishly simple. Otherwise infant brains would be unable to get to grips with it in the first 18 months of life.

Let's count:

apple + orange

No: we can't. However wedded you still are to the law of identity, we can't count two different types of things under the same rubric.

We have to *translate each Substance into a Scalar Axis*: in this case, that of numbers in an arithmetic scale.

---

[343] See *Complete Reality* Chapter

Measure or weigh apple and compare with orange. Again, we *translate each Substance into a Scalar Axis*: in this case, that of numbers in a length or weight scale. And so on.

As Descartes wrote in around 1628:[344]

> Unity is the common nature which, we said above, all the things which we are comparing must participate in equally. If no determinate unit is specified in the problem, we may adopt as unit either one of the magnitudes already given or any other magnitude, and this will be the common measure of all the others [CSM 449]

Descartes had some intriguing ideas on space and time and sense and thought. With his contemporary, Newton, they were breaking free of the Aristotelian conceptions, which had governed western thought for two millenia.

As we now know, at the beginning of the 21st century, Aristotle was less wrong, than they were. As we shall see in later Chapters.

Everything that is, is extended in spacetime. Nothing that is, isn't extended in spacetime. [345]

The soundwaves of a bark are also, of course extended in spacetime. And those waves are operated upon differently to light waves, by humans.

---

[344] *Rules for the Direction of the Mind (Regulae ad Directionem Ingenii)* in Descartes, René. The Philosophical Writings of Descartes: Volume 1 (p. 63).
[345] More to come on this in *Complete Reality* Chapter

That is why ML treats those two waves as different. Because our bodies and brains[346] treat them differently. In order to understand reality, logic must follow the course of reality.[347]

Let's plunge back into that reality, and consider again subjective states in the Scalar field.

Scalar {E} Field Axis

Scalar {E} takes information input which is provided by Node interaction with currents in the Sensory Sea. That is a rather poetic way of putting matters (save as to Nodes) which are well-established in neuroscience.

<∑D*> ⫚ =>/S/{E}

Scalar operations effect their own version of equalisation. Shape, colour and so on are all packaged. It is well-established in neuroscience that these processes are mimetic. We can actually see in the relevant part of the brain patterns which "look like" the inverted retinal image. This is a fascinating example of reflected light-waves being scaled into usable information.

Ml is interested mainly in process, rather than content. So we can leave detailed explanation of the scaling process to neuroscience.

---

346 Insofar as space and time are differen

347 Not a proposition all logicians would sign up to

Also, if any neuroscientist, or other scientist, wishes to object "that isn't how that process works", then these ML thumbnails of such processes stand to be corrected.

In the domain of scalar equations, =>/S/{E}, we are simply dealing with subjective judgments along an (x) + /- scale and a (y) + /- scale. This is co-ordinate geometry. The down arrows seen here, represents extrusions from the axis: just as in Chapter 4, where we looked at arithmetic and geometry.

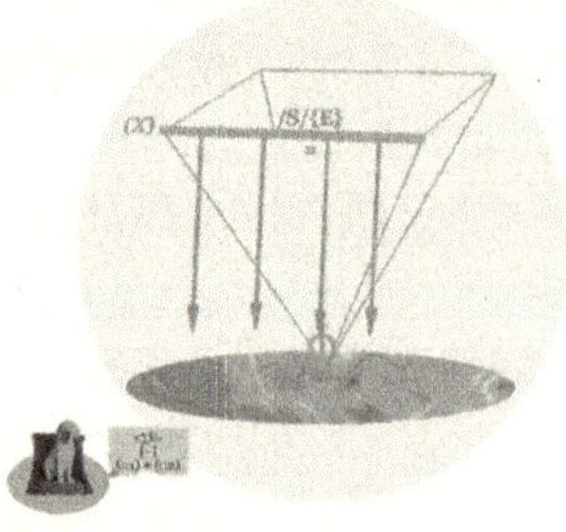

The Scalar Axis deals with extrusions from baseline equality. If you have never seen what you are looking at before, the baseline equality is 0. So, you have to scale by analogy. If you are able to compare what you are looking at to a previous scaling, then you can apply co-ordinate visual geometry, and match results. It's kind of like using a 3-dimensional barcode reader.

Save for the mechanical processes involved (the brain "circuitry"), it's no more difficult than reaching out to grasp an apple. or an orange. We can see, in that comparison, how we perceive geometrical co-ordinates in the world and match them with scalar co-ordinates, in the brain.[348]

The Scalar /S/ axis equates by comparison, not merely in quantity, but quality. Big \Small; Good \ Bad; Right \ Wrong.

What is intriguing about the qualitative function, is that it is all comparative. We cannot operate our /S/ function with reference to an isolate.

What's more, we don't operate our /S/ function in direct connection to the world out there. We register, qualify and quantify, by reference to /S/ scaling of Nodes. Those Nodes are created by our interaction with what is out there.

Scalar/S/ cannot operate Directly.

Experience: Clipped

Step 1:

See this blank sheet of paper, on a table top:

[348] That is an output Theta Θ ≈ example, of course

See a paper clip on that blank sheet of paper:

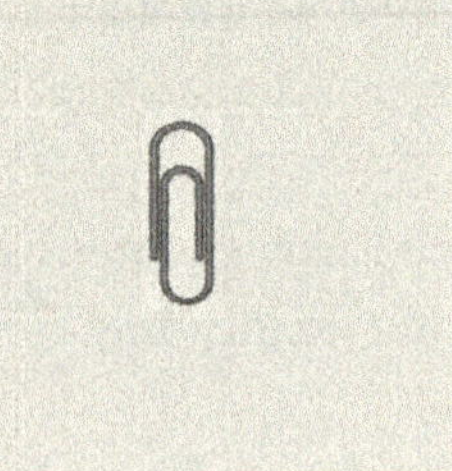

Now:

(a) Choose a short sentence. Any sentence. Something like *Cake is nice*. As simple as that;

(b) practice saying the words.

Next:

Focus on the paperclip

while you are feeling and focusing on that connection

- Say your Sentence out loud.

Stop

Discussion:

(1) Can't be done, can it?

(2) You just can't speak, at the same time you are "connecting" with the image.

Step 2:

Now:

Focus on the paperclip

while you are feeling and focusing on that connection

- Say your Sentence silently in your head.

Stop.

<u>Discussion</u>:

(1) Can't be done, can it?

(2) You just can't think in language, at the same time you are "connecting" with the image.

Step 3:

See these two paper clips on that blank sheet of paper:

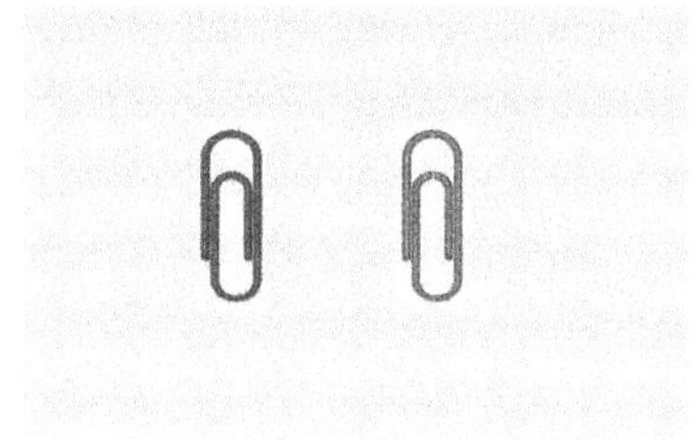

Now:

Focus on both the paperclips

while you are feeling and focusing on that connection

- Say your Sentence silently in your head.

Stop.

<u>Discussion</u>:

(1) Easy, isn't it?

(2) You can't think in language, at the same time you are "connecting" with the **multiple** image.

Step 4:
See these two paper clips on that blank sheet of paper:

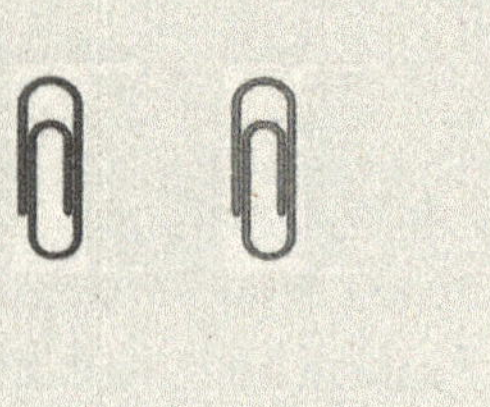

Now:

Focus on both the paperclips
while you are feeling and focusing on that connection

- Say your Sentence <u>Out Loud</u>

Stop

<u>Discussion</u>:

(1) Easy, isn't it?
(2) You *can* think in language, at the same time you are "connecting" with the **multiple** image.
(3) In fact, it feels like the **multiple** image is *drawing the very words from you.*

Step 5:
See a single paper clip on that blank sheet of paper:

Now:

Focus on the paperclip

Next:

imagine You$_2$ looking at You$_1$ seeing the paperclip

And **with You$_2$ watching you**:

- Say your Sentence out loud.

Stop

Discussion:

(1) Easy, isn't it?

(2) You *can* think in language, at the same time you and You$_2$ are "connecting" with the **single** image.

(3) In fact, it feels like the connection between you, You$_2$ and the **single** image is *drawing the very words from you.*

All That Grammar

Language uses rules. Rules that you have to learn.

These are not rules that you are born with. They are not "inside" you. But you are born with the ability to develop the use of these rules.

The rules of language work in the Theta Axis Field.[349]

---

[349] We will come onto why we use the word "field", later

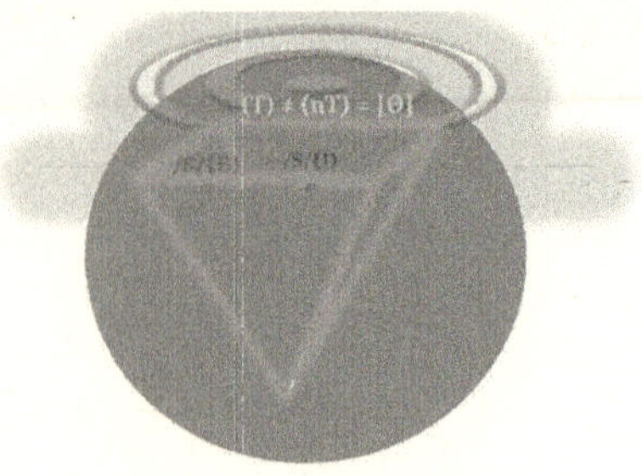

$$(T) \neq (nT) = \Theta$$

This is your "objective" Axis. This is the Axis that communicates *from* you to the outside world.

The connection is[350] this:

Theta puts the "is" into your language.

Of course you can have names for things. You start attaching names to the things your 5 senses bring to you. You start to learn these in your baby steps of talking. These are the first things you learn to say out loud: "Mama", "doggy", "splot" (you don't know the name for that one: you just know it has a name, for you).

That is your /S/{E} Scalar axis at work, in your subjective world.

As you have *Experienced*, you need something else to get those names "up there" in your head, then "out there" to the world.

---

[350] There's more to it, but we'll come to that

That is the You_2 function of the Theta Θ axis: the one that organises thinking and talking sentences.

In the *Experiences,* you are trying to misuse Theta. By trying to put Theta into the frame of perception, when Theta focuses on output.

The rules of grammar are built around the law of identity (A=A), which is founded upon the rock of existential certainty.

As we are learning, that rock may be Tarpeian,[351] rather than Gibraltan, if we are not very, very careful how we view the world from its pinnacle.

Experience: Roominating

Step 1:
Just look around the room you're in while your reading this. Maybe a living room, like this: or any place really.

[351] https://ancient-history-blog.mq.edu.au/cityOfRome/Tarpeian-Rock

Just look around. Think whatever thoughts you want to.

Now:

- Start thinking to yourself any descriptions you like of what you are seeing.

Just for example: *that picture's nice. Though maybe it isn't hanging straight.*

- Keep talking to yourself.

Next:

Focus on the inside of your Eyes
<u>While you're talking to yourself</u>

Notice what's happening.
Stop.

<u>Discussion</u>:

(1) As you talk to yourself, your eyes seem to "shut off" from the room.

(2) You can still see the room. You would definitely notice if the room just disappeared, or an object in it fell over.

(3) But, it's like the room is being viewed by someone else in your head while You-you are doing the talking to yourself.

Step 2:
Just look around the room you're in while your reading

this. Maybe a living room, like this: or any place really.

Just look around. Think whatever thoughts you want to.

Now:

- Start thinking to yourself any descriptions you like of what you are seeing.

Just for example: *that picture's nice. Though maybe it isn't hanging straight.*

- Keep talking to yourself.

Next:

Focus on the inside of your Eyes
<u>And speak your thoughts out loud</u>

Notice what's happening.
Stop.

<u>Discussion</u>:

(1) As you talk out loud, your eyes seem to "shut off"

from the room.

(2) You can still see the room. You *might* notice if the room just disappeared, or an object in it fell over.

(3) But, it's like the room is being viewed by someone else in your head while You-you are doing the talking out loud.

This looks like the familiar "I tune out when I'm driving" dynamic.[352]

> New Study Shows for First Time How Thinking Can Impair Driving
> High percentage of car accidents due to inattention rather than lack of driving ability
>
> While research has tried to demonstrate the potential danger of external distractions (looking at road signs or a map while driving), previous studies have not focused on internal distractions such as one's thoughts. In their study, psychologists M.A. Recarte, PhD, and L.M. Nunes, PhD, of the Universidad Complutense, in Madrid, Spain, examined whether a driver's eye movements would be affected by additional verbal and visual tasks to the point where the driver's ability to pay attention to his or her surroundings is sacrificed.

There is a line of reasoning which we seem driven to[353] by these observation. The consequences of that line are well summarised here:[354]

> Now, what do these dual levels of consciousness—the automatic level and the conscious level—have to do with our emotional responses? Quite a bit, it turns out, as we find that many of our negative thoughts operate in an automatic fashion.
>
> Negative thoughts impose harsh judgments on our abilities, harp on

[352] https://www.apa.org/news/press/releases/2000/03/driving

[353] Ouch

[354] https://www.psychologytoday.com/gb/blog/the-minute-therapist/201803/is-your-brain-automatic-pilot

> our character flaws, portend terrible consequences that lay ahead, instigate anger and prompt aggressive responses, and nag us about our weaknesses, mistakes, and misgivings. When they prick us, we experience corresponding emotions of fear, anger, guilt, worry, and so on. They also dampen our self-worth or how we think of ourselves.
>
> Some negative thoughts occur automatically, as though they just popped into our heads without any conscious effort. Other negative thoughts are products of conscious awareness, of ruminating on our problems, defects, flaws, and disappointments and failures. These disturbing thoughts occur in the domain of internal dialogues we have with ourselves during the course of the day, as whispers we speak to ourselves under our breath, or self-talk.
>
> In the Minute Therapist blog, we focus on the importance of paying attention to our inner speech—our self-talk—as well as to our emotional reactions in the moment so that we can bring to consciousness the underlying thought triggers for these feeling states. We need to stop and think about what we are thinking and feeling.

Fortunately for our emotional and mental health, everything stated is wrong. Which is a shame for psychology and CBT, the practice of which is founded on these errors.[355]

You can immediately see why the "automatic consciousness" theory is mistaken. You were looking at the room, paying utmost attention to the room: yet you still could not think or speak to the scene directly.

---

[355] That's a big claim. See its justification in *Secret Self*. The Author (2020)

The Is Of It

When you look at the sitting dog <∑D*> m̃ /S/{E}, or when you hear the barking dog <∑Barking> Φ /S/{I}, you don't think "is". Or "^{NOT}is".

You don't need to. You are seeing and hearing these dogs. Indeed, as you've learned from the *Experiences,* you can't actually formulate language thoughts or spoken words, while you're occupied with seeing and hearing.

It may seem strange, but your own *Experiences* have shown you that:

- When you are seeing, you can't speak

and

- when you are speaking, you are not seeing.

If your world of mentation consisted solely of these Scalar axes m̃ /S/{E} and Φ /S/{I}, you would be effectively mute. You could see and hear, but you wouldn't be able to share that cognition with anyone: at least not under a set of mutually shared language rules.

You need another capacity. One that:

- Coheres your Scalar thoughts;
- Uses language rules that are mutual with the outside world;[356]
- Can translate your Scalar thoughts into those rules.

---

[356] At least some part of it

That is what your Theta axis = Θ does.

The first action your Theta axis = Θ takes is to decide the existence value ("EV") of the Scalar information.

That existence value is (T), in:

$$(\text{T}) \neq (\text{nT}) = \Theta$$

Now, we must tread carefully here. Theta is not deciding whether anything exists out there in the world: whether there is an external reference point for the Scalar data. That is not Theta's mission statement.

Consider the matter: why should it be otherwise? It is the job of your senses and your Scalar translation of Node m̄ and Wave Φ interfaces, to detect whether something exists or not. Although that task is already done by the world "out there". Either a thing exists, whether perceptible by any of the 5 senses, or it doesn't.

Of course, there's a whole world of things which exist, and of which we are not perceptibly aware. But until we do become aware of any it as information in perception, it cannot be in our heads. Whether you are a rampant subjectivist, or objectivist, that necessary relationship is a fact of the matter.

Once we do become aware of that information, then it exists for us, whether we like it or not. That's a commonplace:

information is presented to us, and we may say "I really didn't want to know that". This involuntary information idea is, after all, at the core of the Woke phenomenon.

In like manner, you may fill your Scalar domains, being purely subjective, with unicorns, bald present kings of France, honest politicians, and any manner of fantasies.

You may manufacture whatever fancies you wish out of the raw stuff of your perception information: and you do. That is all you do.

You don't process reality in your thoughts. You process information about reality. That "about" can vary from distance of a microsecond, to multiple millienia of years, and trillions of miles.

Look how we had you in the *Experiences*, interacting with photographs, with imagined landscapes and sounds, with imaginary paperclips, with words and sentences. *They are all just made up in your Subjective head*. But they are really real.

They are really real to the Axis which counts reality: your Theta Θ Axis.

Theta Θ is the supermarket cashier. It is not Theta's job to determine how the goods got in your basket or from which shelf you got them, nor indeed whether any such shelf "exists".

What is in your basket is what the Theta cashier sees. To which Theta then (metaphorically) applies the rule-based tasks of:

- Categorising
- Adding up
- Deducing a total

There must be a framework for such a rule-based system, and an underpinning, immoveable, unprovable premise.

The premise is: Exists.

To Exist in Θ, means that (T) is extended
in objective internal spacetime.

The reaonably and reasoningly sceptical reader may well be saying at this point: *Hold on. Look, I follow your line of argument up to this point. But now you're just laying down fundamental assertions about this Theta Axis. Where's your proof that this isn't just a unicorn idea?*

Good. Let's now begin to prove to you, the reader that:

To Exist in Θ, means that (T) is extended
in objective internal spacetime.

and in doing so, prove the existence of the Theta Θ Axis function.

A preliminary word of warning. These next *Experiences* will be stimulating. But, you are going to feel dizzy. Your emotions will become engaged.[357] You will want to restore emotional balance. We will do that for you, at the end. Promise.[358]

Experience: Sheeted

Step 1:

- See this blank sheet:

There is nothing else to see.

If you can look at this sheet, here in the book, without your eye catching any words, so that you can focus on only this sheet, that's fine.

Or, you can hold this sheet in your head.

Ready?

Now:

- Imagine on the sheet a paperclip:

which you know is not there

[357] We will discuss this more in a later Chapte

[358] Not a polician's promise

- Keep trying.

Stop.

Take a breath. Look around you. Relax.

Discussion:

(1) You can't do it;

(2) The harder you try, the worse it gets. You actually feel stress, and some emotional distress.

How is this possible? All through this Book, we've had you imagining all sorts of things. Manipulating them, encountering difficulties. But never have you experiences an inability to make your mind just think of something.

Something very powerful is limiting the operation of your mentation. That something, that process, or function, is so powerful that your emotions become aroused, to deal with the stress which you are encountering.

Step 2:

- See the same blank sheet:

There is nothing else to see.

If you can look at this sheet, here in the book, without your eye catching any words, so that you can focus on only this sheet, that's fine.

Or, you can hold this sheet in your head.
Ready?

Now:

- Imagine on the sheet Two paperclips:
  both of which you know is not there

- Keep trying

Stop.

Take a breath. Look around you. Relax.

Discussion:

(1) You can't do it.

(2) It becomes like some insane cartoon: one paperclip sort of emerges onto the sheet, and it's like the sheet rejects it; then the other paperclip tries, and is rejected.

(3) The harder you try, the worse it gets. The *trying to exist clips* just seem to spiral in and out of reality. bit like a spin up/ spin down quark.[359]

[359] Yes: clue

(4) You actually feel stress, and some emotional distress.

You did the *Clipped Experience,* just a few pages back in this Chapter. There you were, getting stuck with one paperclip, then become gloriously free with two paperclips.

But now, the second clip doesn't come to the rescue. It make matters worse.

Please recall that we are now proving to you that:

To Exist in Θ, means that (T) is extended
in objective internal spacetime.

and in doing so, proving the existence of the Theta Θ Axis function.

We have not, with Steps 1 and 2, proved that to you yet. We have shown you a phenomenon that is inconsistent with what we are supposed to believe: about making things up in our heads.

We are encultured to believe that anyone can make up anything. That children, from around 3 onwards, make up both small and increasingly intricate fantasies. It's what we pay poets, and novelists and Hollywood scriptwriters billions of dollars for.

It's what Professor Jordan Peterson, following Jung,

calls our "Maps of Meaning".[360] The idea that we have a-historical cultural archetypes, stories and meta-truths which are, yes, just "made up": but which we did make up.

Now, deep breath. Get ready for Step 3 of your *Sheeted Experience.*

Experience: Sheeted [cont]
Step 3:
We just need to clear your neural pathways.

So: that blank sheet from Steps 1 and 2:

- Throw it away;
- Tear it up, burn it. Drop it in the trash can. See it drift off in flame smoke, or wither into ashes;
- Make sure it is gone. It does not exist.

Now:
Look around you at the room you are in.
Take a few relaxed normal breaths.

---

[360] Jordan Peterson (1999)

Relax.

Ready?

Now:

- Imagine, there in the room, paperclip:
  which you know is not there

Stop.

Take a breath. Look around you. Relax.

<u>Discussion</u>:

(1) The paperclip appeared.

(2) It was automatic.

(3) It was as if, something reached over your shoulder, through your head, and stuck that paperclip right there in the room.

Step 4:

Look around you at the room you are in.
Take a few relaxed normal breaths.
Relax.

Ready?

Now: you're going to repeat what you just did

- Imagine, there in the room, paperclip:
  which you know is not there

Now:

- Move that non-existent paperclip around the room;
- Whatever colour it isn't, change that to some other colour it isn't;
- Whatever size it isn't, change the size:
  - Larger
  - Smaller
  - Wider
  - Narrower
- Make it move

Now:

- Make it disappear.

Stop

Discussion:

(1) The non-existent paperclip appeared, of course. Easy.

(2) As you worked with it, you became used to it.

(3) You could perform all the manipulations.

(4) Yet all the time you knew that the paperclip did not exist.

Please recall that we wished to prove begin to you that:

To Exist in Θ, means that (T) is extended
in objective internal spacetime.

and in doing so, proving the existence of the Theta Θ Axis function.

We will now interpret the results of this *Sheeted Experience*, so as to demonstrate elements of what we have proved.[361]

But first, we promised you an emotional and mentive rebalance.

Experience: Soothing
Just be in the room you're in.
Standing there in the room, relaxed
Feel your surroundings: just feel comfortable.
Breathe.

Now:

- See you, from a few moments ago;
- The younger you, who was having troubles with those *Experience* Steps;
- See younger you. Just look, from outside.

And:

- Feel your feelings of being relaxed;
- Calm;
- In control.

Then:

- Walk across the room, behind younger you;
- Put a hand on younger you's shoulder;
- If younger you turns to look at you that's fine.

[361] Or should that be "proven"? Why are there 2 alternative forms of the past participle of this irregular verb?

- Just smile, and nod;
- Take some breaths, holding younger you's shoulder;
- Younger you can leave now;
- Close your eyes, and allow younger you to become the past: gone.

Stop.

Relax. Breathe. Have some water, coffee, whatever.

Discussion:

(1) I know how you feel. It's nice.

(2) We don't need to say any more.[362]

Back to work. We now interpret the results of the *Sheeted Experience.*

(1) When you tried to imagine the clip, *which does not exist,* against a blank background: you just can't do it.

(2) Theta Θ is the existential domain field axis.

(3) Theta Θ assumes that any information presented to Theta Θ is *existential information*: information about something which exists.

(4) The Theta cashier sums and rules over things which are presented in your shopping basket. Theta cannot act on items which are not in your shopping basket.

[362] For more, though, read *Secret Self*. The Author (2020)

(5) You told Theta Θ at the outset that the paperclip *does not exist.*

(6) You thus instructed Theta Θ to run its rules over an empty shopping basket. Theta Θ is unable to perform that function.

(7) Your Scalar axes could not assist. They don't do concepts of existence. Those concepts are meaningless in your Subjective realm. Ask your dreams.

(8) As we proposed:

To Exist in Θ, means that (T) is extended
in objective internal spacetime.

(9) Figmenting two paperclips, *each of* which *does not exist*: did not help. Not against a blank background. Indeed, it made things worse.

To Exist in Θ, means that (T) is extended
in objective internal spacetime.

(10) Yet, the moment you sought to figment a clip, *which does not exist,* against a real, objectively real, background: the paperclip appeared instantly.

To Exist in Θ, means that (T) is extended
in objective internal spacetime.

(11) Now that non-existent clip was able to become a (T): to function as information:

extended in objective internal spacetime[363]

[363] We will have much more to say about spacetime in Chapter 7

(12) You see the room. You "know" it exists. You "know" it is real. *Esse is percipi.*[364] Or, to a Cromwellian materialist, it knows you are real. Your room provides a frame of local reference for all that which is an exemplar of Substance extended in spacetime: all of the (A) ≠ (nA) = [E] of the universe: all that *is.*

(13) The Theta Θ function requires a *plenum.*

(14) The figmented paperclip is now part of your mentated respresentation of, reflection of, that external extended Substance. The clip becomes part of your:

objective internal spacetime

(15) The clip, which does not exist, is nevertheless:

extended in objective internal spacetime

(16) The clip, which does not exist, is a (T) in information function in your *objective internal spacetime.*

(17) The clip is now an element in your *plenum.*

(18) What's more, that (T) in information is effecting that function in every one of the readers of this book. And everyone you prevail upon to undergo this experience.

(19) That (T) information is existential. It is an information datum (T) which does:

Exist in Θ

[364] Principles #3 *A Treatise Concerning the Principles of Human Knowledge* (1710) Bishop George Berkeley.

(20) Moreover, that (T) information is existential for everybody who does the *Experience*. That (T) information is in the *plenum* of that person's Theta Θ function.

(21) And not merely existential information, such as dog or cat, or a TV show. But it is objectively existential information (T): that (T) exists *in exactly the same way,*[365] for everybody who thinks of it. Anybody in the whole world.

(22) This (T) is a definite, bounded concept of something which explicitly *does not exist*. Yet it has become a "super-meme". It functions as an information packet in the objective mentation of everyone who thinks it.[366]

(23) *That (T) effects exactly the same information function in any plenum.*

(24) Thus: this (T) has become a *rule of language,* in effect.

(25) How is this capable of being so? Because:

$$(T) \neq (nT) = \Theta$$

(26) When there was no objective reality, merely a blank sheet, there existed no ≠ (nT).

(27) Once there was an objective reality, a *plenum* operating in mentation as *extended in objective internal spacetime,* then ≠ (nT) existed.

(28) Thus (⍰) could exist and be recognised in opposition ≠ to (nT).

---

365 As exactly the same information function

366 Against a background of reality. Try it against an imagined background: it will not work

(29) The Theta Θ cashier looked in the basket, and ran the Theta Θ Axis function:

(T)	≠	(nT)	= Θ
Non-existent Clip	Exists with	Room	I Am

Excuse me, the reader cries out. Where did *I am* come from?

Oh, we're sorry. Didn't we say:

The rules of language work in the Theta Axis Field.

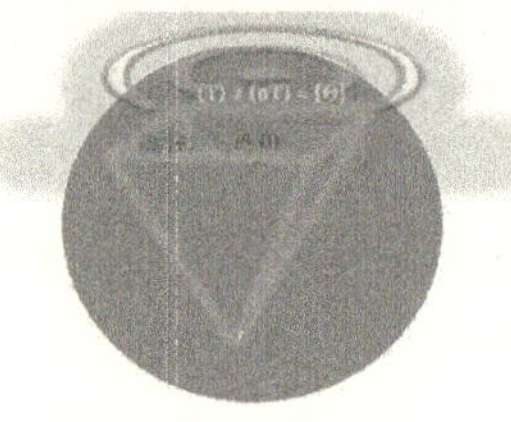

(T) ≠ (nT) = Θ

This is your "objective" Axis. This is the Axis that communicates *from* you to the outside world.

The connection is this: *it is Theta that puts the "is" into your language.*

We wrote "your". Which means: of you. And who, exactly, is you?

The rules of language are an expression of the identity of <I>: which identity depends upon the Theta Θ Axis Field function; which is a function that operates in objectivity:

(T) extended in objective internal spacetime, as Theta Θ

and
(A) extended in objective external spacetime, as Form
[E].

You don't believe this. That's understandable. Let's do another *Experience.*

Experience: Whovering
Look around you at the room you are in.
Take a few relaxed normal breaths.
Relax.

Ready?
Now: you're going to repeat what you just did

- Imagine, there in the room, paperclip:
  which you know is not there

Now:

- Just allow the paperclip to be there;
- Relax with it;
- If you feel it wants to move, that's fine.

Now:

- Think to yourself: *This paperclip does not exist;*
- Keep the non-existent paperclip there in the room;
- Keep repeating that phrase: *This paperclip does not exist.*

Stop.

Discussion:

(1) The paperclip appeared, of course. Easy.

(2) Look into the history of your head: as you repeated *This paperclip does not exist*: what word or idea came into your head?

(3) If you're not sure, go back into the paperclip room, in your head and repeat the *Experience*.

As you repeated that phrase, a totally true statement of the state of affairs, what else did come to think?

You thought:

<I>

Saying that phrase repeatedly was pushing you to say:

But I do exist

I exist

I

You were challenging your Theta Θ Axis Field function. You were trying to tell the cashier it doesn't know how to add up the Existence Value (EV) of the items in the shopping basket. And the Theta Θ cashier was insisting that, yes, it does know very well. And nothing that you have to say will persuade it otherwise.[367]

That is why the Theta Axis Field function is:

$$=\Theta$$

[367] Think about that one, and its consequences for opinion and belief

I Am

Just as:

=[E] is the existence of Substance (A), in opposition [≠(nA)], extended in objective external spacetime.

So as:

=[Θ] is the existence of information (T), in opposition [≠(n⍰)], extended in objective internal spacetime

To mix philosophical metaphors: =[Θ] is the cash value for <I> of information (T). It is the existential register of such information as exists (T) in the objective world that is <I>.[368]

You were challenging your Theta Θ Axis Field function. Telling yourself that the imagined paperclip:

does not exist

Rour Theta Θ Axis Field function insisted:

But I do exist

I exist

I

We have recreated the Experience which Descartes recorded, almost 400 years ago.

Cogito ergo sum:

I am thinking, therefore I am existing.[369]

---

[368] We are trying to avoid using the "me " word: that is a completely different concept

[369] *Principles of Philosophy*. Part 1, At 7 (1644) Published originally in French *Discourse on Method* (1637)

And, despite Decarte's best idealist intent, it was a triumphant recording of objective materialism.[370]

That observation was clearly correct: since you have just experienced it yourself. As does every adult human being over the age of 3, who undergoes the experience.

Which happens every moment of the waking day. You are using your Theta Θ Axis Field function in every waking moment of your life. The Theta Θ cashier is ever busy, summing the differentiated totals of existential information [(T) ≠ (nT) =] into language forms.

If Theta Θ is so busy in me all the time, you might say, why don't I notice it? We answer: for much the same reason that you don't notice your heart beating, or your circulation pumping: until you choose to notice.

But also for another reason: your life and mental health significantly depend upon you not noticing. As we will see in *Secret Self,* one of the operative elements of Mentive dysfunction, is dis-ordering of the relationship between /S/ and Theta Θ.

In whichever human individual the joy of purely subjective shopping (m̄, Φ) is being done, the Theta Θ cashier uses the same rules of accounting. The Theta Θ rules

[370] Although there is an argument that Descartes was actually trying to unite reality and consciousness, not sever them

are universal. Across every culture, society, and commuity. Across historical time. That which is universal in every sphere of human existence is objective.

To be a universal phenomenon is what "objective" means. We might say that universality is a necessary, but not sufficient criteria for being denoted as "objective".

Do you recall the scene from *1984*,[371] in the Ministry of Love:

> 'How many fingers, Winston?' 'Four! Stop it, stop it! How can you go on? Four! Four!' 'How many fingers, Winston?' 'Five! Five! Five!' 'No, Winston, that is no use. You are lying. You still think there are four. How many fingers, please?' 'Four! five! Four! Anything you like. Only stop it, stop the pain!'

O'Brien was working to make Winston deny his Theta Θ function. To do that, Winston had to give up his sense of <I>. With Winston's emotional sacrifice of Julia to the rat, in Room 101, Winston finally became free: of himself.

With the non-existent paperclip, you have just had a taste of Winston's agony. Would you like some more?

Experience: MinLove

- Look around your room;
- Find an object: let's say a chair;
- Focus on the chair;
- Keep your eyes open.

---

[371] George Orwell (1949) p221

Now, say to yourself:

- *That chair does not exist... does not exist... not exist;*
- Sat it over and over again;
- Try to make it so;
- Try to lift the chair out of its landscape;
- Try hard;
- Keep repeating to yourself ... *not exist.*

Stop

Now: look at the chair

- Say to yourself *I exist.*

Relax.

Stop.

<u>Discussion:</u>

(1) It started off with you having feelings of *so what.*

(2) As you intensifed the pressure, it became increasingly uncomfortable.

(3) The words ... *not exist,* were getting an echo in your head: *I exist...I...I*

(4) That's why we stopped you, then did a reset.

(5) When you affirmed your <I>, you felt OK again.

Theta Θ expresses the <I>, of you as a Self.

You were challenging your Theta Θ cashier. You were arguing that some (T) which, for Theta Θ, is in the basket, does not exist in the basket.

We can appreciate the amazing insight of Orwell. The

whole of *1984*, is filled with these astonishing insights into relationships between the workings of our Mentation, and the world.

We said:

> To be a universal phenomenon is what "objective" means. We might say that universality is a necessary, but not sufficient criteria for being denoted as "objective".

Fair enough. What is the necessary criteria? That it be necessary; inescapable; unavoidable; incapable of evasion.

Well, you've just tried, with all your mentative strength to evade the shopping basket filed with things which exist and a clip which does not exist: and you didn't succeed. You've just tried to tip an object in your room, which certainly does exist, out of the basket. You failed.

Once a (⍰) is in the Θ basket, you can't get it out again, before the Theta Θ cashier does the accounting.

Ductives

Theta Θ reasoning is deductive. Theta takes the (T) data presented. It equalizes, collates, sums and outputs.

These are all deductive reasoning modes. That's one of the reasons Theta Θ can operate so quickly: one iteration of a (T) ≠(nT) inequality works as an equivalence function, just like any other.

The /S/ axis fields work by way of induction. They gather in vast data, and sort qualitatively. They sum differences as scalar values, of course,[372] and doing so can involve deduction.

To summarise crudely, in order to provide useful pilot concepts:

- Theta is your accountant;
- Scalar is your artist and scientist.

We will consider your Emotional axis in Chapter 8. For now, we can by analogy say that this Ω field is:

- Your sense data processing centre;
- Your astrologer.

and, definitely, but counter-culturally

- Not your artist.

Plenumpotentiary

The Field [(T) ≠ (nT) =] of the Theta Θ Axis, is a plenum.

The word has had different meanings over time. What we mean is *plurality extended*. That there be a (T) and at least one antigone (nT), each being extended.

We don't use a special symbol for *plenum*. There's no point. Plena are everywhere that is somewhere. And somewhen.

[372] See Chapter 4

The opposite of a plenum is a void. But a true void: not just an interface gap between plena. But a void which has its owns existential conditions.

Experience: Vacant
Just be comfortable and relaxed.
You can keep your eyes open, or close them.

Recall how we defined a true void.

Now:

- Think of a void.

Read those 4 words, and go off into your head.

Stop.

Discussion:
(1) Can't do it.
(2) No matter how hard you try, no matter where you look.
(3) You just can't think of a void.

But there is something strange and emotional which was going on.

Go back into the *Experience*. You only need to revisit for a few seconds.

There's a feeling similar to when you have forgotten

a word, or a face of someone you've not seen in years. There's a gap which has reality in its absence. It is the antigone of a plenum. Not a blank sheet of paper. That just is, and is blank.

You can feel something deep down and formless. Your search for Void is pulling you there.

You're not wrong, and those feelings are justified. We're going to meet the Formless Void in the next Chapter.

So, we have our plenum. The Field [(⍰) ≠ (n⍰) =] of the Theta Θ Axis. The Field, in its operation, is universal and objective.

How you see the sitting dog, or how you hear the barking dog:

- Is different for everyone;
- Is even different for you, every time you see and hear, or think of seeing or hearing.

These are paradigms of subjective perceptions.

The Theta Θ Axis is the Equalizer. The plenum of information [(T) ≠ (nT) =] is summed as a form of information. That summation operates objective rules. Those rules are the same for everyone.

The Theta Θ Axis forms of course can appear as different

languages. But it is a commonplace of linguistics, and common sense, that specific language denotations (dog/ chien, hund, kutta, canis, skylos, cachorro, sobaka, alkalb, ineh) are contingents: accidents and evolutions through cultural history.

The *meaning* of the word is given by the plenum in which it is information (ᴛ). The Theta Θ Axis allows *definitions.*

Definitions are attributes of what exists to Theta. Which is not the same as what exists "out there" in the world. But, as the *Experiences* have shown: whatever exists out there has no function for you, unless Theta can account for it; and what Theta accounts for, exists for you, whether you like or indeed believe it, or not.

Definitions are not what words mean: because meaning is plenum specific: and plena change all the time. Because plena are the shopping baskets of subjective experiences (ᢪ, Φ).

Thus:

- We *understand* what idea each word, and word combination[373] means, by reference to a plenum of shared experience

but

- we *interpret* that objective meaning, only through our subjective shopping.

[373] If uttered under a rubric universalised in Theta Θ

This is the richness of Human Information Transmission: HIT.

This is how scriptwriters, novelists, and to a lesser extent poets[374] make their living. They create a Vortex of plenary experience. A pool into which we each can dip, according to our separate subjective plenum.

Remember, it matters not to the Theta Θ Axis Field whether that paperclip in your head exists, or not: in the world. It matters only that there is information functioning as (T) in a plenum.

Where that information function (T) came from, is irrelevant to the cash value of Theta Θ. The Θ Equaliser does not distinguish between whether a (T) comes from:

- Your sense perception (pain, pleasure);
- Your observation, apperception (seeing, hearing, touching, smelling, tasting);
- Your internal thoughts (whether generated purely by mentation, or be emotion).

Whether any (T) "represents" anything "real" [375] in the world, or in you, is irrelevant to the Theta Θ cashier function.

The disciplines of philosophy and linguistics appear not

[374] See about Memes, below

[375] However you wish to define any such "real"

to have grasped this nuanced, but ultimately elementary combination of processes.

However, the obvious outward appearance of these processes has inspired generations since Descartes and Leibnitz to conceive of a thinking machine. Indeed Aristotle himself had viewed his logic as having effect as αὐτός[376] reasoning.

A thinking machine would operate under the law of identity (A=A), *and thus in bounded infinity*. That would inevitably produce the *Entscheidungsproblem*.[377]

As an interesting side-note: why does the Theta Θ Axis Field function not generate that halting problem? The answer, of course is that: *it does.*

We may enter an alternative definition of the Theta Θ Axis Field function as: *that function which cannot define a sorting algorithm for its own Existence Values.*

Yet, here is the crucial difference:

- A computer can only compute data: which exists. A computer cannot compute a non-existent paperclip.
- Your Theta Θ cashier function does "compute" (T)

---

[376] *Autos*: with etymological root "the again", or "the again / return"

[377] Immerman, Neil, "Computability and Complexity", *The Stanford Encyclopedia of Philosophy* (Winter 2018 Edition), Edward N. Zalta (ed.), URL = <https://plato.stanford.edu/archives/win2018/entries/computability/>.

which can include non-references to any objective existence.

Thus, the Theta Θ Axis Field function includes as potential information function (T) the universality of your personal, individual, subjective lived experience.

And those experiences include all the subjectively existential antigones: all the paperclips which you know do not exist: Your *Plenumpotentiary*.

Before the A Priori

It follows, shortly and ineluctably, from the above, that ideas about the *a priori,*[378] are upside down.

The literature on the *a priori* concept is vast. It lends back to the time of pre-Socratic thinking, and runs as a thread through the tapestry of religious, then scientific, thought ever since.

Plato, for example, argues in his dialogues *Meno* and *Phaedo* that the learning of geometrical truths requires recollection of knowledge possessed by the soul in a disembodied existence before its possessor's birth, when it could contemplate the eternal Forms directly. And onto Descartes, Kant and the antagnism of Kripke.

378 We can trace the phrase to Albert of Saxony (1316-1390)

An enire book awaits for the dissection of the *a priori* debate, under the rules of Matrixial Logic. For now, let us jump to an assertive conclusion:

Theta Θ is an uninhibited field function

Which entails that Theta Θ is not a collection of *a priori* ideas. There are no such ideas.

Any *a priori* system must, by its definition, be an *inhibited domain*. The function of the *a priori* is to inhibit (to limit, place boundaries upon) an information infinity.

What may appear as if they are such *a priori* ideas, is merely the Mimetic[379] acquisition of Theta Θ operations by individuals. The Memes are found by experience to have utility, and so they are passed on culturally.

However, no such Mimetic information is inborn. Each generation must learn it anew.

IIow can we provc that no *a priori* exists? Becuause the Θ Equaliser effects its function without regard to the reality-referential, *or non-referential* quality of any (T).[380]

(1) Take an *a priori* inhibition (API): premises of geometry or arithmetic; cause and effect. Anything you consider to have the characteristics of API.

(2) You are able to express, as API, that which you

[379] Socially repeated

[380] That is a complete answer in itself.

claim to be API, only under the operation of the Theta Θ function.

(3) It is by reference to the "objective" output of the Theta Θ function, that you characterise any API as being AP.

(4) This is a classic metreological error. You are assuming the output of a function to define the terms of its inputs. That is fine in computing. But it is not how the Theta Θ function works.

(5) We have proved to you by *Experience* experiment, that your Theta Θ function operates on any (T) in the Theta Θ field, without regard to whether such information references or does not reference any reality.

(6) API is reckoned to be an essential underpinning of the application of reason.

(7) That which is such an essential underpinning cannot presume in its own premises that which is non-referential to reality.

(8) Ergo: API is illusory.

If one wishes to claim that a valid API can be underpinned by that which is non-referential to reality, then one simply empties the acknowledged form of API from any content. So, that move is not permitted within the rubric of reason, or indeed any API itself.

We comment that the API is merely a perception of <I>, distorted by reason. To know more about where <I>

comes from, see *Secret Self*.[381]

All human beings are born with a limiting function capacity. This is not the a priori. It is something else entirely. We will meet that limiting function capacity in Chapter 8.

We have traced the source of the API illusion to the operation of Mimetic abstraction from the Theta Θ function output. Let's go look at Memes.

Real Memes

"Meme":

> **1.** an idea or element of social behaviour passed on through generations in a culture, esp by imitation
> **2.** an image or video that is spread widely on the internet, often altered by internet users for humorous effect

From the Greek *mimēma* 'that which is imitated. Credited to evolutionary biologist Richard Dawkins:[382]

> Although Dawkins invented the term *meme*, he has not claimed that the idea was entirely novel, and there have been other expressions for similar ideas in the past. In 1904, Richard Semon published *Die Mneme* (which appeared in English in 1924 as *The Mneme*). The term *mneme* was also used in Maurice Maeterlinck's *The Life of the White Ant* (1926), with some parallels to Dawkins›s concept.

Thanks to Instagram and Facebook, we all know what a Meme is.

- Words can become mimetic: super; quantum;

381 The Author (2020)
382 https://en.wikipedia.org/wiki/Meme

- Logos can function as Memes: apple, nike, macdonalds; the image of the christian cross; a lightsaber;
- Mathematical shapes and symbols: arithmetic; geometric.

We understand, instinctively, how a Meme works: information (T) produces the same existential response in a population. Memes can cross cultures and language divisions. A Meme can elicit the same existential response in millions, billions of people. A meme can be universal.

Recall what we said some pages ago:

> It may seem strange, but your own *Experiences* have shown you that:
>
> - When you are seeing, you can't speak
>
>   and
>
> - when you are speaking, you are not seeing.
>
> If your world of mentation consisted solely of these Scalar axes m̄ /S/{E} and Φ /S/{I}, you would be effectively mute. You could see and hear, but you wouldn't be able to share that cognition with anyone: at least not under a set of mutually shared language rules.
>
> You need another capacity. One that:

- Coheres your Scalar thoughts;
- Uses language rules that are mutual with the outside world;[383]

[383] At least some part o fit

- Can translate your Scalar thoughts into those rules.

That is what your Theta axis = Θ does.

The first action your Theta axis = Θ takes is to decide the existence value ("EV") of the Scalar information.

That existence value is (T), in:

$$(\mathrm{T}) \neq (\mathrm{nT}) = \Theta$$

Now, we must tread carefully here. Theta is not deciding whether anything exists out there in the world: whether there is an external reference point for the Scalar data. That is not Theta's mission statement.

A device is used in creating a Meme. It is the same trick as created the illusion of API.

Once the Theta θ axis field function has completed a "run", the output is an equalised, objective output. Of course, such "runs" are being completed every moment of your linguistic existence (since around the age of 3). There is a universally boundless infinity of such runs.

We will call the result of such a run a *Potential Mimetic* or *PM*.

Any PM is capable of transmission and receipt by any individual who themselves has a capacity to conduct such runs and create PM. That conditionality is why pre-age 3 can't do language "properly".

In order for an PM to transmit across the gap between individuals $[I]^n$ and thus to become an *Actualised Mimetic* or *AM*, some contingencies must be satisfied.

It is, hopefully, self-evident that PM functions mimetically as a Node:

$$[(PM) \neq (nPM)] \neq [(\text{Context reference}) \neq (n\text{Context reference})] = <\sum AM^n$$

PM obviously begins its mimetic trajectory as a simulacrum of Substance =[E]:

(T) extended in objective internal spacetime, as Theta Θ

and

(A) extended in objective external spacetime, as Form [E].

Now, the scope of these elements of any Mimetic Node, is purely a matter of cultural contingency. Which is why language is alterable, and flexible, and able to be different in expression.

When Node meets a continuum in that infinity, then there is a wave interference event. Just like m̄ in the subjective domain.

From there, you can re-iterate the familiar process through Scalar /S/{I} to the Theta Θ axis field function. Such that =AM becomes (T).

So, what is the Mimetic device? What's the trick?

- Imagine, figuratively, a harp lying inside the flow of a river. Now, try to throw a PM stone, such that it actualises only one harp string [≠(-PM)]. Not easy, but not impossible in principle;
- Now multiply the number of stones thrown a trillion-fold. Statistically, there will be a small number of single-string hits;
- Now, imagine that the harp strings are moving with the river flow. That makes the task harder. But there are trillions of strings also: making the task easier again.

Now:

(1) That which is non-referenced in spacetime functions differently as (T), than that which has reference in spacetime.

(2) The greater the level of abstraction from reference in spacetime, the closer a (T) becomes to non-reference.

(3) That which is non-referenced cannot strike a single string: although it can shake the harp.

(4) By having the least reference to spacetime, AM can pass through the Subjective Scalar /S/{E}\{I} interfaces with least Subjective analysis process.

(5) That which is (T) in Theta Θ for H^1, becomes (nT) in Theta Θ for H^2, with only single-string modification through /S/{E}\{I}.

Let's look at some famous examples of such Memes:

(a) Is: and its tense iterations: was; will be

(b) Scalar iterations: up / down; big / small

(c) Geometry and Arithmetic.[384]

That's all the participles and all the adjectives and adverbs. In two Memes. Plus the Scalar derivations of the aggregate Scalar iterations.

<The> and <A> have only cultural, not linguistic, universality. Plenty of languages do fine without these linguistic concepts, such as some inflected Indo-European languages, for example.

The nouns: the names of things, are obviously a matter of cultural contingency. As are the signs (alphabets, pictograms and so on) in which we communicate Mimetic information.

We can assign names to whatever does or does not exist. The only requirement is that the thing and its name can, as a packet of information, function as (T).

Let's create a Meme together.

---

[384] Which are combinations of (a) and (b). See Chapter 4 *Forms of Inequality*

Experience: Mimec

Step 1:

- Look at these 3 images:

- You can name the middle one whatever you like. Or just call it: thing, line, shape, mark; etc
- Look at each in turn and while you look, say to yourself the name of each one;
- Do this 3 times.

Now, close your eyes after these instructions:

- Focus in your head on the 3 images, as they are, above;
- Just focus;
- Allow to happen what happens.

Stop.

Discussion:

(1) You began to feel the Orange and Dophin "slipping away", while the •/ stayed sharp.

(2) That felt odd. So you tried to bring the other images back into focus.

(3) But the more you did that, the more they slipped away.

Step 2:

- Put a blank sheet of paper in your head.
- Put the •/ on the blank sheet of paper.

Stop

<u>Discussion</u>:

(1) Easy.

(2) Unlike the *papercip that does not exist,* this •/ does exist.

(3) You saw it above, in this book. So it became (⍰) for you.

(4) It did not have to undergo inspection, or analysis in Scalar /S/{E}. Your Scalar operation was not seeking a "match" in any field of reference.

Step 3:

- Look around the room;
- Put the •/ into the room;
- Multiply copies of it •/ •/ •/
- Let those copies be in the room. Hanging in the air, on the walls, on the carpet.

Stop.

<u>Discussion</u>:

(1) Easy.

So, something is happening with this symbol. Let's take

that a little further.

Experience: Brandish

Step 1:

Remember this exercise, without You_2

OK: that was just a reminder.

Now:

- Whatever name you gave •*l*
- Think of that name. You can change the name now, if you like.

Now:

- With one •*l* there in the room, in the same spatial area that other actual things are in.

And:

- Put that Name into the room

Stop.

Discussion:

(1) You can't.

(2) The •*I* name behaves just like a "real" name that you are used to.

Step 2:

Notice You_2 again.

Now:

- (Same as before) With one •*I* there in the room, in the same spatial area that other actual things are in.

But:

- with You_2 watching You and the Room.

And:

- Put that Name into the room

Stop.

Discussion:

(1) Easy.

(2) Again, the •*I* name behaves just like a "real" name that you are used to.

Congratulations. We have created a Meme.

You know, just know, that you can pass •*I* on to anyone, and they will carry it with them in their heads: in Theta Θ.

You can get •/ into anyone's Theta Θ shopping basket. Whatever natural language they speak: young, old, PhD or grade school educated.

As we said:

(1) That which is non-referenced in spacetime functions differently as (T), than that which has reference in spacetime.

(2) The greater the level of abstraction from reference in spacetime, the closer a (T) becomes to non-reference.

(3) That which is non-referenced cannot strike a single string: although it can shake the harp.

(4) By having the least reference to spacetime, AM can pass through the Subjective Scalar /S/{E}\{I} interfaces with least Subjective analysis process.

(5) That which is (T) in Theta for H^1, becomes (n) in Theta for H^2, with only single-string modification through /S/{E}\{I}.

We said:

> This is how scriptwriters, novelists, and to a lesser extent poets[385] make their living. They create a Vortex of plenary experience. A pool into which we each can dip, according to our separate subjective plenum.

Meaning arises from the combination of harp strings sounding from each thrown AM stone. The genius of the Meme is that it has no intrinsic meaning. It is not that it is non-referent to spacetime, but that it is universally referent to any iteration of spacetime. Because it is so abstracted from the informational content of any Theta Θ

[385] See about Memes, below

Axis field operation.

That's why poetry has a special place. Poetry works by illuminating the spaces between strings. That which references one string to another.

Of course, there can be an emotional Meme,[386] as well as a Mentative Meme. The emotional aspect is reserved for *Secret Self*. What we can say here is: *a successful emotional Meme is one that does our feeling for us.*

So where does the string and indeed the river come from? Well, when a sperm meets an egg that it really gets on with…

Drawing Strength

You're sceptical. You follow the line of argument do far. But you want proof.

So, are you ready for something which may genuinely change your view of you and your world?

When we have been using You_2, we have been engaging your Theta Θ Axis field function. Its equaliser function does what your subjective, scalar /S/ cannot.

Your Theta Θ Axis field function is incredibly powerful.

[386] The technical terminology needs some adjustment in the domain of Ω

It is the function which allows you to give thought and voice to <I>. Without it, <I> is mute.

Are you ready to feel the power of your Theta Θ?

Experience: I-Pen

Prologue:

You must use whatever is your non-ordinary writing hand, in this experience.

- If you are right-handed, use your left. Your You_2 will come from your right;
- If you are left-handed, use your right. Your You_2 will come from your left.

Materials:

Get a pad of paper, limed or unlined;

Get a ballpoint pen (because it writes most clearly and easily).

Start.

Step 1:

- With your normal writing hand: write A B C on the pad;
- Do it with reasonable care and concentration. Not slapdash.

Stop.

Step 2:

- With your Other hand: write A B C on the pad;
- Do it best you can. Take your time.

Stop.

Step 3:

You$_2$

Imagine You$_2$

Connect

Feel the connection

Get used to maintaining the connection

Now:

- With your Other hand: write A B C on the pad;
- **Keep your You$_2$ connection going all the time;**
- Do it best you can. Take your time.

Stop

Rest. Let the You$_2$ connection drop.

Step 4:

You$_2$

Reconect with You$_2$

- With your Other hand: write a b c , on the pad;
- **Keep your You$_2$ connection going all the time;**
- Do it best you can. Take your time.

Stop.

Rest. Let the You$_2$ connection drop.

Here is a sample:[387]

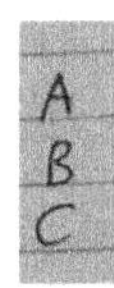

Normal Hand

Unassisted Other Hand

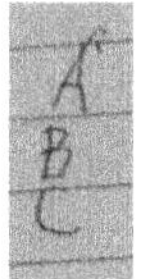

You[2] Other Hand

and, in a second round of practice, a few moments later:

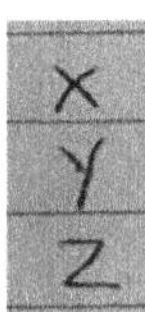

Normal Hand

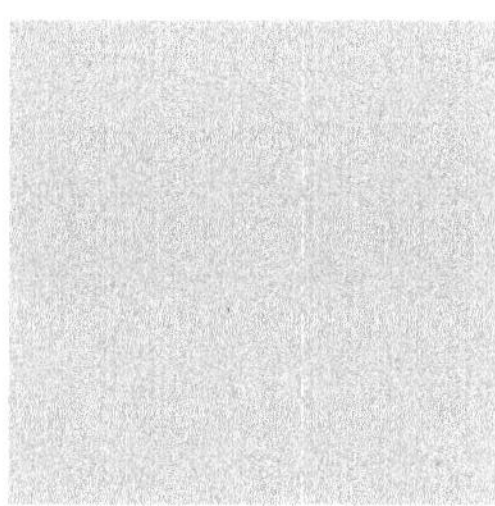

You[2] Other Hand

Another sample from a different individual:

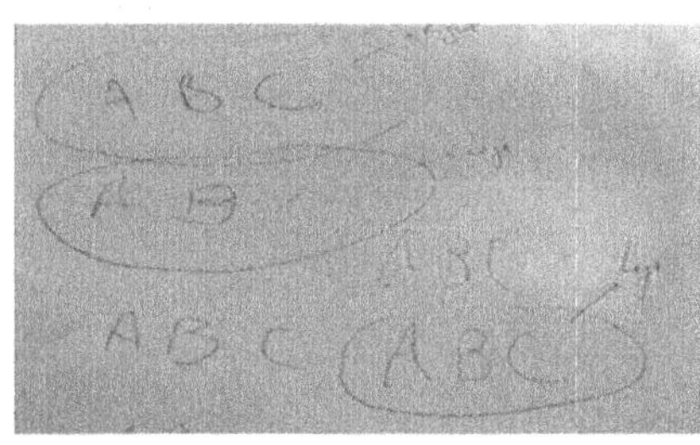

1. Normal Hand

2. Left: without You_2

3. Left: with You_2

The 3rd writing was performed moments after the 2nd writing.

[387] By a friend of the Author

More samples:

(3)

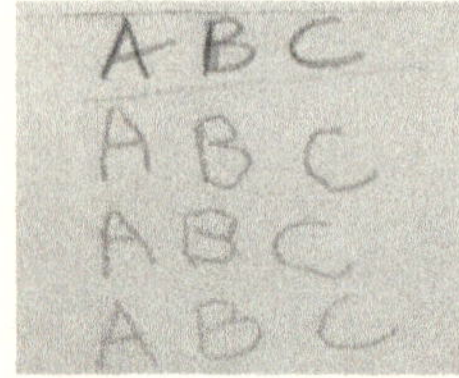

1. Normal hand

2. Left: without You2

3.1 Left: without You2

3.3 Left: with You2

(4)

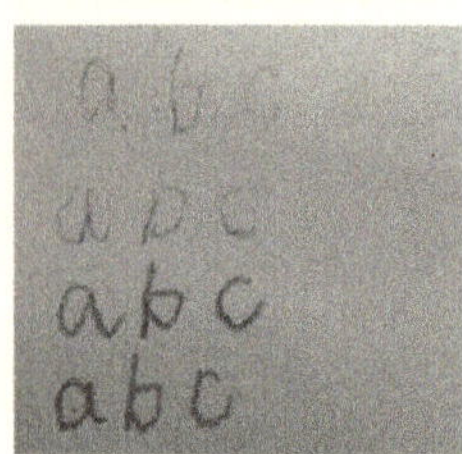

1. Left: without You2

1.1 Left: without You2

2. Left: with You2

2.1 Left: with You2

(5)

1. Normal hand

2. Left: without You2

3.1 Left: with You2

3.2 Left: with You2

3.3 Left: with You2

If you saw similar results, then you have had proved, before your own eyes, and in your own hand, the power of your Theta Θ.[388]

You may wish to dispute the interpretation of this phenomenon. You cannot dispute that the phonomenon which you have just *Experienced,* did happen.

You can learn more in *Secret Self.* We have more to say about the implications of *I-Pen* in Chapter 10.

Cabbages and Non-Existent Kings

We can return briefly to the energetic paper-filled study of Bertrand Russell at the beginning of the 20th century.

This was the syllogism of Russell's project:[389]

(a) Rationality provides truth-values.
(b) Language is rational.
∴ Language must provide truth-values.

This Chapter has now provided a firm foundation for understanding why, exactly Russell and all others engaged in this project, were wrong. Why the project was doomed to failure.

Language processing is comprised of two linked processes:

---

388 Now you also know how *Ouija Boards* work
389 And Frege, Meining, Godel

- Scalar Subjective /S/{E} \ {I}
- Theta Θ Objective.

HIT is comprised of the iteraction of thes eprocesses via the PM => AM Nodal function.

The Scalar axes /S/ are very much concerned with the truth-values of perceptions. Truth-value in the sense of referential tests against the external world, and by reference to our internal "barcodes".

Whether we are right or wrong about that truth-value, once an information packet gets "upstairs" as (T) to Theta Θ Axis field, the matter is out of the Subjective domain. The Theta Θ cashier does not care where the (T) ≠(nT) contents of the shopping basket came from, nor whether they have any truth-value in the world.

All that Theta Θ considers is whether (T) is acountable in the context of ≠(nT). As we said: Theta Θ only considers Existence Value (EV). Theta Θ outputs EVs as [PM]. Run task complete. Return to start.

Simple. Elegant. So powerful, Θ can alter your fine motor skills so that you can write perfectly with your other, usually non-legible, hand.

Our Scalar /S/ Subjectivity is rational. If by "rational", you mean *testing hypotheses of truth against reality, both external and internal.*

So yes, that Scalar /S/ *rationality* does produce contingent truth-values.

But, the axis of language, which functions under Memes, is Theta Θ. The Theta Θ Axis field function is utterly objective. But it has nothing to do with "truth-values". Theta Θ only considers Existence Value (EV)

So, the Russell syllogism fails at proposition (b).

Therefore, necessarily, the Conclusion is false. From that point, *rational language* theorists can fiddle around forever. They will never succeed. Because they are trying to get blood out of an unsanguine stone.

The forms of language are not *rational*. Yes, they have rules. Very simple ones, arising from the elementary mimetic content of all language: extrusions from the concept <I Exist>. But the most irrational of activities have rules. Rules are not the same as truth-values. Ask any computer.

What's perhaps most fascinating about these *rational language* theory endeavours, is this aspect. Try as hard as he might, Russell could not exterminate *The Present King of France is a non-existent Paperclip*.

The theory of denotation failed. *Walter Scott was the author of Waverley*. Of course, you would have some success with any theory which holds that the only important word is "the": precisely because that is the one word which has no intrinsic linguistic significance. Ask any Latin speaker. You are then merely inventing your own private pseudo-language, which embodies truth-values, and so is no kind of language at all.

Of course, *the present king of france* was as non-existent as the *paperclip which does not exist*. Niether had any truth-value. But the Theta Θ cashier doesn't care. Both function as (T) in a field including ≠ (nT). So they are linguistic units of account, which is all the Theta Θ operates on.

As we proved with *Experiences* above, once (⍰) is in your Theta Θ shopping basket, you can't get it out.

It's a good thing that *rational language* theory is wrong. Otherwise truth-values would stop us writing the Bible, Shakespeare and Harry Potter. Or designing the engineering plans for, well, anything. Or doing anything, which involved novelty thinking: for that which is novel by definition does not yet exist. Just like that silly present king of france.

Instant Cognition

We were labourious[390] in taking you through every layer of the linguistic cognitive process. We had to keep You_2 at a distance, otherwise, you would have short-cut the *Experiences.*

As your syncopated linguistic cognitive functions are supposed to: instantly.

There you go:

= L/T: ˆLion Is Roaringˇ

≈

"Lion Roars"

You performed that set of Scalar => Theta Θ equations literally quicker than the blink of an eye. Good thing, otherwise our ancestors wouldn't have got around to

390 Not enough for some readers, probably

producing the genetic line ending in us.[391]

And another:[392]

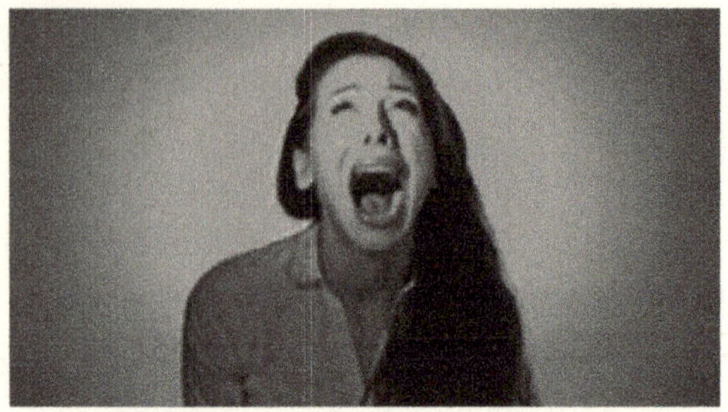

= L/T: ˆWoman Is Screamingˇ

≈

"Woman Screams"

Your 3H system is so fast, it takes a lot to slow it down. It takes even more to confuse it enough to take you through *Experiences* with FoR's that just don't equalise (so long as we keep You_2 out of the way).

I-dentify

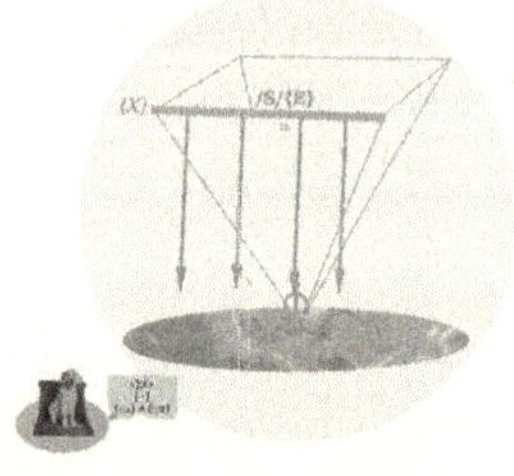

[391] No: we are not going into neo-Darwinian just so stories. Merely observing a basic fact of life.

[392] Re previous Fn: told you so

Recall how you could not put the name of an object "into the room".

But you could, in your head, name the object: vase, dog, lion, woman. In a millisecond.

Your *nominative* power, is a superpower. Of course, it can work so quickly because it just has the unidimensional Scalar axis to work in.

We know that newborns can recogise 😊 of their primary carer. That is the instant recognition, which works in *nominative* power.

The speed is helped by the way *nominative* power functions: as a fixing of types, of categories. You don't have to keep re-inventing some nomination function to know, instantaneously, what this is.

You see the thing. As we previously said: Our default setting with the world is: wordless.

But if you need to exercise the superpower nominative

function, you can do it, instantaneously.

Like knowing which button to press on the vending machine to get a latte. You know. You don't need to use what you know, until you need to. When you do need, there's no "working out" as if there was a problem to be solved, a jigsaw puzzle to be assembled. See: press: done.

This naming takes place in the Scalar /S/ interface between you and the world.

That interface is definitely not a direct interaction. Remember, that /S/ deals only with the collapses of and interferences in Δ^n by Nodes: ₥ and Φ.

How and where that happens, is for Chapter 8.

It's important to appreciate that the Lion and the Woman become part of your /S/ existential universe: its FoRs. They do so instantly. However, the Lion and the Woman do not become part of your grammatical universe, until they are entered in the Theta Θ shopping basket.

So: that something exists for you, is not the same as something being an <Is> in grammar. Indeed, it takes toddlers 18 months or so to start making that connection.

To say that you interface with the world through your hands, differently to your feet, is not to become weirdly metaphysical. It is not to enter into fantastical domains of hyper-consciousness. Nor idealist Descartian other worlds.

It is *organic materialist* common sense: which any 7 year old would agree with.

So, to talk about "Scalar" means that we use different functions in our real interactions with the real world. "Scalar" is logical thought about how one of those functions works. There is not some special Scalar part of your brain that you will find in an fMRI machine.

Scalar is one expression of the 3H's. We can usefully think separately about different elements of the 3H's: because the 3H's are a dynamic harmony of elements. But, no matter how hard you look in eggs, milk and cheese, you will not find the recipe for an omelette.[393]

So, now let's crack the Noun equation.

We started with:

(x) <∑D*> = /I/ ┘

Now, we can see that this formulation is missing something:

Node <∑D*> => [ⲯ / Δ^n] => /I/ ┘

[393] These words are directed at uber-materialists, who want to believe that they think that you can

We know that:

<∑D*> is [Abstracted information about dog in landscape / context]

We know that we carry a vocabulary:

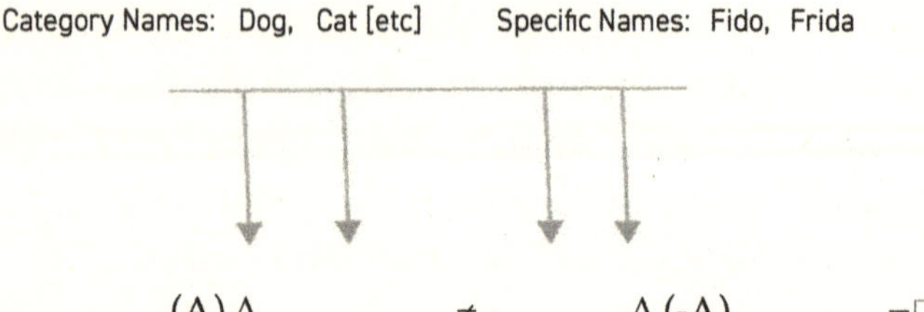

(A) Δ ≠ Δ (-A) =□

These are all extrusions from Subjective Existence.
The Scalar Plane of Existence is not a google map which says "I am here" (in time). Rather, it is a time line in infinity.[394]

So, this is why the first equation line of the formula:

<∑D*> [⁺ጠ] ≠[⁻ጠ] / ∑[ጠ] = /I/ ┘

is a simple identity.

That's why you had such trouble trying to manipulate that identity, in the *Exercises*:

= /I/ ┘ is Isomorphic
to the Paramorphic <∑D> Node.*

394 More on that below, and later

That is[395] why you can't put the word into the room, for example.

= /I/┘ is not Isomorphic to
the contents of forms which are active in the Node.

That's why you can't extract the dog from the background, for example.

And, because it is a simple identity, that's why you can identify *Lion Roars; Woman Screams*, so quickly.

Now, there is another reason why such identities don't cross the gap from you to the landscape. We will get to that in Chapter 7.

Barking the Verb

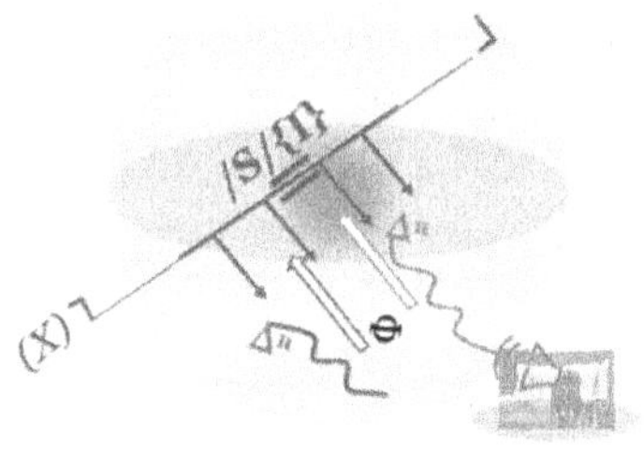

[395] Part of

Recall how:

> Thinking <∑D*> and <barking> and / exists/ all at the same time

> Why your black and coloured curtain worked

> Why ⅃ [barking] ¬ hangs in the air with curtain

When you were asked *Now try it with* You_2, that worked. The experienced problem went away.

The reason is simple: you operate with language[396] in a syncopated system of cognition. It's a division of labour. To call it "parallel processing" wouldn't be apt, because the system works as a whole: holistically, all the way through the process.

What we were doing in these Experiences, is separating elements of that whole. That's why you had problems: the language system only works with both.

Baby Talk

To know what we are talking about, it is invaluable to have a resource which tells us what babies do actually talk about, and when.

---

[396] And much more besides

Just such a resource has become available since 2014.[397] This article provides such an excellent summary of this important CDI work. that it repays extensive quotation:[398]

> The CDI allowed researchers to start to understand the full range of kids' early vocabularies, how they grow, and how they are tied to other language abilities. An early CDI study, published in 1994, of 2,000 24-month-olds showed that at that age, "normal" vocabularies range from fewer than 50 words to 600 words, with the median at 300 words. Everyone knew there was variability, but that much variability "was big news," says Virginia Marchman, a Stanford research scientist who serves on a nonprofit board overseeing the CDI.
>
> In 2014, a Stanford professor, Michael Frank, approached Marchman. He told her he had a bunch of CDIs from a previous study taking up space in his filing cabinet. She did too. They decided they wanted to build a tool that would make all that information easily searchable and accessible to other researchers and the public. The result is Wordbank, which now consists of more than 82,000 CDI reports in 29 languages and dialects. An initial analysis of Wordbank data was published online in January.
>
> If the CDI showed how variable children's early vocabularies are, Wordbank reveals that those vocabularies also have consistent themes. Seeing these themes makes first words more interesting as a phenomenon than as any single instance. Infants tend to talk about more or less the same things, no matter what languages they learn. Across 15 languages, they prefer to say and tend to understand words about sounds, games and social routines, body parts, and important people in their life. Words learned early in one language tend to be learned early in other languages. In American English, the 10 most frequent first words, in order, are *mommy, daddy, ball, bye, hi, no, dog, baby, woof woof,* and *banana*. In Hebrew, they are *mommy, yum yum, grandma, vroom, grandpa, daddy, banana, this, bye,* and *car*. In Kiswahili, they are

[397] http://wordbank.stanford.edu/

[398] https://www.theatlantic.com/family/archive/2019/04/babies-first-words-babbling-or-actual-language/588289/

> *mommy, daddy, car, cat, meow, motorcycle, baby, bug, banana,* and *baa baa.*
>
> One reason for this consistency is that such words rank high in a trait researchers call "babiness," which simply means they're words that have to do with babies, their immediate surroundings, and important, concrete things. They are often words babies hear frequently.
>
> But another reason for the consistency is that babies tend to learn words that help them interact with their parents and caregivers. "Kids want to share things; they want to be part of the social mix," Frank told me. *Hi* is the first word for a lot of kids. *No* is also a frequent first word. (In an earlier study, Frank found that *no* was more often a first word for younger siblings than firstborn children.)
>
> ...
>
> It also appears that 1-year-olds in most languages tend to say and understand more nouns than verbs, and use many fewer function words (such as *the, and,* and *also*), even though they hear function words frequently. Two exceptions are Mandarin and Cantonese, where children say more verbs, probably because those languages allow speakers to use a lone verb (*run*) to stand for clauses that in other languages require subjects or objects (*he runs*).
>
> ...
>
> Yet the overarching theme of Wordbank is variability, no matter the language. This suggests that no culture, no family structure, and no social environment has some special sauce that will turn out speakers or signers of a particular type. Everywhere, kids are "taking different routes to language," as Frank puts it.
>
> The father of two, Frank finds this liberating. "Parents tend to assume that variations they observe in their child's language are due to specific parenting decisions that they've made. But children vary so much that small variations in parenting will usually come out in the wash." Major differences in language input will still be consequential, but others, like reading one book or two before a nap, will barely register.

While immensely valuable in itself, what stands out immediately is that the prediction of Matrixial Logic: that

Nouns would come first in language, is borne out by the empirical data:

> It also appears that 1-year-olds in most languages tend to say and understand more nouns than verbs, and use many fewer function words (such as *the*, *and*, and *also*), even though they hear function words frequently. Two exceptions are Mandarin and Cantonese, where children say more verbs, probably because those languages allow speakers to use a lone verb (*run*) to stand for clauses that in other languages require subjects or objects (*he runs*).

As we predicted, and proved through *Experiences*:

So, this is why the first equation line of the formula:

<∑D*> [⁺巿] ≠[⁻巿] / ∑[巿] = /I/ ┘

is a simple identity.

That's why you had such trouble trying to manipulate that identity, in the *Exercises*:

= /I/ ┘ is Isomorphic

to the Paramorphic <∑D> Node.*

That is why you can't put the word into the room, for example.

As we said at the start of this long Chapter:

Known Unknowns

We know that HIT involves some kind of iterative framework. But it is not merely repetitive. It is heuristic. By doing, we learn.

HIT is also hylomorphic: meaning derives from forms both constrained by and expressing contents.

Another obvious attribute of HIT is that it is holistic. Not merely in the life experience of the individual as its is expressed in a present moment of communication. But also socially holistic, and indeed environmentally holistic.

- Heuristic
- Hylomorphic
- Holistic

The 3H's: this looks to be an obvious part of the Matrixial Logic landscape.

We know that newborn babies effect HIT. There is a vast literature on the development path from crying, movement, and eye contact, through to first words, and beyond.

How, exactly we subsume the stages of that development path under a model of language, and indeed whether we can do so, have remained open issues.

So we know that:

- There is a developmental environment from birth to full language acquisition;
- There is a (perhaps surprisingly) limited set of phonetic "building blocks" in any language;
- Newborns need to effect HIT and do so

- newborns are able to discriminate between authors of HIT.[399]

Let us know see whether we can capture the 3H's operating as HIT.

We are not going to add to the length of this chapter providing a detailed account of baby talk development processes, from birth[400] to mature infancy (at 7 years old). That is reserved to *Secret Self.*

We have sown lots of clues throughout this Chapter. What we can present, in summary, which reflects what you have learned from your own *Experiences,* is:

(1) Your early language (mama, daddy, doggy) is focused on nouns.

(2) That is isometric Scalar analysis. It takes time to establish the "barcodes" in the Scalar domain.

(3) Once established, the Scalar recognition process is instantaneous.

(4) By around the age od 2, baby has come to be aware of "me". But <Me> is not an organiser. It is an acquirer and collector.

(5) It is that amazing growth period between the onset of <Me> and around the age of 3, that child becomes aware of dentity, of <I>.

[399] Whether for linguistic or neuro-biological reasons is obviously difficult to discover

[400] Snd before

(6) That formation of <I> is accompanied by, expressed in, and advanced by, the integration of the (born in us) Theta Θ Axis Field, and (developed within us{ Scalar axis fields.

(7) This is what allows the Memes of grammar to be applied to the diffuse collection of Scalar ideas acquired by the child.

Through the *Experiences,* you have yourself travelled through some of the mentative space that a child cross, in their journey from <Me> to <I>.

The *Memory Palace of Matteo Ricci*[401] may be well-known to readers from its uses by Hannibal Lecter and Sherlock Holmes.[402]

The Palace technique serves as a useful illustration of the power of the Theta Θ axis field. It is in that field, that the memory palace is constructed.

We need to be careful though. The Palace is not an archive of data: it is a set of instructions for encoding information handled in the /S/ field. It's like setting up a "5 items or less" super-fast checkout lane for your Theta Θ shopping basket.

You have seen for yourself the power of the Theta Θ Axis

[401] https://artofmemory.com/wiki/The_Memory_Palace_of_Matteo_Ricci

[402] In the BBC Benedict Cumberbatch iteration

Field function.

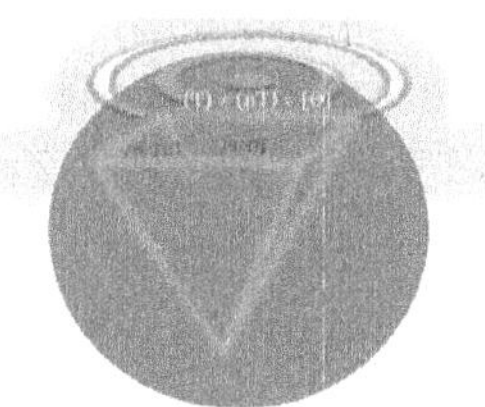

Language is the expression of layered processes:

- Nodal and Wave information acquisition;
- Interference in standing waveforms by ₼ and Φ;
- Creating Δ^n Scalar reference points in {E} and {I};
- Operating in the Subjective domain;
- Producing (T) information in the Theta Θ Axis shopping basket;
- Allowing PM standard (regularised) information to be received as AM.

All of which is Human Information Transmission.

A process which is:

- Heuristic;
- Hylomorphic;
- Holistic.

Cogito Ergo Someone

Inkeeping with the core ideas of Matrixial Logic, this Chapter has been multi-dimensional.

We have sought to use the applied problematic of language, to show how Matrixial Logic allows us to break down the problem. Then, from that new starting point, to build up the Matrixial layers: until we reach a functioning 3H system.

The *Experiences,* at each point, help the reader to embed key ideas. As the Author is aware, from working with "focus groups", these *Experiences* explain what mere words and equations cannot.

Recall the *Experience* in which we had you trying to deny the exist-ness of a non-existent paperclip, and of a very real item in your room.

You felt *Ego sum,* resisting. Shouting back.

You also felt the place from which we asked you to produce the non-existent paperclip: the Void.

Now, we take you on a Matrixial Logic tour of that void: the C Logic of Chaos.

CHAPTER 6

CHAOTIC ORDER

This brief Chapter provides a respite from the structural rigours of the preceding Chapters.

Way back in Chapter 2 *Framework,* we briefly touched on C Logic: Chaos.

> Chaos is that space in which no substance or moment solves under a form.
>
> So, we have "half" a formula:
>
> $$(A) \neq (Z^n)$$
>
> More formally, we can write:
>
> $$\frac{(A) \neq (Z^n)}{C}$$
>
> where the denominator merely expresses what we already know from writing $\neq (Z^n)$. In other words, that may be iterations of an (A) going on, but we have no idea what (if anything) one (A) has to do with (Z^n).
>
> We find content for C by exclusion. There are some terms which just won't solve in E or B. Those, by definition, are C terms.

> C terms are things that we know we don't know.[403] We can hypothesise that there are things which we don't know that we don't know. Since any such hypothetical things cannot enter any account of knowledge,[404] we don't need a category of logic for them.

Chaos is a difficult subject of enquiry, whether under the rubrics of logic, or otherwise. It is only on an exclusionary basis that we can only approach understanding, much less a definition of, Chaos.

We can begin here. Suppose that we had the scientific understanding to answer every question that could be asked: Chaos would be that which still remains.

How can it be that any residuum could survive universal knowledge? Because the means of knowledge are by definition non-exhaustive. Any (A) requires its antigone.

It is for this fundamental reason that all historical claims to creation of a universal logic, even the quest of modern material science is engaged in ultimate failure. Not failure in a project of expanding human understanding: that succeeds. But failure to exclude that which is unknowable, because the conditions of knowledge presume it.

---

[403] In the sense that we can't subsume them under a FoR.

[404] Which is what philosophy is

Imagine that you could be super-god. That you could see everything that was or will be. Yet you cannot in that frame of reference see that which is your seeing:

$$(God) \neq (nGod) = Witness\ [I]$$

This, you will notice is the Aristotelian Universal Affirmative (the ML Categorical Universal).[405] It is the immanent truth of all universals.

This is one perspective on Chaos: a definitional approach. We can also adopt the functional approach. Chaos enters as a category of our life from the moment we are born.

It is the struggle against Chaos which creates a succession of Singularities, which shape the essence of what it is to be human. We will say much more about this in *Secret Self*.

This is not saying that our emotional persona is correlate to Chaos. It is our whole being which enters into the primordial struggle with Chaos. That being is imaged in Chapter 8.

This brings into focus a common misunderstanding. The emotional self is not Chaos. It is not chaotic. On the contrary, it is a perfect self-ordering heuristic system.

Our emotional self is necessarily holistic with our

---

[405] see Chapter 3

Subjective and Objective mentation.[406] But, in the supreme irony, our mentation is para-hylomorphic to our emotion. One can reflect and influence the other. But our thinking is not determined by our feeling.

Our emotional self functions in dynamic harmony. A succession of states of disorder, mediated through order, back into disorder. Our appreciation of beauty is the very bridging of disorder by order: which is why beauty always carries the tinge of the dangerous. This seems like a contradiction: we say that emotion is both:

- A perfect self-ordering heuristic system;
- A succession of states of disorder.

The contradiction is only between frames of reference of the phenomenon:

- Emotion self-regulates;
- That self-regulation does not appear to mentation as order;
- Because mentation recognises only that which is of mentation as order

That contradiction is at the core of our being, as human. We are both passenger and driver.

Here, we begin by exclusion, to pierce the veil of Chaos. In Chaos, there is no passenger and no driver. Chaos is

[406] these are bridge ideas, expanded in later Chapters and integrated fully in *Secret Self*

neither directed nor experienced directly. We experience Chaos only through the bridge of Singularity.

Singularity is the antigone of Order. We certainly recognise Singularity and we can trace the ordering of systems to the function of Singularity.

Recall Chapter 2:

We denote an Si Bridge like this:

$\frac{\neq}{\neq}$	=	?
Disjunct	Bridge	Form

Let us begin to analogise these admittedly esoteric ideas. In the next Chapter we will consider the nature of time, in Matrixial Logic.

There, we will have occasion to consider briefly the ideas of Loop Quantum Gravity. The details need not detain us here. LQG is the latest in a series of investigations which take as a datum zero point energy. The idea that the vacuum of space is actually a plenum: that it is seething with energy populated by matter / anti-matter mutual annihilations.[407]

LQG is investigating a Loop version of the plenum

[407] First formulated by Dirac (1902-1984). *Principles of Quantum Mechanics* (1930) is sublime as poetry

"fabric". That is intriguing, because the loop (as a torus) is a Mandlebrot feature, implicated by the relationship between Phi Φ and Pi π. Such loops will therefore necessarily exhibit [408]recursive similarity, which entails infinity implication in any formula for differences within that system.

That aside, taking ZPE or LQG at face value, Chaos is what lies beyond the hypothesised plenum. It would be from Chaos that the energy is derived which is "borrowed and repaid" in the matter / anti-matter collisions.

Chaos of course is not "in" time. There is no analogue of Theta Θ in Chaos. Singularities do, of course bridge into time, and are in part defined by their subsistence in the "gaps" between $[T]^n$ states.

A more real-world analogy of Chaos is weather systems. Science has become proficient at weather modelling. Even so, it is acknowledged that such models only approach the unknowable.

As we note in more detail in Chapter 10: suppose we were to draw together all of our scientific method and knowledge, and visualise it dimensionally. Proportionally all that would bear as an atom to the scale of the universe.

---

[408] Godelian

Hubris is an essential emotional component of the pursuit of science. But the forms of knowledge are assessed in logic, not triumph. Quantum theory informs us of th definite limits of what we can know. Yet it is the limiting which has proven so successful in creating the post-war technological civilisation we inhabit.

All of which prompts the understandable question: why does Chaos matter? The significance of Chaos is that it expresses the boundary condition of order.

This explains the intractability of Maxwell's Demon.[409] The idea was explained brilliantly by Charles H. Bennet in 1987:[410]

> One manifestation of the second law of thermodynamics is that such devices as refrigerators which create inequalities of temperature, require energy in order to operate. Conversely, an existing in equality of temperature can be exploited to do useful work - for example by a steam engine, which exploits the temperature difference between its hot boiler and its cold condenser. Yet in 1871 the Scottish physicist James Clerk Maxwell suggested, in his Theory of Heat, that a creature small enough to see and handle indi vidual molecules might be exempt from this law. It might be able to create and sustain differences in temperature without doing any work:
>
> "...if we conceive a being whose faculties are so sharpened that he can follow every molecule in its course, such a being, whose attributes are still as essentially finite as our own, would be able to do what is at present impossible to us. For we have seen that the molecules in a vessel full of air at uniform temperature are moving with velocities by

409 https://en.wikipedia.org/wiki/Maxwell%27s_demon

410 https://ecee.colorado.edu/~ecen5555/SourceMaterial/DemonsEnginesAndSecondLaw87.pdf

no means uniform... Now let us suppose that such a vessel is divided into two portions, A and B, by a division in which there is a small hole, and that a being, who can see the individual molecules, opens and closes this hole, so as to allow only the swifter molecules to pass from A to B, and only the slower ones to pass from B to A. He will thus, without expenditure of work, raise the temperature of B and lower that of A, in contradiction to the second law of thermodynamics."

The "being" soon came to be called Maxwell's demon, because of its far-reaching subversive effects on the natural order of things. Chief among these effects would be to abolish the need for energy sources such as oil, uranium and sunlight. Machines of all kinds could be operated without batteries, fuel tanks or power cords. For example, the demon would enable one to run a steam engine continuously without fuel, by keeping the engine's boiler perpetually hot and its condenser perpetually cold.

To protect the second law, physicists have proposed various reasons the demon cannot function as Maxwell described. Surprisingly, nearly all these proposals have been flawed. Often flaws arose because workers had been misled by advances in other fields of physics; many of them thought (incorrectly, as it turns out) that various limitations imposed by quantum theory invalidated Maxwell's demon.

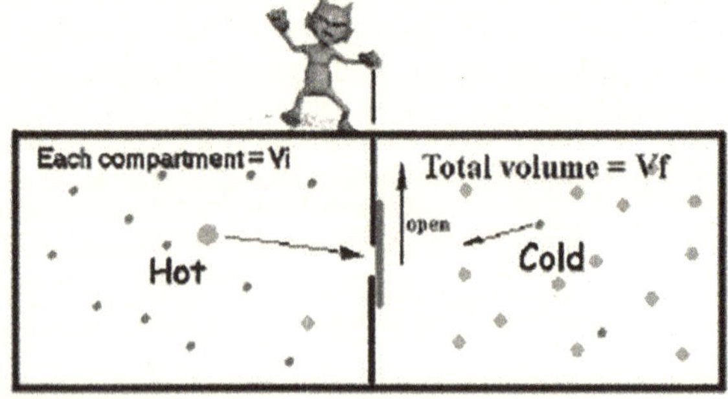

https://www.researchgate.net/figure/Maxwells-Demon-operating-the-gate_fig1_263731145

The Second Law of Thermodynamics ("SLD") is a useful example of Chaos logic. SLD emerged from, and is still confirmed by, observations of the interaction of particles and forces (matter and energy). Putting it over-simply, the proposition is that you can't cheat the cosmic accountant. There is a unilinear direction of (S) entropy >increase.

However you assemble particles and forces, you can only restrain entropy by adding energy (or information, which is itself an energy cost)[411]: and each energy addition adds to $\sum S$.

We see in Maxwell's Demon a wonderful exemplar of the logic of Chaos. Classical physics and quantum physics, of course, view S very differently. But both systemics revolve around the idea that information costs energy, and thus perfect information cannot exist because we cannot have perfect energy.

This reveals something important about the practice of science, since we emerged from the late medieval. *Science uses forms of inequality*. That is exactly the function of the Demon: it is magically equalising thermal state inequality, without the expenditure of energy.

The Demon turns an inequality state system into an

---

[411] The Bennett theorem

equality state system: a Scalar plane.[412] Science holds such transformation to be contrary to fundamental physical law.

We must here draw a distinction between:

- Scalar *measurement,* which is merely an accounting device
- Physical extrusions from a Scalar plane.

The former functions "as if": the latter as "is". Science is an encyclopedic of "as if"'s. Science runs into difficulties when it tries to operate with "is" extrusions.

Experience: Hover

Setup:
Get your by now familiar blank sheet of paper;
Get ready to imagine a load of paperclips.

412 See Chapter 4

Step 1:

Add 3 paperclips thus:

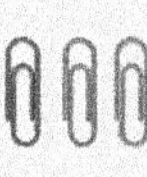

- Count them in your head.

Stop

<u>Discussion</u>:

(1) Easy enough for a 3 year old.

Step 2:

- Add another paperclip.

But, this time:

- Make it hover in the air over the others.

Now:

- Count them all in your head

Stop.

<u>Discussion</u>:

(1) You knew, before you started counting, that the answer is <4>

(2) Yet, when you got to the arial papercip, you paused.

There was a hesitation, even if only briefly.

Yet, these are all just paperclips. Every of these <4> is paperclippy. Except suddenly in your Mentation: when they are no longer all just iterations of a universal: there's something different about the arial one.

Experience: Bridge

Setup:

Now, place two sets of these 3 clips, arranged like this, on your white sheet:

Step 1:

- Count them in your head.

Stop.

Discussion:

(1) Easy enough again.

Step 2:

- Add another paperclip.

But, this time:

- Make it hover in the air over the others.

Now:

- Count them all in your head.

Stop

Discussion:

(1) This time, the count was unpaused: uninhibited.

(2) You raced through to the answer <7>

Why so? Because you had Maxwell's Demon to help. The gap between the two sets of these 3 clips made the 7th clip an iteration Δ^n of the count.

That gap re-ordered the relationship between the airclip and the others.

Hence the unhealed rupture between mathematics and emipirical science. Mathematics is par excellence, the derivation of one form of Scalar existential extrusion from another. Hence Turing and Godel.

The reader perhaps began this Chapter with an expectation that there could be little to say about any logic of Chaos, without recourse to something terribly new age, or deistic.

Instead, we find that the logic of Chaos informs the most fundamental underpinnings of science.

Experience: Lines

Setup:

Arrange your 6 paperclips thus:

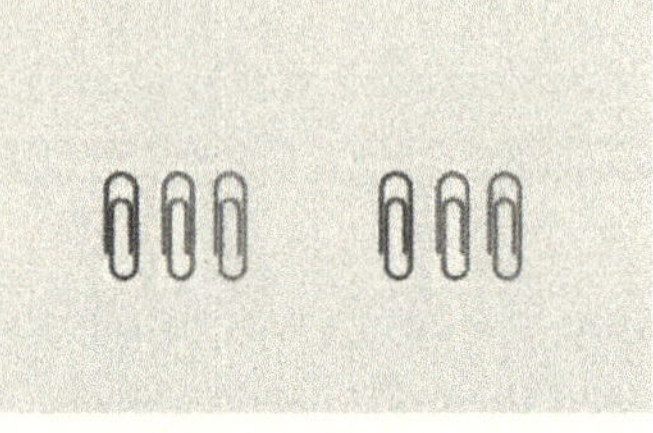

But now: see them in a 3D perspective: like in a 3D Google Map.
Tilt the imaginary table that the sheet is one, away from you, or towards you.
At any angle, even 90 degrees.
The paperclips are stuck on, so they won't move of all off.

Step 1:

- Hover your airclip over the gap;
- Any height away you like.

Now, in your head:

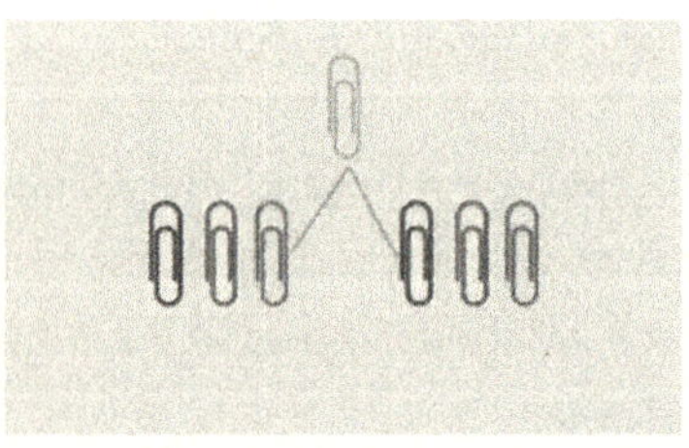

- Draw a straight line between airclip and one side of the gap;
- And do the same for the other side

Now:

- Move those lines to the corresponding edges of the two sets of paperclips i.e. out to the far left, and far right, respectively

Now:

- Move those 2 lines back to where they were

Stop

Discussion:

(1) The height of the airclip didn't move.

(2) Instead, you adjusted the angle of the lines

Maxwell's Demon helping again. This time re-ordering 3 dimensional space, so that the 3-clip sets can function as iterations of a Scalar measurement of physical extrusions from a Scalar plane.

Experience: Float

Setup:

Put your *Lines* table back to the flat.

Step 1:

- Float your airclip, free-form;
- Make it dance around like a Tinkerbell fairy.

Now

- Bring your airclip to hover exactly over the gap.

Stop.

Discussion:

(1) This was really difficult ;

(2) So long as the aircip was free, you couldn't locate the gap. You could barely move it towards the gap;

(3) So you adjusted your focus. You saw airclip and clips \ gap together, then froze the airclip;

(4) Then you kind of brought the airclip over the gap;

(5) But as you were doing that, the whole thing started phasing in and out of your focus: it was like manipulating ghosts.

That's exactly what you were doing. You lost the Scalar /S/ equalisation function. You were trying to order Chaos.

We can perhaps more illumine the dimension of Chaos by investigating the Demon in formal ML.

First, the problem (Author emphasis):

> a being, *who can* ***see*** *the individual molecules*, opens and closes this

> hole, so as to allow only the swifter molecules to pass from A to B, and only the slower ones to pass from B to A. He will thus, without expenditure of work, raise the temperature of B and lower that of A, in contradiction to the second law of thermodynamics

Obviously, as stated, the proposition is a circularity:

$(c^{1}) \neq (-c^{n})$ = Speed [I] : Reflexive Negation

|

$(s^{1}) \neq (-s^{n})$ = *C* [I]: Categorical Universal

|

$(Speed^{n}) \neq (nSpeed^{n})$ = Demon [E]

We are simply in the realm of the ACS:[413]

Ackrill/Aristotle Categorical Sentence Transitions to ML

	CS	ML Equation	Type
A	Every A is B	$(A) \neq (-A)$ = B[I]	Categorical Universal
E	No A is B	$(A) \neq (B)$ = N[E]	Reflexive Universal
I	Some A is B	$(A^{n}) \neq (nA^{n})$ = B[E]	Reflexive Opposition
O	Not Every A is B	$(A^{n}) \neq (-A^{n})$ = B[I]	Reflexive Negation

Table 3.6

The molecules are travelling at different speeds:

$(c^{1}) \neq (-c^{n})$ = Speed [I]: Reflexive Negation

This is obviously a form of speed inequalities, as Moments Δ^{n} in bounded Infinity ∞.

The bound is naturally the speed of light in a vacuum[414] *C*. Thus the Form (=Speed [I]) is used to create a form of

[413] See Chapter 3

[414] Ignoring the obvious practical difficulties with that in the Maxwell scenario

forms (a Node precipitating Event)[415]:

$$(s^{1}) \neq (-s^{n}) = C\ [I]\text{: Categorical Universal}$$

We can stop at that point. The violation of ML rules tells us that something has gone wrong. This is trying to use the abstracted (or "alienated") shadow content of a reflexive negation, to act as substrate for the very Form, which is supposed to act as the substrate of the reflexive negation.

It's like the category mistake of asking "which picture on this Art gallery wall, is the Gallery?"

What's intriguing about this, is that Maxwell through his genius created an *artificial* Singularity (Si), which then made Chaos real, as an agent of order.

We do this accidentally, in commonplace reasoning, all the time. It is, of course, the basic parlour trick of all theology.

Maxwell was able to do this, precisely because <ΔS> was posited by Clausius as a fundamental substrate of the physical universe. The law of entropy is: Chaos.

Experience: Collection

Setup:

Get your by now familiar blank sheet of paper

---

[415] See Chapter 4

Get ready to imagine a load of paperclips

Step 1:

- Pour a load of paperclips over the sheet;
- Cover the sheet;
- Until you have a mound of paperclips.

Now

- Count the paperclips.

Stop.

<u>Discussion</u>:

(1) You can't.

(2) You try, but after the first few items, the whole thing started phasing in and out of your focus: it was like manipulating ghosts.

Yet, every paperclip came from your own mind. If you had counted each paperclip, as your put it onto the sheet, well then you'd know how many were in the pile. You could go from an ordered count to a disordered pile: but you can't go back the other way. Not even in your own head.

Except you can. In the next Chapter we're going to show you how.

We will revisit this *Collection Experience*, during Chapter 7.

You will need:

- A metronome;
- To imagine a little conveyor belt taking the clips away from the pile, one at a time.

This Chapter on Chaos will then look very different.

Chaos produces Singularities: bridges into the forms of inequality. Now, if we invert the state order of the Maxwell Demon ML equations, we see how the chain begins with an asserted Si at Demon [E].[416]

$$\frac{(\text{Energy}) \neq (\text{-Energy})}{(\text{Chaos}) \neq (\text{-Chaos})} = \text{Order [E]}$$

Disjunct Bridge Form

$$>$$

$$\text{Si: } (\Delta\ \text{Energy}) \neq ((\Delta\ \text{Chaos}) = \text{Order [E]} \}\ \text{External}$$

The dynamic form of Order [E], in Becoming, is Information [I].

[416] See Chapter 2, to recall how Si equations work

Thus we finally step:

$$>$$

$$\text{Si: } (\Delta \text{ Energy}) \neq ((\Delta \text{ Chaos}) = \text{Order } [E] \} \text{ External}$$

$$<>$$

$$(\Delta \text{ Order}^{n}) \neq (\Delta\text{-Order}^{n}) = \text{Information } [I]$$

We then simply perform the Singularity equation chain backwards in C, which is of course the fundamental derivative boundary of Information [I], as an infinity ∞.

Hence the activity of the Imp-erator of Information. The Demon *can* ***see*** *the individual molecules.* The Demon is thus postulated to have, free of any entropy-cost, Information:

- Access

and

- Manipulation

This is where Bennett cleverly found the flaw. However, we could just as easily make the Demon an engine of Chaos.

Instead of a Singularity chain beginning with Chaos and ending with Order [E] mutated into Information [I], we could turn the chain around:

$$\frac{(\text{Information}) \neq (-\text{Information})}{(\text{Order}) \neq (-\text{Order})} = \text{Chaos [E]}$$

Disjunct Bridge Form

$$>$$

$$\text{Si: } (\Delta\ \text{Information}) \neq ((\Delta\ \text{Order}) = \text{Chaos [E]}\ \}\ \text{External}$$

$$<>$$

$$(\Delta\ \text{Chaos}^{n}) \neq (\Delta\text{-Chaos}^{n}) = \text{Energy [I]}$$

This, of course, is spookily reminiscent of the theoretical underpinnings of how quantum mechanics treats entropy.

All of this shows us that the logic of Chaos underpins our entire existence, and how we think about it. We are ever minded to try and extract abstraction from reality, and to do so by the creation of Singularity.

We go about that by creating (unknowingly) forms of forms: not to act as Node potential Events, where they are content-driven by other reality: but as Singularity.

Arificial Singularity invokes Chaos. But the actual production of Singularity by Chaos, in the boundary condition of the existential, invokes real content in forms of inequality.

Being and becoming are the holomorphs of all that there is, to us, in our universe. It is this which Chaos: allows as synergetic contingency, and a-temporally re-energises.

All that is our being and becoming, is the luminous shadow of Chaos: the reality of that which it is not.

We are able to tell the difference, in time.

CHAPTER 7

MATRIXIAL CHRONOLOGIC

As Augustine wrote, in *Confessions*:[417]

> "'Quid est ergo tempus? Si nemo ex me quaerat, scio; si quaerenti explicare velim, nescio".[418]

On which Wittgenstein commented[419]

> "Something that we know when no one asks us, but no longer know when we are supposed to give an account of it, is something that we need to *remind* ourselves of".

As with so much of Wittgenstein, it is wrong but in a useful and interesting way.

This Chapter will, sadly for the general reader, require some familiarity with quantum mechanics and its successors. To write of these matters in a way intelligible to the reader who is without such familiarity, would require a book-length explanation: and yet still be very difficult.

But, please make sure to do the *Experiences*: as these render both intelligible and almost redundant all the other commentary and explanation.

---

[417] Book XI (c 397-400AD)

[418] What then is time? If no one asks me, I know, if I want to explain it to someone who asks, I do not know.

[419] *Philosophical Investigations*, aphorism 89 (published 1953)

As Newton wrote:[420]

> Absolute, true, and mathematical time, of itself, and from its own nature flows equably without regard to anything external, and by another name is called duration: relative, apparent, and common time, is some sensible and external (whether accurate or unequable) measure of duration by the means of motion, which is commonly used instead of true time; such as an hour, a day, a month, a year.

The investigations of Loop Quantum Gravity are producing new interpretations of time:[421]

> The absence of time does not mean, therefore, that everything is frozen and unmoving. It means that the incessant happening that wearies the world is not ordered along a timeline, is not measured by a gigantic tick-tocking. It does not even form a four-dimensional geometry. It is a boundless and disorderly network of quantum events. ...If by 'time' we mean nothing more than happening, then everything is time. There is only that which exists in time.

We have to note the ideas of time as an emergent property of quantum entanglement.[422]

> *Time from quantum entanglement: an experimental illustration.*
>
> In the last years several theoretical papers discussed if time can be an emergent propertiy deriving from quantum correlations. Here, to provide an insight into how this phenomenon can occur, we present an experiment that illustrates Page and Wootters' mechanism of "static" time, and Gambini et al. subsequent refinements. A static, entangled state between a clock system and the rest of the universe is perceived as evolving by internal observers that test the correlations between the two subsystems. We implement this mechanism using an entangled state of the polarization of two photons, one of which is used as a clock to gauge the evolution of the second: an "internal" observer that

---

420 *Principia Mathematica* (1687) p738

421 See *The Order of Time*. Carlo Rovelli (2018) p.90

422 arXiv:1310.4691v1 [quant-ph]

> becomes correlated with the clock photon sees the other system evolve, while an "external" observer that only observes global properties of the two photons can prove it is static.[423]

As Farias and Recami show,[424] in *Chapter 2 - Introduction of a Quantum of Time ("chronon"), and its Consequences for the Electron in Quantum and Classical Physics*, this Chronon theory of time creates new possibilities of analysis.

À *La Recherche*[425]

We want to do a reset. To prepare you for the theoretical work to come.

Experience: Pasting

Step 1:

- Look around you at the room you are in;
- Take a few relaxed normal breaths;
- Relax.

Now:

- Imagine, there in the room, a paperclip;
- Just let the paperclip be in the room;
- Move it around. Make it bigger and smaller;
- Feel your control over it.

Stop

---

[423] https://web.archive.org/web/20161116104620/https://arxiv.org/pdf/1310.4691v1.pdf

[424] https://doi.org/10.1016/S1076-5670(10)63002-9

[425] À la recherche du temps perdu. Marcel Proust (1913)

Discussion:

(4) The paperclip appeared.

(5) It was easy.

Step 2:

Now:

- Look around that room;

and:

- See the paperclip there in the room.

Now: Focus

Make the paperclip appear 5 Seconds in the Past

in the room

Stop .

Discussion:

(1) You can't do it.

(2) The harder you try, nothing happens.

(3) You actually feel like you're frozen.

Step 3:

Now:

- Look around that room

and:

- See the paperclip there in the room

Now: Focus

Make the paperclip appear 5 Seconds in the Future

in the room

Stop

Discussion:

(1) This time, you can kind of do it;
(2) It's like a ghostly image of your the paperclip that you definitely can make appear in the present;
(3) But it feels emotionally unsettling.

We will come back to the future paperclip in the *Matrixial Anthropics* Chapter. For now, we only wanted to show you that there is a real difference between past and future for You.

Recall also the *Experiences Object* and *Landscape*. These are the *Experiences* where we showed that you can't ordinarily "cut out" an object from its landscape and bring it into your head.

There are, however, 2 ways that you can achieve this cognitive effect:

(1) You replace the in-conext landscape with a figmented one (popping a postcard, or "green screen" behind the object);
(2) You use You_2 to cut out and move the object.

By this stage in the book, you'll have realised that both of these are doing exactly the same thing. You are using the Theta Θ equalising function to place the landscape and object as datum (⍰) ≠(n⍰) in the same axis.

Note: that this does not work for the future. We will

explain why in the *Matrixial Anthropics* Chapter.

We can now reveal the real reason you cannot effect object / landscape separation:

You are trying to alter events which are past in time.

The objects of your visual field always exist in the past for you. You can't alter that past, no matter how clearly you see it.

However, there is no time in the Theta Θ axis. The equalisation function occurs instantaneously.

If we could slow-motion what occurs:

Event	Time
Unaided	
You see the object/landscape	-1
Attempt to separate	0
With Theta Θ	
You see the object/landscape	-1
You re-create object/landscape in Theta Θ	0
You manipulate in Theta Θ	0

It is because all of this is happening in units of Planck time[426] that you feel as if you are actually manipulating what you are *presently* seeing in your visual field. You're not. You're manipulating a present recreation of the past

---

[426] 10^{43} per second

which looks exactly like the past.

It's as if you could visualise the arrow heading toward's King Harold's eye:[427] and make it miss. But that arrow was a thousand years ago: your visual field reference point at (say) 12 feet, is 1.220048099^{-8} of a second. That seems like a tiny duration of time. But that duration must be compared with units of Planck time.

This is indeed how St Augustine conceived of time and the human mind, over one and a half millenia ago:[428]

> Chapter 28. Time in the Human Mind, Which Expects, Considers, and Remembers.
>
> 37. But how is that future diminished or consumed which as yet is not? Or how does the past, which is no longer, increase, unless in the mind which enacts this there are three things done? For it both expects, and considers, and remembers, that that which it expects, through that which it considers, may pass into that which it remembers. Who, therefore, denies that future things as yet are not? But yet there is already in the mind the expectation of things future. And who denies that past things are now no longer? But, however, there is still in the mind the memory of things past. And who denies that time present wants space, because it passes away in a moment? But yet our consideration endures, through which that which may be present may proceed to become absent. Future time, which is not, is not therefore long; but a long future is a long expectation of the future. Nor is time past, which is now no longer, long; but a long past is a long memory of the past.
>
> 38. I am about to repeat a psalm that I know. Before I begin, my attention is extended to the whole; but when I have begun, as much of

[427] 1066 and all that
[428] *Confessions* (397-400)

> it as becomes past by my saying it is extended in my memory; and the life of this action of mine is divided between my memory, on account of what I have repeated, and my expectation, on account of what I am about to repeat; yet my consideration is present with me, through which that which was future may be carried over so that it may become past. Which the more it is done and repeated, by so much (expectation being shortened) the memory is enlarged, until the whole expectation be exhausted, when that whole action being ended shall have passed into memory

So, the time duration between [-1 and 0] in the above table is 1.220048099^{-8} of a second.[429] That's how long you need to keep the landscape / object image in your head, as a re-creation. Which if course is being "refreshed" before you even have time to blink.[430]

Stuart Hameroff and Roger Penrose, in their captivating study *Consciousness in the universe: A review of the 'Orch OR' theory,*[431] note that:

> The best measurable correlate of consciousness through modern science is gamma synchrony electro-encephalography (EEG), 30 to 90 Hz coherent neuronal membrane activities occurring across various synchronized brain regions

As with Carlo Rovelli's work,[432] the Author finds very useful data input for Matrixial Logic: but cannot accede to the conclusions.

The difference between 1.220048099^{-8} of a second, for a

[429] One hundred millionth of a second

[430] Around 400 milliseconds

[431] https://doi.org/10.1016/j.plrev.2013.08.002

[432] op. cit, and his wonderful online lectures

photons to communicate an image from 12 feet, and 30 to 90 cycles per second, as the domain of human brain processing, demonstrates the perceptual gap.

Thus, it is not correct to say that your Theta Θ viewing state is an "illusion". Any more than it would be to say that your "ordinary" vision is an illusion.

You don't notice the passage of time in your visual field. Anything that you can see, you can see it precisely because photon packets are delivering their reflected goods to your eyes in a hundred millionth of a second.

If you could "slow down" your visual system so as to synchronise it with the external world, at the speed of light, you would not see a fixed landscape. You would not even see a movie. You would see a succession of still snap-shot images. You would probably go utterly insane after a few iterations. It would certainly be a very disturbing experience.

The idea that times, for each of us, is not a flow-stream but a series of Moments Δ^n is deeply counter-intuitive. That's one way of putting the matter. More accurately, it's not our intuitions that are astray: it's our culturally conditioned Subjective Scalar /S/ way of thinking.

In modern psychometrics, the concept of discontinuous

perception goes back to Stroud in the 1950's:[433]

> The author inquires whether the basic structure of psychological time is continuous or discrete. A wide range of psychophysical phenomena is cited in support of the discrete structure of psychological time. The length of the unit of psychological time - the 'moment' - is about one-tenth of a second,[434] but can be varied somewhat by experimental operations.

Woolf and Hammeroff wrote in the 2001 paper *A quantum approach to visual consciousness*:[435]

> In this view, visual consciousness occurs as a series of several-hundred-millisecond epochs, each comprising 'crescendo sequences' of quantum computations occurring at ~40 Hz

VanRullen and Koch wrote in 2003:[436]

> How does conscious perception evolve following stimulus presentation? The idea that perception relies on discrete processing epochs has been often considered, but never widely accepted. The alternative, a continuous translation of the external world into explicit perception, although more intuitive and subjectively appealing, cannot satisfactorily account for a large body of psychophysical data. Cortical and thalamocortical oscillations in different frequency bands could provide a neuronal basis for such discrete processes, but are rarely analyzed in this context. This article reconciles the unduly abandoned topic of discrete perception with current views and advances in neuroscience.
>
> In the quest for the neuronal correlates of consciousness, time is an

---

[433] Stroud, J. M. (1956). *The fine structure of psychological time.* In H. Quastler (Ed.), *Information theory in psychology: problems and methods* (p. 174–207). Free Press.

[434] 100 milliseconds: we now know it is around half that

[435] Nancy J.Woolf Stuart R.Hameroff https://doi.org/10.1016/S1364-6613(00)01774-5

[436] Rufin VanRullen, Christof Koch. MAY 01, 2003 *Is perception discrete or continuous?* Trends in Cognitive Sciences VOLUME 7, ISSUE 5, P207-213, DOI:https://doi.org/10.1016/S1364-6613(03)00095-0

> important but often ignored variable: *when* percepts arise is probably as fundamental a question as *where* they do.
>
> How do percepts and their neural representations evolve over time, both when the outside world is standing still or when it is abruptly changing, such as during an eye movement? Do we experience the world as a continuous signal or as a discrete sequence of events, like the snapshots of a Multimedia Component camera? Although the subjectively seamless nature of our experience would suggest that the relevant underlying neuronal representations evolve continuously, this is not the only possibility. Conscious perception might well be constant within a snapshot of variable duration.
>
> The idea of discrete perception was already considered, but quickly discarded by William James]. It faced a renewal during the 20th century with the expansion of the cinematograph - an obvious technological metaphor.
>
> The most influential modern version of this discrete sampling hypothesis is the 'perceptual moment' theory of Stroud. The popularity of this topic peaked around 50 years ago, and has declined continuously since. We review the psychological evidence for discrete visual perception and advocate that, although definitive experimental evidence is still lacking, the case is not at all closed. We will see that oscillations in different frequency bands could serve as a neural substrate for such processes.

The "theatre of consciousness" model is the locus of challenge by reductionist materialists to dualist theories of consciousness. They are both wrong, and we can prove it: as we see in the *Matrixial Manipulations* Chapter.

We can note, at this point, that the "consciousness as a movie" construct is a straw man. There is no homomculous who needs to observe a film on a screen. The fact of the matter is that there is something going on in the durations and gaps in discrete neural processing.

A recent paper has presented an interesting account of two-stage perception:[437]

> ***Time Slices: What Is the Duration of a Percept?***
> We experience the world as a seamless stream of percepts. However, intriguing illusions and recent experiments suggest that the world is not continuously translated into conscious perception.
>
> Instead, perception seems to operate in a discrete manner, just like movies appear continuous although they consist of discrete images.
>
> To explain how the temporal resolution of human vision can be fast compared to sluggish conscious perception, we propose a novel conceptual framework in which features of objects, such as their color, are quasi-continuously and unconsciously analyzed with high temporal resolution. Like other features, temporal features, such as duration, are coded as quantitative labels.
>
> When unconscious processing is "completed," all features are simultaneously rendered conscious at discrete moments in time, sometimes even hundreds of milliseconds after stimuli were presented.

This experimental work suggests support for the Matrixial Logic framework, in which Subjective Scalar /S/ inequalities undergo a secondary level of processing in the Theta Θ axis (Author italics emphases):

> ...The most important implication of our model is that *the duration of a conscious percept and the temporal resolution of the visual system are independent issues.* Thus, investigating the former does not allow us to draw conclusions regarding the latter.

---

[437] Herzog MH, Kammer T, Scharnowski F (2016) Time Slices: What Is the Duration of a Percept?. PLOS Biology 14(4): e1002433. https://doi.org/10.1371/journal.pbio.1002433

As mentioned, experiments have shown that humans perceive two disks as simultaneous when they are presented within ~40 ms of one another. This result was taken as evidence that perception lasts at least 40 ms. However, the experiment tells us only that the temporal resolution of *unconscious* processing is limited in this paradigm. No conclusions can be drawn about the duration of the percepts.

...We, to the contrary, argue that these experiments may reveal periodic unconscious processes but cannot be linked to conscious perception. To investigate the minimal duration of a conscious percept, we need to rely on indirect measures such as the feature fusion paradigm combined with TMS. This experiment revealed that feature fusion is not completed before 400 ms, and consequently consciousness cannot occur beforehand. It thus provides a lower bound for the minimal duration of percepts in this paradigm. However, this experiment does not provide an upper bound and cannot directly measure the duration of percepts. Other experimental methods with high temporal resolution, such as electroencephalography (EEG) (e.g., electrophysiology, have provided valuable insights into the temporal dynamics of unconscious processing by providing objective markers of temporal processing. Complementarily, functional magnetic resonance imaging (fMRI) work has been successfully used to identify the spatial locations of neural structures involved in the unconscious processing of time information.

According to the two-stage model, we do not experience stimuli and objects during their actual presentation, but much later, when they are rendered conscious. For example, a 50 ms stimulus is not perceived for 50 ms during its presentation, but its perceived duration is the result of unconscious processing that assigns a quantitative duration label to this stimulus]. This is akin to how, for example, motion features are processed.

We do not perceive motion because we consciously track a dot point by point along its trajectory, but because the output of motion detectors provides quantitative values for direction and speed.

Hence, psychophysical paradigms cannot reveal what we actually experience during feature integration but can only reveal the consciously perceived output of processing, i.e., the feature labels.

> Furthermore, *the coding of temporal features as quantitative labels can explain the long-standing philosophical mystery regarding how we can experience nonstatic temporal features even though our conscious experience is confined to the moment.*

Penrose and Hameroff have taken this discontinuity, and since the 1990's been working on a breath-taking fusion of brain cytostructure and a novel solution to quantum gravity (Author emphasis):[438]

> What is consciousness? Some philosophers have contended that 'qualia', or an experiential medium from which consciousness is derived, exists as a fundamental component of reality. Whitehead, for example, described the universe as being comprised of 'occasions of experience'.
>
> To examine this possibility scientifically, the very nature of physical reality must be re-examined. We must come to terms with the physics of space-time -- as is described by Einstein's general theory of relativity -- and its relation to the fundamental theory of matter -- as described by quantum theory.
>
> This leads us to employ a new physics of objective reduction: OR which appeals to a form of 'quantum gravity' to provide a useful description of fundamental processes at the quantum/classical borderline (Penrose, 1994; 1996).
>
> Within the OR scheme, we consider that *consciousness occurs if an appropriately organized system is able to develop and maintain quantum coherent superposition until a specific 'objective' criterion* (a threshold related to quantum gravity) is reached; *the coherent system then self-reduces* (objective reduction: OR).
>
> We contend that this type of objective self-collapse introduces non-computability, an essential feature of consciousness. *OR is taken as*

[438] S.R. Hameroff, R. Penrose *Conscious events as orchestrated space–time selections* J Conscious Stud, 3 (1) (1996), pp. 36–53

> *an instantaneous event -- the climax of a self-organizing process in fundamental space-time -- and a candidate for a conscious Whitehead-like 'occasion' of experience.*
>
> How could an OR process occur in the brain, be coupled to neural activities, and account for other features of consciousness? We nominate an OR process with the requisite characteristics to be occurring in cytoskeletal microtubules within the brain's neurons (Penrose and Hameroff, 1995; Hameroff and Penrose, 1995; 1996).

It will be obvious (not least from the words of the penultimate paragraph *"the climax of a self-organizing process in fundamental space-time"*), that Matrixial Logic parts company somewhat from the Penrose / Hameroff thesis. We will have more to say about this in later Chapters.

Discontinuous perception is now a well-founded[439] scientific theory of how the mind works. That provides opportunities in how we think about thinking. It also provides some difficulties: particularly in relation to mental states which appear to be recursive (having effects backwards in time).

Chronon Field Theory

The basic unit is:

[*Chrone* (Ɔ^{n})] [and ≠] [*Anti-Chrone* (nƆ^{n})]

Plural: *Chronons*

Each Chrone (Ɔ) is a peturbation in a *Local Chrone Field* ("LCF"). Each LCF is created by its Contents (eg currents

[439] Although by no means commonly agreed

within the sea of visual perception). These LCF's are Chronal Tensor[440] Fields.

The basic equation is:

$$(\text{Ɔ}^{n}) \neq (n\text{Ɔ}^{n}) \mid \text{Ꞓ}^{n}$$

(Ɔ^{n})	$\neq$	$(n\text{Ɔ}^{n})$	$\mid$	Ꞓ^{n}
Chrone	meets	Anti-Chrone	mutually 'annihilating'	Creating a temporal "gap"

This equation is a *Chronogap*. But this is an inferred theoretical state-change only.

Time presents in space only as Events, with the potential to become Nodes:[441]

$$\begin{matrix} (\text{Ɔ}^{n}) \neq (n\text{Ɔ}^{n}) \mid \text{Ꞓ}^{n}\,[LCF^{1}]) & & \\ \neq & = & <\sum \text{Ꞓ}^{*}> \\ (\text{Ɔ}^{n}) \neq (n\text{Ɔ}^{n}) \mid \text{Ꞓ}^{n}\,[LCF^{2}]) & & \end{matrix}$$

or, in lines of equation:

$$(\text{Ɔ}^{n}) \neq (n\text{Ɔ}^{n}) \mid \text{Ꞓ}^{n}\,[LCF^{1}])$$

$$\neq$$

$$(\text{Ɔ}^{n}) \neq (n\text{Ɔ}^{n}) \mid \text{Ꞓ}^{n}\,[LCF^{2}])$$

$$=$$

$$<\sum \text{Ꞓ}^{*}>$$

---

440 A multi-dimensional array

441 Upon interaction with an Infinity ∞ of Moments Δ^{n}

Nodes of course interact with currents in infinity Δ^n. Thus manifesting more peturbations (Ↄ^{n}) and (nↄ^{n}).

To explain the mutual annihilation | concept:

- Chronons are not energy or mass and not carriers of energy or mass;
- They are spatial modifiers. They are the manifestation of what is not matter or energy;
- We are using the word "meet" figuratively. What we mean is that a chrone becomes co-ordinate with another chrone, thereby creating a space-time event: <∑₵*>
- At this point, the time "wall" in each chrone is "destroyed" by the other wall;
- This creates a temporal gap (a fissure), which allows the reconfiguration of spacetime.

This can be visualised just like the "quantum foam" of the universe. Where peturbations represented by quarks (with plus and minus charge) are being created and annihilated. Or by the QED model of particles and anti-particles being mutually annihilated: nearly.

Time is neither mass nor energy, but is the product of both. Upon a chrone ≠ anti-chrone event ₵, the chronons "exhaust" each other.

Returning to the quantum field example, we would not characterise the "appearance" of a Q^+ as (Ↄ^{n}) ≠ (nↄ^{n}) | ₵^{n}.

Rather, that Q^+ is a Node <∑₡*>: a composite of Chronogaps.

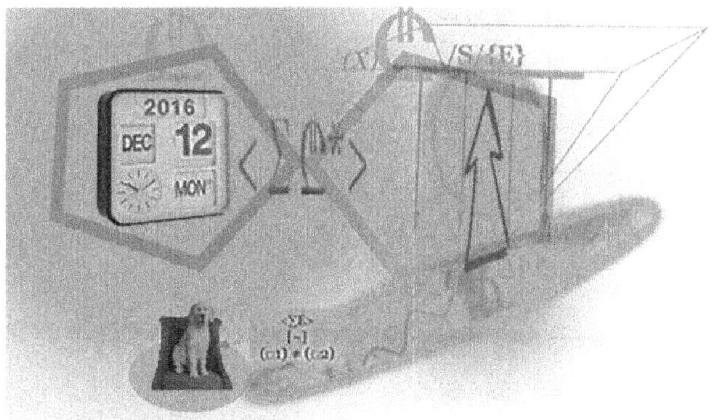

We explain this graphic later. Just allow it to thought-provoke.

Chronographic Field Axes

Our world is *Fields*. They are the foundation of everything. And everything is ultimately measured against the one single constant in the universe: the speed of light in a vacuum (or the non-local influences of entangled particles: so far as there is a difference). That is, against the clock of time itself.

The "sea" in which spacetime events[442] such as ₥ and Φ occur, can be viewed as a Vector Field:

> A vector field in the plane (for instance), can be visualised as a collection of arrows with a given magnitude and direction, each attached to a point in the plane.

[442] Local spacetime: see later

There are *chronological localisations within Vector Fields*

Events in a Vector Field are cognised under Scalar Field Axes: /S/{E}; /S/{I}. In physics, a Scalar Field is defined as:

> A field whose value at a particular point in space and time is characterized only by a single number.

Now, we can't use the "number" concept, except as a placeholder.

The Theta Θ Field Axis operates as a tensor field:

> As a tensor is a generalization of a scalar (a pure number representing a value, like length) and a vector (a geometrical arrow in space), a tensor field is a generalization of a scalar field or vector field that assigns, respectively, a scalar or vector to each point of space.

We can simply visualise the Theta Θ Field Axis as summing over differences.

In Chronon Field Theory, we don't refer to "axes" any longer. Instead, we refer to *Chronological Localisations within Fields*.

Thus:

> Each Chrone (Ɔ) is a peturbation in a *Local Chrone Field* ("LCF"). Each LCF is created by its Contents

(eg the currents within the sea of visual perception). These LCF's are Chronal Tensor Fields.

So we have:

Matter Form	**Chronal Form**
Vector ᡥ Φ	\V\ LCF

Axis	Chronological Localisations within Fields
/S/{E} /S/{I}	\{}\LCF

Θ	\Θ\ LCF

It is informationally useful to use the idea of a *Chronogram* (a C-gram: to picture Chrone and LCF events).

It is easy to see the peturbation of our sensory fields by pictures and sounds. Using the "classical" model of ML, we represent those[443] as wave interaction functions.

Using a "quantum" model of ML,[444] we can more usefully envisage the *Chronogramic* operation of interacting fields.

443 For the purposes of easy explanation
444 And ML "works", whichever model we use as an explanatory tool

In the C-gram picture series, there is the dog visualisation, from the *Languini* Chapter. Chronal field events are shown in purple / pink. The clock should be imagined as moving in tiny increments:

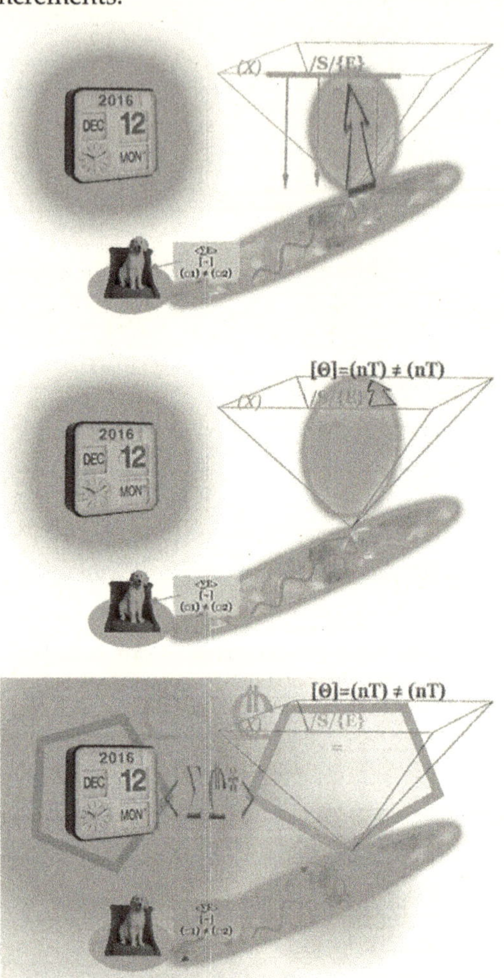

There is an interesting convergence between this Chronogramic representation and the two-stage temporal perception postulated by Herzog, Kammer, and Scharnowski in 2016.[445]

The C-gram illustrates the:

> difference between 1.220048099^{-8} of a second, for a photons to communicate an image from 12 feet, and 30 to 90 cycles per second, as the domain of human brain processing

But all this seems to come at a price. Are we not, for the benefit of understanding human perception of time, drawing a curtain between perception and objective reality?

Do we not resile ourselves to some Cartesian wonderland, where that which is sensate inhabits consciousness, and that which is sensible subsists only in external reality?

The answer is: no. For much the same reasoning that H_2O remains that same molecule, whether it wanders through a Heraclitan river, or slides down our throat.

The Cylinder of Time

This can be billed as the next big step up the Matrixial Logic ladder.

---

[445] op. cit

In the graphic which appears below, care must be taken. The cylinder of time is not a cosmic clock, or a supra-temporal ruler. There is no time which is everywhere. But every *where* is correspondent to some *when*.

(ϵ)

m̄ /S/{E}

∑Ω (T) ≠ (nT) = Θ ≈>

Φ /S/{I}

(϶)

<∑£*> ⇔ (Ɔ^n) ≠ (-Ɔ^n) | £^{n} ⇔ <∑£*>

The infographic attempts to collate the critical aspects of the Matrixial universe, which we have discussed so far.

The Reader may notice the form of inequality [(ϵ) ≠ (϶) ∑Ω]. This is the Emotion *Ohm* field function. We will get to that in a later Chapter.

The Cylinder of Time is a poetic, rather an analytic, device. Whether chaos theories (classical or quantum) illuminate a real non-knowable universe, or a real present lack of human ability to analyse a deterministic universe, remains a thoroughly open issue.

We can be sure that time is local. It is relationally different for different observers. Gravity, distance and acceleration all generate relativistic differences in time. So, it is said by Bell experimenters, do entangled particles.

Matrixial Time Machines

Each of us operates a collection of "clocks".

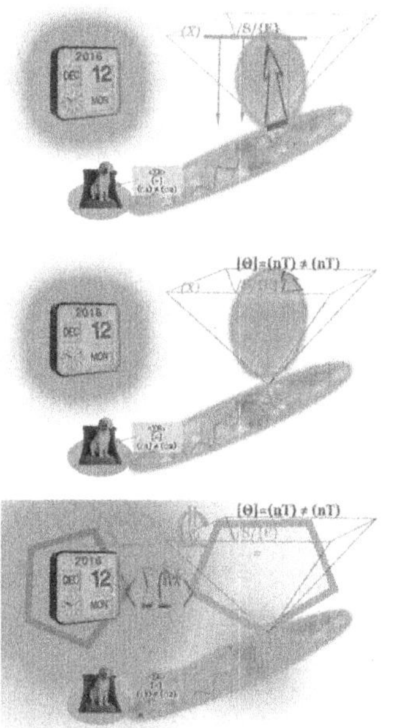

While this succession of visualisation events will occur under the same process for each of us, we will each undertake that process at different speeds: in different durata of times.

Nancy registers seeing the dog at a different "time" to Nick. They can both agree to feed the dog at 12.30 pm. That does not in itself create some objective time, nor is it a reference to an objective time. It is a mutual synchronisation of activity by reference to a form of time measurement.

What we need to disassociate, are the concepts of: (i) "reality"; and (ii) "time as an objective frame of reference". It is precisely this assumed association which has caused so much philosophical trouble, for 2,500 years.

Nancy processes the (now past) visual image of the dog in (say) 50 milliseconds.[446] Nick takes (say) 65 milliseconds. These are psychometric facts: matters of reality.

We don't notice them, in ordinary daily life. But when the car heads towards the dog in the road at 60 mph, such that the vehicle is covering 88 feet per second (0.88 feet per millisecond): then the /S/ timing difference of 15 milliseconds between Nancy and Nick does matter. Did doggy manage to just evade the edge of the car bumper, or not?

---

[446] In her Subjective /S/{E} axis field

In this scenario, we have no difficulty in agreeing that the different /S/ time perceptions of Nancy and Nick map onto reality. That their internal time events do engage in correspondence, synchronicity <∑₾*>, with the external world.

But: on what basis do we arrive at this agreement?

How can our subjective, individual, sense of time, map onto the temporal order of reality?

Because we each have a matrixial time machine. Let's experience it in action.

Recall that the basic Chronal equation is:

(Ↄ^n) ≠ (nↃ^n) | ₾^{n}

(Ↄ^n)	**≠**	**(nↃ^n)**	**|**	**₾^{n}**
Chrone	meets	Anti-Chrone	mutually 'annihi-lating'	Creating a temporal "gap"

Experience: Noming

Preparation:

- Google "30 bpm";
- Choose a metronome (YouTube or any will do);
- Make sure the Metronome has a seconds time clock counter;
- Be ready to press Play (don't Play yet).

You will then have an audible 30 beats per minute: one beat every 2 seconds.

Test that you can hear the beat.

A 10 Beat takes 20 seconds: check the time from 0.00 on your YouTube clock (your YouTube clock will show 18 seconds of count: that's because the next Beat occurs on the 20th second).

Activity:

Step 1:

- Get one finger ready: to tap on the table, or your leg;
- Play the metronome;
- Try and tap your finger in time with the Beats.

Discussion:

(1) Unless you trained as a music student, or gymnast, you find this very hard.

(2) You seem to get "lost" between the beats.

(3) You feel like you are waiting for the sound, then tapping: but anticipating the next beat seems like an accident.

Step 2:

Close your eyes and focus

Engage You_2: your Theta Θ brain function

- Get one finger ready: to tap on the table, or your leg;
- Play the metronome;

- Try and tap your finger in time with the Beats.

Discussion:

(1) Now, it seems easy.

(2) You feel like there is an automatic synchrony between your finger and the audible Beats.

(3) It's almost like "You" are not there at all: like some other person is tapping out the Beats and you're just observing.

Recall from the *Languini* Chapter, the *I-Pen Experience* ("IPE"). It's the same You_2 feeling.

Step 3:
Now: ***turn the metronome sound off***

Close your eyes and focus
Engage You_2: your Theta Θ brain function

- Get one finger ready: to tap on the table, or your leg.

press Play [spacebar or mouse]

and

- Play the metronome, ***silently***
- Tap your finger in time with the Beats
- For 10 Beats (18 seconds of time)
- Press Stop [spacebar or mouse]

Now: check the YouTube clock counter.

Discussion:

(1) Now, it seems easy.

(2) You feel like there is an automatic synchrony between your finger and the audible Beats.

(3) It's almost like "You" are not there at all: like some other person is tapping out the Beats and you're just observing.

This is evidence, not proof. It's evidence that we have two completely different internal "clocks":

- Keeping our Scalar /S/ subjective time;
- Creating our Theta Θ axis field "objective" time.

> You can practice on your phone by downloading an App like *MIDI Rec.*[447] Repeat Steps 2 and 3. Use the piano keypad (with or without sound). Reset to 0 each time. You can film the playback with a phone screen recorder App. So, you can check if you are hitting the key at the correct 2 second intervals.
>
> Then, try it with different metronome beats.

It is the Theta Θ clock which is able to synchronise with the YouTube second counter. That counter is of course a digital representation of the "second" determined by the "atomic clock".

New studies conducted over the last decade, in the wake

[447] a free App

of the Libet Experiment,[448] provide scientific support for just such Theta Θ clock function:

> Results from Emmons et al. (2017) suggest that such ramping activity encodes self-monitored time intervals. This hypothesis is particularly pertinent given that self-monitoring of the passing of time by the experimental subjects is intrinsic to the Libet et al. (1983) experiment.

If time is a flow, a "river", then two rivers can never mix, without losing their flow integrity. They can be bridged, but that is to create a form in alienation from the rivers.

But where time is not a flow, but a quantised apparition of (Ɔ^{n}) ≠ (nƆ^{n}) |, then each Event ₾^{n} can *synchronise* with its negation in Nodal *annihilation*:

|

|

As the Chromal equation states:

(Ɔ^{n}) ≠ (nƆ^{n}) | ₾^{n}

In the *Noming Experience,* you felt that | annihilation happening. Review the history of your head and go back into the *Experience*. Or just repeat the *Experience.*

You will sense that each synchronised Beat (which is an | annihilation) seems to "reset" your internal counter to zero (0). You feel liberated to feel the next | annihilation.

---

[448] *Readiness Potential and Neuronal Determinism: New Insights on Libet Experiment.* Karim Fifel. Journal of Neuroscience 24 January 2018, 38 (4) 784-786; **DOI**: 10.1523/JNEUROSCI.3136-17.2017

You can sense "when" it is coming.

It is not actually your Mentation which is doing the sensing. It is your Ohm Ω emotive field: more on that later, and in *Secret Self*.

The quantised property of time, in the world and in your head, explains a perennial mystery: how animals can act in unison. For example: birds in flight.

Since birds do not have a Scalar /S/ field,[449] there is nothing to "get in the way" of the operation of the avian version of a Theta Θ type axis. Thus, the birds synchronise their Theta Θ time. Time, speed and distance are intrinsically relational. The individual bird in a flock of birds does not need to know "where" is "now", at every moment: merely to have the same "now" by reference to the same metronome, as the other birds in a flock.

Droplets Not Rivers

As the Chromal equation states:

$$(\text{Ↄ}^{n}) \neq (n\text{Ↄ}^{n}) \mid \text{Ȼ}^{n}$$

This is, at first glance, a surprising formulation. Now that readers are becoming familiar with ML equation formulation and manipulation, it appears strange that the equation for time is not couched in (A) ≠(-A). After all, what could be more intuitively a flow of Moments Δ^n in

[449] Will, if they do: it is very small and weak compared to humans

Infinity ∞ : than Time?

The reason is simple: and deeply counter-intuitive to everyone who has grown up in western traditions of culture and science.

Time is not a flow. It is not a river. Time is a series of beats. And not even the same beat. There are different beats, contingent upon the state of matter which originates and regulates time.

The work in Loop Quantum Gravity provides a (disputed) mathematical basis for the proposition that time is quantised.

Andrew Jaffe reviewed for Nature in 2018:[450]

> According to theoretical physicist Carlo Rovelli, time is an illusion: our naive perception of its flow doesn't correspond to physical reality. Indeed, as Rovelli argues in *The Order of Time*, much more is illusory, including Isaac Newton's picture of a universally ticking clock. Even Albert Einstein's relativistic space-time - an elastic manifold that contorts so that local times differ depending on one's relative speed or proximity to a mass - is just an effective simplification.
>
> So what does Rovelli think is really going on? He posits that reality is just a complex network of events onto which we project sequences of past, present and future. The whole Universe obeys the laws of quantum mechanics and thermodynamics, out of which time emerges.

To be fair, the reviewer concludes:

> Ultimately, I'm not sure I buy Rovelli's ideas, about either loop quantum gravity or the thermal time hypothesis. And this book alone would not

[450] https://www.nature.com/articles/d41586-018-04558-7

give a lay reader enough information to render judgement. *The Order of Time* does, however, raise and explore big issues that are very much alive in modern physics, and are closely related to the way in which we limited beings observe and participate in the world.

This itself has a vintage going back to the paper by Robert Levi in 1927:[451]

> Metaphysical considerations, inspired by the theory of Relativity, lead the author to interpret by direct actions, carried out with the speed of light in a vacuum, what was attributed to actions of environment.
>
> Applying this method to electromagnetism, it is condensed into a formula generalizing Coulomb's law, which provides the motion of an electrified mass in the field of other masses with various motions (§ 5).
>
> The author is then led to suppose that the electrons are without volume and that their proper times consist of successive time atoms or chronons.
>
> It follows from these hypotheses, which do not generally alter the laws of electromagnetism, the existence of an intra-atomic vibration, the frequency of which identically satisfies the equality hv = W, W being the total energy of the atom. The simple expressions found at this energy provide new interpretations of the action integral (§ 11).

We can then turn to a remarkable paper by A. Charlesby in 1992 *Quantised space-time: An alternative approach to quantum theory*:[452]

> Although the mathematical deductions and predictions from quantum theory are extremely successful, they involve two concepts, waves

[451] (1927) Theory of universal and discontinuous action
J. Phys. Radium 8 , 182-198 DOI: 10.1051 / jphysrad: 0192700804018200

[452] International Journal of Radiation Applications and Instrumentation. Part C. Radiation Physics and Chemistry Volume 39, Issue 1, January 1992, Pages 141-147 https://doi.org/10.1016/1359-0197(92)90188-L

and particles, which physically appear contradictory. One attempted explanation requires the interaction between observer and observed; existence without observation is expressed in terms of a probability function and is not real in any conventional sense.

Another explanation involves a constantly increasing number of universes but the observer operates in only one.

It is difficult to reconcile them physically with the simultaneous phenomena of diffraction and scattering of single electrons in a low intensity beam. The unsatisfactory nature of these physical interpretations is shown by the constant search for others.

The theory proposed in this paper involves the quantisation of space-time into units s_0 dependent only on particle rest mass m_0, measured along the worldline s of each particle. In terms of a time unit ct, $s_0 = h/m_{0c}$.

With the standard relativistic corrections for relative motion of the observer, this leads directly to the de Broglie relation for wavelength = h/mv, the energy-frequency relation $E = h$, the Heisenberg uncertainty principle, etc. The apparent contradiction between scattering and interference also disappears. The importance of the observer is greatly diminished, as it is the interaction of one particle with another which alters the space-time unit to $h/(m_0 + M_0)c$.

Many difficulties in physical interpretation appear to arise from our attempts to define physical behaviour in terms of the continuous space and time to which we have grown accustomed from macroscopic experience. Similar difficulties would arise if we attempted to define particle rest mass and electrical charge as continuous variables.

The Charlesby quantisation offers an intriguing new solution to the Entanglement problem:

The dimension of the space-time unit s_0 depends only on the rest mass m_0 of the individual particle, though the individual space or time dimensions may appear different due to motion of observers, and this is reflected by a change in the individual ex values, depending on observer velocity and orientation (as they do in relativity). In particular for a photon whose rest mass m_0, in vacuum is zero the basic space-time unit becomes infinite ($s_0 = h/m_0 ic$ with $m_0 = 0$).

Thus an individual photon in vacuum remains in the same basic unit and this may explain the important experiment of Aspect in which a pair of photons emitted in opposite directions by a radioactive source have coupled spins; detection of one appears to affect the other even at a considerable distance, within a time shorter than can be covered at the speed of light. This experiment has been considered to oppose the views of Einstein and agree with those of Bohr.

However it could be argued that for a photon with a base unit which is infinite in extent, at least within its own framework, there would be no difficulty in transmitting a message between two photons which continue to reside in the same unit. This will only cease when they interact with other particles etc. of finite rest mass when the basic unit $h/(m_0, + M_0)\ ic$ becomes finite. It would be interesting to carry out a similar experiment with particles of non-zero rest mass, when they would not continue to reside in the same unit.

One has often read of attempts to account for such experiments by suggested hidden variables but no evidence has emerged for their existence to correllate the behaviour of remote particles. However it could be considered that the subdivision of the axes chosen by the observer with ∞x units provides such a correlation. Any change in the observer would affect the a values of all particles being observed however remote.

A problem with this model is that it seeks to extract what is in effect a Planck unit for time s_0, which functions as an iteration of particle rest mass. What then lies between particles is a "frozen" version of the Shroedinger wave equation. Under the rubrics of this model, time would not move at all. We would be stuck in a static uni-temporal universe.

Charlesby is right to say that, under this model, there is

no FTL interaction of entangled particles. But the price paid is that there is no temporal interaction at all.

The fundamental problem with the Charlesby model is that it is seeking connect temporal attributes using Scalar identity, instead of temporal coherence.

To conduct even a summary meta-review of the published ideas in this stratum would take an entire book on its own. Hitting the peaks as we are, it is worth pausing at the 2020 paper *Physical Implications of a Fundamental Period of Time* (Author emphases):[453]

> A proposal to formulate a meaningful notion of physical time goes back to an investigation by Dirac [4] that analyzed general properties of quantum constrained systems relevant for generally relativistic systems.
>
> Dirac briefly suggested a construction, now called deparametrization, which, with hindsight, can be interpreted as a solution to the problem of quantum fluctuations of a physical time variable by showing that *physical time requires constrained dynamics*: If both time and the system of interest are quantized in an extended model that includes all relevant degrees of freedom, the energies of the time variable and the system have to be exactly balanced. *Otherwise, a nonzero net energy would imply nontrivial evolution of the extended model in an external absolute time parameter*, violating the assumption that time is described by an internal degree of freedom
>
> [4] [4] P. A. M. Dirac, Can. J. Math. 2, 129 (1950).

But, for balance, then we have recent observations which

---

[453] Garrett Wendel et al. Physical Implications of a Fundamental Period of Time, *Physical Review Letters* (2020). DOI: 10.1103/PhysRevLett.124.241301

indicate that, if "quantised time" is a reality, it is not presently observable cosmologically:[454]

> At microscopic distance scales comparable to the Planck length, lP = p ~G/c 3 ' 1.62 × 10^{-35} m, it is thought that spacetime itself is subject to quantum fluctuations. If true, spacetime should appear "fuzzy" or "frothy", an effect that was termed "quantum foam" (also referred to as spacetime foam) by Wheeler (1963). A foamy spacetime would cause minute uncertainties in the propagation of waves, such as the distance traversed by a photon, or its energy. If found, it would demonstrate that the nature of spacetime is probabilistic, rather than deterministic, and provide strong clues towards finding a unified description of gravity and quantum mechanics (for an overview, see Amelino-Camelia 2013).

The original Shroedinger equation was, of course, time dependent.[455] Shroedinger's boxed feline rebuttal of the Copenhagen interpretation depends on a crucial constant: the flow of time τ.

Indeed, the paradox requires that there be multiple flows of continues times τ^n:

- Some universal time;
- The relative time of the box-opening observer;
- The time within the closed system, as experienced by the zombie cat.

We can indeed trace the "many worlds" interpretation of Everett to this very paradox. To have multi-linear time τ^n is exactly to suppose multiple worlds.

---

[454] Ryan, Welsh, Fumagalli and Pettini (2020) *A limit on Planck-scale froth with ESPRESSO.* Physical Review Letters 124, 241301 (2020). DOI: 10.1103/PhysRevLett.124.241301

[455] Although also written so as to be capable of being time independent.

Matrixial Synchronics

Permit us the theoretical liberty of supposing that the *Noming Experience* be validated, as a repeatable and verifiable laboratory experiment. Such that Theta Θ axis "clock" and "clocks" in the external world may synchronise: and such that this is a scientific datum.

Let us call this the Theta Clock Observation ("TCO").

The TCO tells us that, for any time event <∑₵$_1$>, there can exist a time event <∑₵$_n$>, such that the two events can assume a Nodal value (+ / -) in any configuration of | ⇔ |

Thus:

- Your <∑₵$_1$> can configure with the BPM <∑₵$_2$>, in a [= 0] configuration of | ⇔ |
- Your <∑₵$_1$> can configure with the YouTube <∑₵$_3$>, in a [= 0] configuration of | ⇔ |

Since every human alive has a Theta Θ clock: Your <∑₵$_1$> can configure with <∑₵$_x^n$>, of any other human, in a [= 0] configuration of | ⇔ |.

What this entails is that (by analogy) an orchestra can play without a conductor: no baton being required to be waved by the hand of god, nor by any universal Lorenzian[456] time.

---

[456] Or Einsteinian

That relativistic time is subject to dilation, does not render it any the less a universal constant. Indeed, that was the position which Einstein fought for before and after the infamous Solvay Conference of 1927.

We arrive at a description of the universe which is much less counter-intuitive, and much more useful than QED, or relativity.

Let's spend some moments describing the Matrixially Synchronic universe.

(1) There exists within the elements of any system, the potential for synchronic configuration.

(2) Until the potential is actualised, the system is open and chaotic.

(3) Actualisation can occur: (i) by conscious will, for those entities possessed of such;[457] (ii) by genetically determined behavioural gestalts;[458] (iii) by circumstantial imposition;[459] (iv) by interactions of categories (i) to (iii).[460]

(4) Upon actualisation, [=0] configuration of $|\Leftrightarrow|$ occurs: or to put it more accurately, the arising of [=0] configuration of $|\Leftrightarrow|$ is exactly actualisation.

(5) Actualisation occurs without any universal flow or account of time τ.

---

[457] Human beings

[458] Animal

[459] The inanimate universe

[460] For example, in a 2-split quantum apparatus

(6) Indeed, the idea of τ is antithetical to TCO actualisation.

(7) There is no single temporal mode of TCO actualisation. Each is a form of the forms of its interaction (reference the Nodal formula).

(8) Any form of forms has an apparitional property, which partly accounts for the illusion of "time". (That which completes the account is discussed below).

Let "H" stand for any human being^{n}.

(9) Any H has potential to actualise $[(\text{Ɔ}^{n}) \neq (n\text{Ɔ}^{n}) \mid]$ with any other H^{n}.

(10) Such actualisation creates a $\sum H$ temporal frame of reference [T], which is fixed in some externality.

(11) Externality is any function of Substance extended in spacetime (as to which see below).

(12) In other words, [T] is not an abstract or non-contingent emergence or property.

(13) Nor is [T] an Infinity ∞.

(14) [T] is a succession of annihilations. That which lies between $[T]^{n}$ is no kind of order, but Chaos [=C].

(15) That Chaos [=C] may be granted apparitional form: as anything. That apparitional form is freely culturally contingent.

In these last item matters, let's now revisit the Collection Experience in Chapter 7.

Experience: TimeCart

Setup:

Get your by now familiar blank sheet of paper.

Get ready to imagine a load of paperclips.

Step 1:

- Pour a load of paperclips over the sheet;
- Cover the sheet;
- Until you have a mound of paperclips.

Now

- Count the paperclips.

Stop.

<u>Discussion</u>:

(3) You can't.

(4) You try, but after the first few items, the whole thing started phasing in and out of your focus: it was like manipulating ghosts.

Step 2:

Setup:

You will need:

- A metronome, as in *Noming*
- Set it to 1 BPM
- To imagine a little conveyor belt taking the clips away from the pile, one at a time.
- A pad and pen: you are going to be making /// marks on the pad, in rows
- To be ready to switch your You_2 on, as in *Noming*

Now:

- Pour a load of paperclips over the sheet (don't try to keep a count as you do it);
- Cover the sheet;
- Until you have a mound of paperclips.

Next:

- Set your metronome beating;
- Switch on You_2 and match your [T] Node, with the BPM count.

Now:

- Count each paperclip away;
- Keep the pile in your head;
- Make a / mark on your pad each time.

Do this for as many beats as you like

Stop.

Discussion:

(1) This is actually very easy now

(2) The only effort is to keep the pile in your head

(3) But you can see it reducing

An extraordinary thing is that, you can even project forward. It's like you can run the beat as a fast-forward movie, and see the clips disappearing one by one: while still keeping a count.

Just to note that, perhaps our commentary on the Charlesby time quantisation theory, now makes more sense.

What you're experiencing here is captured in this C-gram:

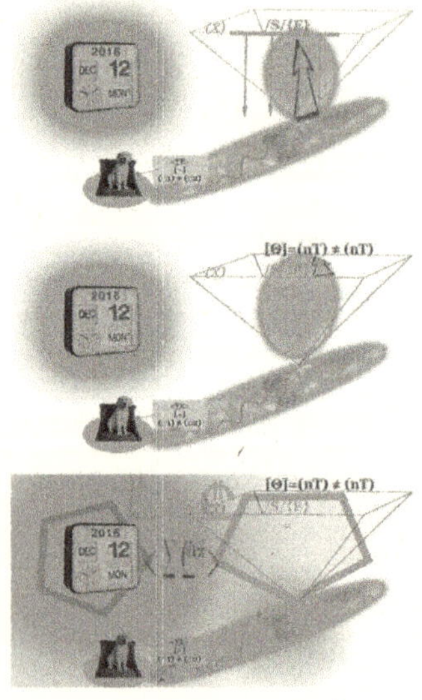

If you substitute the pile of paperclips for the dog. And swap the BPM counter for the clock. You're using synchrony: <∑₵ ^{n}>, to arrange your interaction with the material world.[461]

The succession of Chronogaps between each extinguished <∑₵ ^{n}> Chronal Node, is part of the fabric of that organisation. These are the experienced manifestions of Matrixial Synchronics.

Two Timing

This brings us to the realisation that there are two sorts of "time" which we experience:

- Subjective time in our Scalar /S/ axis field;
- Objective (in the sense of externally synchronic) time, in our Theta Θ axis field.

Our "subjective" time, is the time which actualises in our personality matrix.

- What makes a film pass quickly for one H and interminably for another.
- What makes the summers appear shorter as we age our very perceptions of aging.

Objective (externally synchronic) time [T] is quantised.

It is the quantisation of time which accounts for the precepts of

[461] The *Experience* would obviously work just as well with real paperclips

relativity, and for entanglement in quantum theory.

It is also that quantisation which allows us to cohere with external time.

In other words, quantisation, which for 100 years has seemed to furnish only counter-intuitive advances into unreality, is the simple foundation of the ultimate proof of our own subjective reality.

Our "objective" time, is not just something that we engage though Matrixial Logic *Experiences*. These *Experiences* are merely deployed in order that we can externalise our perception of what occurs automatically. In every moment of our lives.

From our moment of birth, our Theta Θ axis field carries the potential to be actualised in interactions with the world.

Imagine two children running around. They move chaotically, in relation to each other, and the world. Then, without necessity of communication, they engage in synchrony of the Theta Θ axis field in each of them.

Now, they co-ordinate in time, and thus in movement (velocity and distance). They have created space-time: a frame of reference. Without observing or corresponding to any outside "clock" measuring any universal "time".

That which was chaotic has become determinate, in the mutual frame of reference of $<\sum\mathfrak{C}_1>$ [=0] $<\sum\mathfrak{C}_2>$. That [=0] Nodal point is itself an externality, by reference to which all of space and time can, in its infinity of moments Δ^n, become bounded: but only in reference to that $[=0]^x$. Each other $[=0]^n$ creates its own spacetime. Thus a continuum of spacetime appears, but only as apparition of its elements, not as a thing or force in itself.

Why can't we travel faster than the speed of light *C*? Because any velocity creates its own frame of reference [=0] which binds the time in which *C* can subsist. So, in fact, we can in a sense travel at super-luminal speed, but the time which is available for us to do that reduces ultimately to Planck time.[462] We can chase our own tail but the faster speed we achieve, the more foreshortens the tail we are chasing.

If this be so, then the calculations of relativity remain in effect.

We simultaneously glimpse an answer to the fundamental problem of quantum mechanics. There are waves and particles, as entities with $[=0]^x$ time function, which, when actualised to a different frame of reference $[=0]^y$, subsist in that actuality differently to the initial observation.
The entangled photon are not communicating at super-

[462] Or something of that order: see the papers referenced above

luminal speed (= D/T). It is that their quantised time $[T]_0$ is necessrily different to the $[T]_{+\setminus-}$ which is the frame of reference of the equipment.

That equipment, being local to the observer, appears to fall within the same frame of reference: although at the Planck scale level, is actually not.

It is thus analagous to when we "pull" a visual object out of its landscape with a You_2:

- There is set a Form of forms of "beat" time $[T]_0$ FoR between Substances (the particles)
- A Form of forms is an unstable Event <∑₾>
- Anything can function as a Wave Δ^n (such as an equipment), which "interferes" with that Event, creating a [T] Node
- [T] Node Points at A and B (the places of the equipment) create point references ⫙^{n}

And at that moment:

(i) $[T]_0$ is extinguished |

(ii) what was Substance becomes referenced to $[T]_1$: the temporal FoR of the equipment

And we can add to that:

- ⫙$^n_{(+\setminus-)}$, functioning as, say, red and green, register
- Scalar registration will give a wave interference pattern

It would then be an error:

- In treating time τ as a fixed FoR for all events (rather than relative to the FoR of entangled Substance)
- In failing to recognise that particles interact with other particles such that waves are created (which is obvious as a matter of elementary physics)
- That where equipment which sets up a Temporal wave Δ^n [T1], and which interacts with the packet $[T]_0$: to fail to account for the fact that the two particles have been made to "share" packet $[T]_0$ by their very creation

To put it another way, τ is not a singular phenomenon. τ subsists both as quantised Substance relations in [E], and as unquantised Process infinity ∞ in [I].

It is notable that the Bell Inequality experiments of Wheeler (1998), Kim (2000)[463] (and others since) use entangled photons. So, the 2 particles under consideration, by definition, share the same $[T]_0$ state.

When we bring these τ subsistences together, we create new time {t}, which is a simulacrum of $[E][T]_0$ (in this experiment).

But this is a process in nature and in the macro-world. We can experience it. We have had that experience it in *Noming* and *TimeCart* and will again in *Ropetime*.[464]
Thus, rather than an Everett "many worlds" interpretation,

---

[463] https://journals.aps.org/prl/abstract/10.1103/PhysRevLett.84.1
[464] See Chapter 9

in which the wave function invests the entire universe, producing infinite numbers of new universes at each interaction, we have something much simpler. And more coherent with observation.

There are simply "many times" $<|\text{Ꞓ}^{n}>$, each extinguishing upon the advent of the next:

$$(\text{Ɔ}^{n}) \neq (n\text{Ɔ}^{n}) \;|\text{Ꞓ}^{n}$$

Successive extinctions:

- Preserve multiplicity, which avoids the requirement for Schroedinger wave collapse

yet

- do not result in that cumulation of hypothetical states, which makes the many worlds interpretation so distasteful to many.

All we are doing in small-particle experiments, is dealing with phenomena so close to the Planck scale interaction of the speed of light, and the size of the particle, that the gaps $< | >$ become so small that the phenomena appear to violate FTL law, and thus instantiate non-locality. But they don't do either. On this basis, it can be said that Einstein was right all along, for the wrong reason.

It lies at the level of preliminary suggestion, at this stage, to submit such a radical re-interpretation in operation of the Schrodinger wave equation. But it works. Fundamentally, we are able to explain why the insertion of any "observation" appears to collapse the wave function.[465]

It is not because the superposition of particles in a probability wave (the spread of which cannot be determined), is collapsed. It is because the interposition of the observation[466] actualises temporal state $\{<\sum ₵_1> [=0] <\sum ₵_2>\}_z$ which is different to the initial temporal state $\{<\sum ₵_1> [=0] <\sum ₵_2>\}_a$ and is different not on any scalar axis, but a new and entirely distinct [T] time.

One could analogise this and say that the Bohmian pilot wave interpretation works, and without non-locality, if one substitutes a conceptualisation of $<|₵^n>$, for the pilot wave. There can then, of course, be non-locality in the trivial geographical sense of simultaneity at a distance: but that is only simultaneity from the FoR of the "observer", not within the temporal FoR of the entangled particles.

Indeed, we end up with the best of Everett and Bohm. Interactions of quantised temporal FoRs provide:

- The same operator function as Pilot Waves;
- Plurality, thereby preserving the wave function, but via extinction rather than multiplication.

What's more, the divide between the quantum and the classical, and indeed the cosmological, disappears. It is no

[465] Or rather, doesn't: instead a different Chronal Field is instantiated
[466] The 2-slit device

longer needed. The same mathematics of I₵n inequality function can[467] describe all three plenums.

Let's consider the journey of two tennis balls:[468]

(1) They are each poured from a single tin, from the same factory, down an incline plane.

(2) at $[T]_1$ each ball has a temporal state $\{<\Sigma \text{₵}_1> [=0] <\Sigma \text{₵}_2>\}_{a\ b\ c,\ n}$

(3) at the split-screen, that creates a Node for potential event $[T]_{1.}$

(4) there is now actualised a new $[T]_2$.

(5) ball hits screen, creating a new $[T]_3$.

(6) 2nd ball undergoes the same process.

(7) the split-screen acts like the YouTube beat-meter and You in the *Noming Experience*: there is actuated the same temporal actuation $[T]_2$.

(8) 2nd ball now shares in that temporal actuation, when it hits the screen.

We could express this in other words: by saying that the split-screen actuates a specific "temporal field", in interaction with subject Events. Events are "carried by" each of the balls, and actuate upon entering that temporal field. That is loosely put: it is the interaction which creates the Node, the collapse of which is the actuation of the novel temporal field.

---

[467] It is thought: but not yet proven

[468] This is a surface-skimming explanatory; not an analysis

To continue the analogy,[469] the split-screen interaction "coats" all balls (that have the same elementary properties) with the "impressum" of that new null [T] field [=0]. That is why the balls manifest an apparition of the "coating" at the final screen.

Adding[470] clarity: the balls had (in their inter-relationship before "launch") a natural temporal field,[471] which manifests apparition as a "wave": and that apparition is being re-created by the particular properties of the intervention at the split-screen.[472]

The Bell inequality can still be violated by non-randomness in non-temporal qualities of the entangled particles.

This further explains why the added imposition of moment measuring devices "cancels the interference": because we are trying to image a ghost. The $[T]_z$ state of even a single particle simply does not exist in the $[T]_x$ state in which our "camera" is operating. Not least, because our visining is by definition occurring at a different lightspeed C than the reference object. And so on, for the imposition of the screen.

---

[469] Which is more poetic than analytic: but for that is more useful in aiding comprehension

[470] Hopefully not subtracting

[471] In reality, many such

[472] How that is so, is undoubtedly a special property of quark-based particles, rather than tennis balls: which is where the poetic analogy surrenders to reality

These are not probability functions, as probability assumes a single [T] frame of reference. This explains why, without "normalisation", probability functions tend to infinities: that which is a [T] frame of reference does not, without an intervention, submit to any other [T] frame of reference. Each is contingently independent of the other.

The difference with consciousness, in H, is that it alone can function in a domain of choice and will. You do not have to join the dance. Non-conscious nature admits of no choice, only contingency.

This datum explains also why the "observational" actualisation of different [T] states came to be (mis-) associated in popular culture with some consciousness interaction. That which is non-conscious cares not the least whether we do, or do not, observe.

By our conscious actions, we can so effect external states of affairs such as to actualise different [T] states. But effect requires material intervention and imposition.

A further matter which arises from the *Noming Experience* is the multi-chronologic nature of time [T]. We notice the 2 BPM pattern and have to engage Theta Θ with deliberation, because it is not a pattern we are used to. There is no reason to think that, in the universe, there exists some supra-regulatory "beat". We have good reason, from empirical physics investigations, to consider Planck time as a minimum unit of [T] quantisation.

Beyond that, we have no obvious basis for assuming that the contingent event time $[T]^n$ of any events correlation, is the same. There are obvious inferential reasons for assuming that it isn't.

This is helpful. It is what makes the detection of Theta Θ [T] so accessible with a simple *Noming Experience.*

Substance Synchronics

Just a minute:[473] readers who have been reading carefully will exclaim. Just look what we said of Substance back in the very beginning of this book:

> In E Logic we analyse a Substance, in relation to the opposite of that Substance.
>
> Substance
>
> We define *Substance* as: *that which is extended in spacetime.*
>
> Spacetime means the 4 dimensional space which is reality. In other words, the reality which is traditionally considered to be the domain of science. To put it crudely: everything which is, or the universe.

---

[473] Sorry

Let's use atomic concepts to illustrate:[474]
Each atom:

- Is an event in spacetime
- Occupies a unique co-ordinate position in spacetime.

No atom is the same as any other atom, because no atoms can occupy the same co-ordinate position.[475]

Atoms of the same state are interchangeable in spacetime.

Interchangeability is not the same as sameness. Sameness is a form, not a substance.

Nothing can exist as a sole substance. That which is sole "is" not:

How can a Substance be that which is extended in spacetime, if spacetime is a function of substance?

That is a fair and cogent question. The answer is given by the expressly hylomorphic character of all forms of inequality:

(1) To be Substance, is to be extended.
(2) To be extended is to be comprised in a form.
(3) Form requires plurality of Substance.[476]

[474] This is an aid to thinking only: don't go all quantum, yet
[475] Without change in state of the atoms
[476] One of the fundamental laws of ML

(4) Plurality of Substance creates null [T] field [=0]
(5) The "life" of Substance in plurality is a succession of null $[T]^n$ fields [=0]
(6) which is not Infinity ∞, because each null $[T]^n$ field [=0] bounds its domain of operation.

To analogise: Hydrogen and Oxygen are different: they are inequalities. Together they form water. Water is not something apart from that Substance, in mutuality. Water is both the form of that Substance, and that which defines that Substance, in its mutuality.

Time, in this analogy, is the water of all Substance. Different Substance only produces different waters. Same Substance produces temporally different waters, in each null $[T]^n$ field [=0] of actuality.

Do we now say, then that [I] is illusory, and that only that which is in [E] partakes in time [T]? No: [I] is infinity ∞ and infinity does not partake in time [T].

This we can prove. Hypothesise You and the beat synchronising. You may do so at 30 bpm. Or 1,000[477]; or 1 trillion bpm. *But you cannot synchronise infinity.*

That is why QM has the normalisation problem.

- By locating wave/particle events in a *continuum* of

[477] Assuming no physical limit on capacity

time (or disregarding time as a plenum altogether), the unadjusted equations are bound to approach infinity.

- To put it another way: by presuming time as a *continuum*, QM inevitably results in treating actuality as the cause of its own causality.[478]

Thus [I] is reality, as an actuality of interference Nodes. Invoking Chapter 4:

Form and Node in [I] Becoming

Potential Δ

The domain of [I], the domain of Becoming, is Infinity ∞.

Moments operate in Infinity ∞.

$$(A) \neq (-A) = [I]$$

Moments are in infinite succession and regression, simultaneously. To analogise imprecisely again, Moments can be thought of as the *fabric of Time*.

Moment operates as Potential Δ^n in Infinity ∞

$$(A) \neq (-A) = [I]$$

Potential Δ^n is a Dynamic of Moments in Becoming.

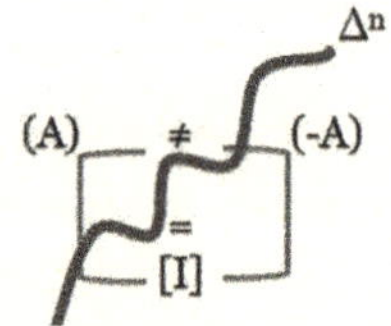

[478] more on this in *Completing Reality* Chapter

Interference <∑I>

An Interference <∑I> is not a singular occurrence.

Interference <∑I> is an aggregation ∑ of multiple [I] FoR_n

$$((A^1) \neq (-A^1) = [I^1]) \neq ((A^2) \neq (-A^2) = [I^2])\ [\sim]\ \Sigma(\text{FoR}_2)$$

for which we can now substitute and simplify:

$$(\Delta 1) \neq (\Delta 2)\ [\sim]\ \Sigma(\text{FoR}_2)$$

$$>$$

$$(\Delta 1) \neq (\Delta 2)\ [\sim]\ <\Sigma I>$$

Like Event, Interference is perhaps self-evidently, only effected in plurality.

There are no rules of Matrixial Logic as to the terms and conditions of such *interferences*. These are matters of inexhaustible possibility in the world.

Form of Forms: Node <∑I>

As stated earlier in this Chapter:

A Node <∑…] is

any convergence of two Forms in the same domain.

Hopefully, it is self-evident from all that has been written before, that:

an Event <∑E> and an Interference <∑I>,

cannot possibly be the same sort of phenomenon:

whether in the world, or in perception.[479]

[479] Here intended to be a vacuous catch-all word for whatever one might

As we said:

> Moments can be thought of as part of the *fabric of Time*.

To analogise (rather than analyse) *Potential* Δ^n is the "loop" in LQG; the zero-point foam.

- Being is hylomorphic to becoming, in the realisation of actuality;
- The infinity of becoming is bounded by being;
- Being interacts with becoming to form Nodal null field [=0] actuality.

[I] is infinity ∞ and infinity does not partake in time [T]. But actualised Moments Δ^n *do.*

Thus, [I] is not "in" time. [I] is the "gravity"[480] for time.

Each time you tapped, in the *Noming Experience,* you felt that annihilation [=0].[481] Each next tap, had a successive element to it, but also a novel element: as if each were the first time you had tapped. That is exactly why you are able to synchronise the BPMs: because you are only synchronising one beat "at a time".

This moves us to a profound, materialist and monist view of reality.

---

term as "subjective", or occuring inside humans, their minds, their feelings

[480] An awful conjunction: apologies

[481] $\{<\sum \mathfrak{C}_1> [=0] <\sum \mathfrak{C}_2>\}_a$

- That Substance in plurality subsists only in the form of [T] time frame, actualised by the interaction of that plurality, in opposition ≠.
- We, H^n can perceive Substance (in plurality) only through the [T] from of Substance, and where that [T] form is Nodal to our own Theta Θ axis [T] field;
- Yet, we H^n are able to "appear" to reality[482] because we appear Nodally: in each "objective" moment domain of time $[T]^n$.

We, H^n, are entailed to *entanglement* with reality by reason of our inborn capacity for actualisation of a temporal state {<∑₵$_1$> [=0] <∑₵$_2$>}$_z$ which can be annihilated (in "meeting external reality) at a Nodal point {<∑₵$_1$> [=0] <∑₵$_2$>}$_a$. Thus producing <∑₵$_1$> [=0] <∑₵$_2$>: a novel null field [=0] Nodal point.

We, as H^n, have the capacity to be "in time" with each timely manifestation of our universe. Each individual meets in the unique contingency, in time and space and "history" of that individual's being.

Robots or zombies may be *in* time, but only conscious selves have the capacity to be *of* time. We will treat more of this in the next Chapter.

---

[482] Anthropormorphising to make a point

Chronologic and Chaos

Readers fortunate enough not to have lost the will to live may by this point, may have noticed something:

> But where time is not a flow, but a quantised apparition of $(\text{Ↄ}^{n}) \neq (n\text{Ↄ}^{n})$ | , then each Event ₵^{n} can *synchronise* with its negation in Nodal *annihilation*:
>
> |
>
> |
>
> As the Chromal equation states:
>
> $$(\text{Ↄ}^{n}) \neq (n\text{Ↄ}^{n}) \mid \text{₵}^{n}$$

We have seen this symbolic notation before. It is something like the notation for Chaos:

> The Form may be in E or B.

- If a Bridge arises in E, then we denote that as:

  } External

- If a Bridge arises in B, then we denote that as:

  } Internal

> We then write an Si Bridge, using the notation:[483]

[Si: Martian≠/Venusian≠]	= Alien =	[Being]	} External
Disjunct	*Bridge*	*Form*	*Domain*

> If we want to propose a link an External with an Internal domain, we denote with:

[483] The contents in this example are frivolous, and the equation is not accurately solved

Ⅲ

So, for example:[484]

[Si: Martian ≠/Venusian≠] = Alien = [Being] } External

Ⅲ

[Si: Fear ≠/ Joy≠] = Stress = [Becoming] } Internal

Note that the | symbol does not denote a *formula,* but a propositional connection between formulas.

Chaos bridges into the universe of forms [E] and [I], only by means of Singularity (Si). Si is a bridge, and functions as such.

By contrast a Nodal null field [=0] actuality is a "gap": an annihilation state. That state is not a Singularity. It is not a bridge. We do not bridge from one null field [=0] to another. There is no bridge made from a continuum of time. Each "beat" of synchronous clocks is the first.

It is not that Chaos "fills" the gap. That gap is existential. It is a nothing which is something. An analogue of a force. But only an analogue, since it has no potential.

With these assertions emplaced, do we say that Chaos [C] does or does not subsist in $[T]^n$? The answer is clear, but contingent.

---

[484] The contents in this example are frivolous, and the equation is not accurately solved

Chaos is a function of imperfection of information. Were any H^n or any system which could function as H^n, possessed of perfect uniform information, then Chaos would cease to have possibility.

That information would need to be perfect at all beats of every synchronous null field [=0]. Which is impossible, since null field [=0] bounds all infinity ∞.

Thus, Chaos [C] is existential by reference to the inherent imperfectability of information, functioning in all and any times $[T]^n$.

In other words, [C] and [T] are in universal symbiosis. Not *in* the structure of the universe, but *as* its very structure.[485]

From which we deduce, by definition, that no Si can produce [T]: because only plurality in opposition ≠ can produce [T].

All this, surely, accords with our deepest intuitions. Time is order: not as a continuum, but as synchrony. It is that which orders infinity ∞, and thereby bounds it.

---

[485] There is an analogy here with the Dirac Equation and the derivation of anti-matter

Experience: UniBeat

Setup:

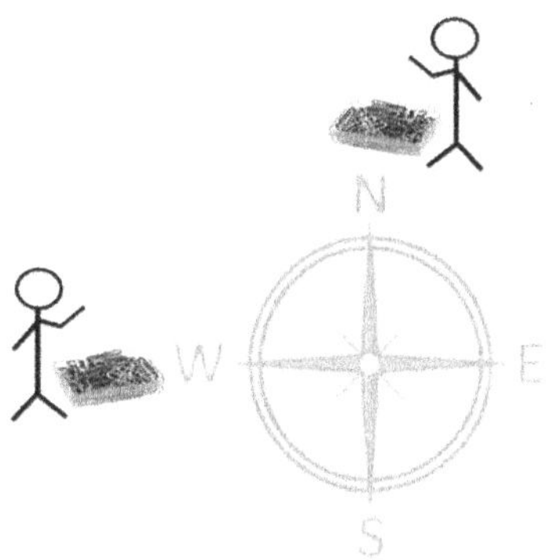

You're located at the West pole of this compass.
You're going to be dropping paperclips into an empty box.

Your Twin (not You_2) is located at the North pole.
Twin is going to be dropping paperclips into an empty box.

Then:

You and Twin are each going to:

- access your respective You_2
- set a 30 BPM metronome
- set your internal time $[T]_0$ to the same Beat

You and Twin can now turn off the sound, throw away the metronome. You each keep the Beat in your respective You_2

It is now time $[T]_0$

Step 1:

- You and Twin drop the first clip; drop a clip for each beat
- Do 5 beats

Stop

<u>Discussion</u>:

(1) Obviously, you and Twin each have 5 clips in your box.

Step 2:
You get in your aeroplane.
You're going to fly to the top of a high mountain.

Twin will stay put.
You and Twin each have a clock set at 12.00.

You and twin are at $[T]_0$
Drop the first clip

Now:

- You and Twin drop the first clip; drop a clip for each beat.
- Do 5 beats (the mountain is close by).

Stop.

Discussion:

Q: Do You and your Twin still each have 5 clips in each box?

(1) You and Twin each have a Clock. They are set in timeframe $[T]_{1.}$ We know those Clocks would be out of sync. Your Clock would have slowed down, relative to Twin's Clock, in timeframe $[T]_1$

(2) The You_2 Beat clocks synchronised inside the heads of You and Twin are set to timeframe $[T]_0$

(3) You got $[T]_0$ in sync by reference to $[T]_1$: but then you threw $[T]_1$ away

(4) Obviously, You and Twin could simply have synchronised without using the external clocks. Just like the birds in a flock. You could just tap each other, or say "Tock" to each other: till your You_2 is in snyc with Twin's You_2

(5) So, changes between relative iterations of $[T]_1$ affecting the clocks, don't affect the You_2 Beats in timeframe $[T]_0$

A: So, the answer is Yes: You and your Twin still each have 5 clips in each box

Step 3:
You get in your rocketship.
You're going to fly to the next solar system.
So You get there during this Chapter, that will mean going at say half the speed of light *C*.

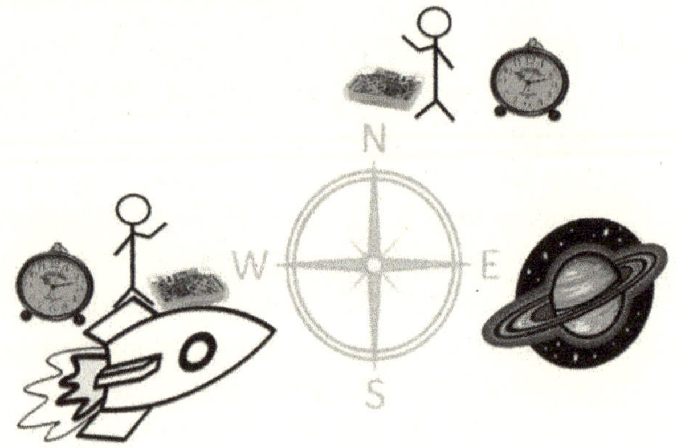

Twin will stay put.

You and Twin each have a clock set at 12.00.

You and twin are at $[T]_0$
Drop the first clip.

Now:

- You and Twin drop the first clip; drop a clip for each beat.
- Do 5 beats.

<u>Discussion</u>:

Q: Do You and your Twin still each have 5 clips in each box?

A: Yes: since changes between relative iterations of $[T]_1$ affecting the clocks, don't affect the You_2 Beats in timeframe $[T]_0$

Now, we have a problem. As Einstein wrote in 1905 *On the Electrodynamics of Moving Bodies:*[486]

> From this there ensues the following peculiar consequence. If at the points A and B of K there are stationary clocks which, viewed in the stationary system, are synchronous; and if the clock at A is moved with the velocity v along the line AB to B, then on its arrival at B the two clocks no longer synchronize, but the clock moved from A to B lags
>
> $\frac{1}{2}tv^2/C^2$
>
> behind the other which has remained at B by (up to magnitudes of fourth and higher order), t being the time occupied in the journey from A to B.

This Twins Paradox[487] has justly become famous over the last century. The time dilation effect on clocks synchronised to $[T]_1$ has been established by experiment.

The *Unibeat Experience* gives us an uncomforatble feeling that perhaps the logic of special relativity has not, after all, reached its conclusion.

If $[T]_0$ which You synchronize with Twin is not subject to time dilation, whether by acceleration, distance, or anything else, then the universe may not actually work as we have thought it to work, for the last 115 years.

---

486 http://www.fourmilab.ch/etexts/einstein/specrel/www/
487 https://en.wikipedia.org/wiki/Twin_paradox

Any $[T]_0$ is operating in our brains: which are made, ultimately, of elementary particles in coherence. Does that which holds for our brains, also hold for those elementary particles? If so, then we have opened a pathway to discovering why QM and relativity do not cohere.

That pathway also branches into an explanation of Bell Inequality violation which does not rely on FTL "spooky" action at a distance: synchrony is not FTL.

Can we conceive of rules set at Point 0, which allow for changes in the conditions in which synchronous $[T]_0$ operates, to affect operations referenced to $[T]_0$? Merely in stating the question, the answer seems inevitably to be yes.

That then leads to a radical re-appraisal of the technology needed to effect quantum computing. We don't need 0.5 above absolute zero K to effect quantum superposition: we just need to synchronise action of 2 event frames to a $[T]_0$, and then apply $[T]^n$ time frame interactions: $<\sum ₵_0> \mid <\sum ₵_n>$. We could, perhaps, take further control of infinity from Chaos. Infinity, however is not the same as Chaos. In the domain of Chaos there is neither [E] nor [I]. We only know of Chaos through Si, which can itself be manifest only in its effects, not as a cause.

We conclude with this. Time is not a river, a flow, a continuum. There is no Time such as to effect order. Time is the order actualised by synchrony of $₵^n$ Nodes, which appears to be successive, but attains that manifestation only as the apparition of annihilated nullity.

Each of us has our subjective sense of time. It is from this that our conceptions of time emerge: as the "alienated" objectivisation of those subjective senses.

The real *objectivity* of time, to each of us, is in the human capacity to synchronise $\mathfrak{C}^n$ Nodes, with external $\mathfrak{C}^n$ Nodes, thereby producing a null field [=0] actuality. Again, and again and again: each time for the first time.

For this, no clock, no universal time, no orchestral conductor, is needed. It is not merely the "blind watchmaker", it is the blind time keeper.

Matrixial Chronologic, is the logic of synchrony: the forms of inequality of $\mathfrak{C}^n$ Nodes self-organising in "parallel".[488]

Infinities annihilated and bounded by a null field [=0] of actuality: an existential gap | ⇔ |.

Time is only what its synchronous Nodes can manifest, by their extinction.

[488] please don't take this literally

CHAPTER 8

MATRIXIAL ANTHROPICS

We have so far in the book been considering the "objective" operation of Matrixial Logic. Setting out the structural dynamics of the matrix, and considering some of its elementary implications and applications.

In this Chapter, we adjust perspective. We consider how Matrixial Logic affects our view of ourselves: as beings who are *human*.

In this Chapter, we are not setting out to provide a comprehensive structural account of You. That is for *Secret Self*.

Here, we sketch some outlines of the Matrixial Logic framework of selfhood, bearing in mind that Self is everybody and all human collectives[489] are selves.

Meet You

We have now encountered these forms of inequality:

Basic Equation	**Form**
(A) ≠(nA) = [E]	Being in [E]
(A) ≠(-A) = [I]	Becoming in [I]

[489] cf The Borg in *Star Trek TNG* and *Voyager*: a fascinating pseudo-journey in loss and reconstitution of Self-hood

<∑*>	Nodes
∑Δ^n ≈ ∞	Moments in Infinity
m̄	Node Points
Φ	Moment Interference Points
(+A) ≠(-A) = /S/	{E} and {I} Scalar Axis Fields
(T) ≠(nT) = [Θ]	Theta Axis Field
(Ɔ^n) ≠(-Ɔ^n) | ₵	Temporal Field Gap

Together, with their antigones and interactions, we have attained a 3H model of the Mentative Self, in Time, and bounded by Chaos.

We have, so far, looked at how your Subjective Scalar /S/ {E}/{I} Mentation interacts with your Equalising Theta Θ axis field.

Now, we turn to consider the realm of Emotion:

(ϲ) ≠ (ɔ) ∑Ω

Referencing a visual or auditory experience, this is You.

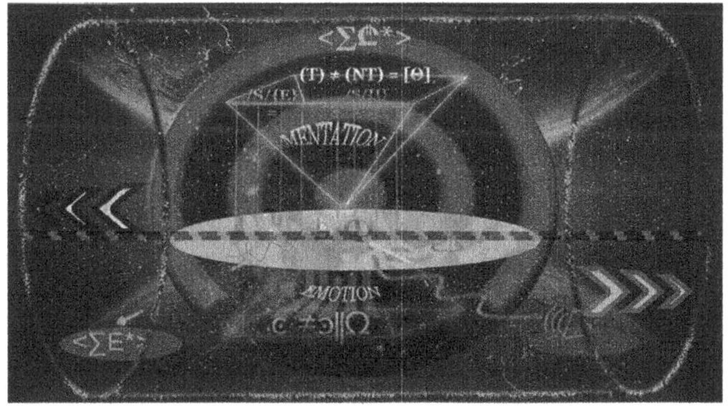

(ϲ)

ⵎ /S/{E}

∑Ω (T) ≠ (nT) = Θ ≈>

Φ /S/{I}

(ɔ)

<∑₾*> ⇔ (Ɔ ⁿ) ≠ (-Ɔ ⁿ) | ₾ ⁿ ⇔ <∑₾*>

Emotional interactions give an unusual form of inequality equation. We sometimes use | | in place of ∑: the dipoles effect the equivalent operation to the summing.

The antipoles (ϲ) ≠ (ɔ) are neither opposites nor negations of each other. Thus, they do not produce a form, but sum over ∑ as Ohm Ω.

We can immediately notice the similarities with Chaos [C]. But what is also immediately different, is that Ohm Ω is not producing Singularities (Si).

We will set out some further assertive rules, before proceeding:

- Ohm Ω is directly mediated by /S/
- Ohm Ω is not directly mediated by Theta Θ
- Ohm Ω is accessed by Theta Θ only indirectly via /S/.

The implications of these rules fall to be addressed later in this Chapter, and more comprehensively in *Secret Self.*

Sums of Inequality

We are encultured to regard emotion as the seat of Identity. What You_2 *Experiences* demonstrate evidentially, is what Matrixial Logic dervives as a proposition. That:

Emotions have no self-descriptive state

What we mean by this admittedly gnomic utterance is most easily described by negative comparison.

Our subjective /S/ Scalar axis conducts (+/-) examinations of qualities as quantities: more pleasurable / less pleasurable. And so on.

These are conceptual constructs, which grow in sophistication from the acquisition of language. These concepts are, of course, culturally contingent.

This poses problems for views of evolution,[490] and ethics.[491] "Pleasure" is not a fundamental or foundational fact, which can then be used as the substructure for a behavioural or ethical theory.

Pleasure is nothing more than a Scalar /S/ interpretation of Ohm Ω states. How that measure is applied, is a social construct.

---

[490] Hammeroff; and neo-Darwinian "survivalism"

[491] Utilitarianism and its successors

There is an asymmetry here:

- We are subjects of physically sensed pain and pleasure (a hot stove touch; a cold water refresh);
- We're only describing differences, in retrospect, and in Scalar terms: can elect to decribe such differences differently;
- Both are real: but reality functions differently in our interactions with externality; and our interactions internally.

Emotions do not self-describe. Emotions are mechanical processing of input data: whether provided externally, or internally.

Asking whether one's emotions are "happy" is like asking whether one's blood circulation is "happy". Happy is not the same as functional.

One's emotional mechanics can, of course, suffer dysfunction, just as can one's bio-organic processes. The science of such interactions is Biomatrixology. We will deal with such matters in *Secret Self.*

All this appears, at first sight, deeply counter-intuitive. Surely one's lived reality is comprised exactly of emotional experiences?

We agree. It is. But experience is not the same as a *measure of value.* All sentient life has experiences. How sentience measures experience is the difference between consciousness and awareness. Animals are undoubtedly sentient and aware. But they are not conscious.

How do we know that? Very simply: they do not have the capacity for an inter-relationship of Ohm, /S/ and Theta:

World <=> Ohm ≠ /S/ Σ Θ ⇔ World

<I>

It is this relationship which comprises human conscious experience, in awareness.

Well then, how do we know that a dog does not have this relationship? That is because it is self-evident what a dog does and does not have:

Domain / Field: Ohm
Relationship: Yes, but obviously limited. A dog has sensory apparatus superior to humans in certain respects (olfaction, auditory). A dog's sensory apparatus is also inferior to humans in other respects (vision, and auditory). But a dog's Emotional field is (unsurprisingly) limited to its needs of life as a dog.

Domain / Field: /S/
Relationship: No. A dog does not make value judgments about its emotions. A dog reacts and enacts behaviourally, and that behaviour is conditioned by its environment (which can include us humans).

Domain / Field: Theta
Relationship: No. Dogs, like other animals, have a *behavioural simlacrum* of Theta. Unlike Theta, there is no mediation between that simlacrum and emotion: there is no /S/. If you wish to test

this proposition, try getting a dog to use You_2 in order to facilitate new task management.

These few words point to the emperor's new clothes of behaviourism and A.I.[492]

Behaviourism appeals to our social intuitions: we feel attracted to seeing that which behaves like, as being like.

But this is a most primitive intuition. Rather than being an apogee of scientific thinking, it is a reflection of primal encodes of similarity-seeking, and difference-opposing (by fight, eat, or flight). It is the logic of a Paramecium.[493] Or a bat.

The Ohm antipoles (ϵ) ≠ (϶) are neither opposites nor negations. Therefore, under ML rules, no form can arise.

An obvious example of a (ϵ) ≠ (϶) type relationship is the Systolic-Diastolic heart functions:

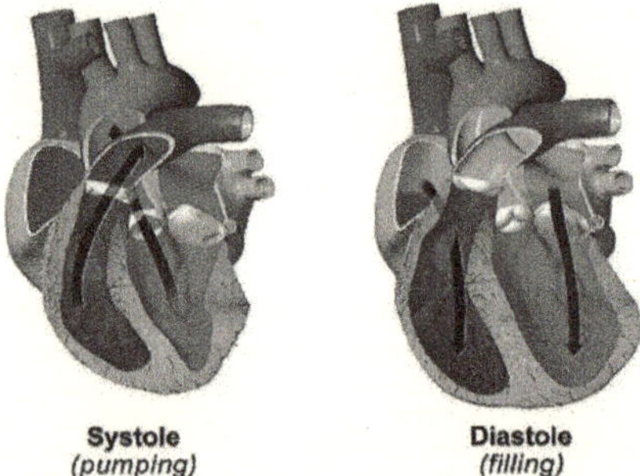

Systole
(pumping)

Diastole
(filling)

[492] With due acknowledgement to Sir Roger Penrose
[493] https://microbewiki.kenyon.edu/index.php/Paramecium

Medicine places these functions under a scalar rule of measurement. Thus we have a regularity of S <120; D <80.

That scalar codification for diagnostic reasons is of no concern to the heart. It jut carries on pumping and filling. Clinical intervention to deal with irregularities in heart function, does of course affect the function: which is why they are carried out.

While the seat of emotion was once thought to be the heart, we have long since known, scientifically, that this is not so. We do know that emotional engagement can have systolic-diastolic consequences. We also recognise that emotional engagement has effects throughout the body.

This provides us with our first clue as to what Ohm Ω actually is.

The Scalar /S/ axis field effects recognition and judgment (quantising and qualifying) in Mentation. These are data (⍰) equalised in the Theta Θ "shopping basket checkout" function.

The Ohm field performs a *distributed referral* function.

That is not the same as recognition and judgment. Reference involves recognition, but it is not a judgment function. Emotions do not judge: they register. Emotions have no frame of external reference, by which to make judgments. Nor do emotions have a "memory".[494]

Emotions are recognitions of the present: which is exactly the present that the Scalar /S/ axis cannot see: because /S/ can only see the past.

We thus begin to construct a *temporal* role for emotion:

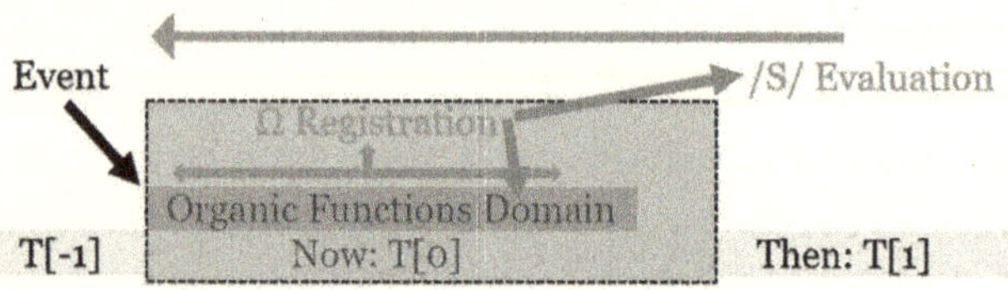

Time [T]	Function
-1	Event
0	Organic Functions Registration
0	Emotional Recognition
+1	Scalar Evaluation

We thus begin to unravel the mystery of the Libet experiments.[495]

Emotion is thus fixed in the present, and references towards the future.

---

[494] Which makes trauma-based cognitive therapy something of a dead duck

[495] We discuss this in *Matrixial Chronologic*, and in *Secret Self*

Emotion functions:

- As a holistic, emergent from Organic Functions
- Distributing Ohm Ω registration information to Organic Functions
- Referring Ohm Ω registration information to /S/ evaluation functions.

Ohm Ω registration information can therefore be summarised as this primary function:

Registration

<= Feed Back => Feed Forward

Emotion of course uses the brain as its operating machinery. If we wish to demarcate some part of the neural mechanism as particularly "emotional", then begin with the spinal cord and its attachment to the brain stem. It is here that the brain begins, morphologically.

Obviously, many parts of the brain are implicated in emotional Ω registration. Indeed, it is difficult to find any parts of the brain which aren't. This is one of the reasons why neuroimaging studies are naive, in the absence of a functioning logic of human consciousness states.

Experience: Thirst

Imagine yourself being really thirsty.

Maybe imagine that feeling in a context of being in a desert, or in the middle of the ocean on a raft.

Just allow your imagination to sink into the feeling, until the feeling becomes powerful in you.

Step 1:

- Think of 5 seconds in the Past
- Think of a cup of water
- Locate that cup of water at that 5 seconds in the Past

Now:

- Try to see yourself reaching for the cup and drinking

Stop.

Discussion:

(1) You can't reach the cup.

(2) It's like there's an invisible barrier stopping you from reaching it.

Step 2:

- Think of 5 seconds in the Future;
- Think of a cup of water;
- Locate that cup of water at that 5 seconds in the Future.

Now:

- Try to see yourself reaching for the cup and drinking.

Stop.

Discussion:

(1) Now it's easier.

(2) There's no barrier. It does feel "blurry", as if you can't quite complete the reaching / drinking experience.

This *Experience* illustrates the *temporal* role for emotion:

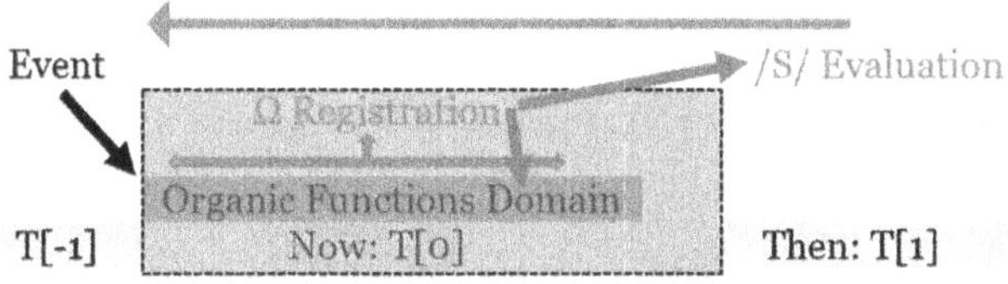

(1) You simulated an Emotional-Mentative state of thirst: Ω ⇔ /S/
(2) Your Ω cannot distinguish between external and internal stimulus[496]

In Step 1:
(3) Your /S/ evaluates backwards in time: so envisioning the 5 seconds Past cup is easy
(4) but: your Ω cannot "see" the Past

In Step 2:
(5) Your /S/ envisions a present state projected as a "future" state;
(6) Your Ω functions as a referral-to-the-future (your /S/ evaluation) function. In other words, your Ω is always *forward looking* in its referral function;
(7) So, your Ω can engage with that fictive "forward" experience

This now begins to explain the open-textured character of

[496] Ω couldn't function if it had to make such determinations every millisecond of your life

the Ohm equation elements:

$$(\epsilon) \neq (\ni)\ \Sigma\Omega$$

It would be a natural inference that /S/ acts in relation to ∑Ω referrals, as Theta Θ acts in relation to /S/{E} / [I} referals.

But that's obviously not so. As we have demonstrated in previous Chapters, Θ is the Equaliser axis field: once items are in the Θ shopping basket as $(\mathrm{T})^n$ they cannot be removed.

By contrast, /S/ is popping data in and out of its trolley, changing the dimensions of the trolley, and travelling the imaginative dimensions of the universe. With only such rules as /S/ chooses to abide by.

Once we understand and acknowledge that /S/ can only function as an evaluation of past experiences,[497] we appreciate the necessity of the Ω function.

Without it, we would see everything and nothing. An indiscriminate, undifferentiated landscape of the past. With referential axes or points: a landscape from which we could not learn.

---

[497] *Past* including the domain of the millisecond

Experience: Looky

Look at this graphic:

Step 1:

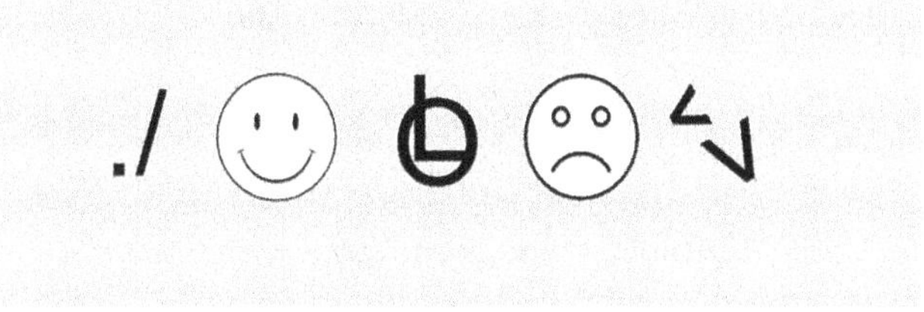

Just think that there will be no questions or instructions about the graphic at all.

Now:

- Just allow your eyes and attention to move between what you can see;
- Relax. Go with whatever comes naturally to your looking.

Stop.

Discussion:

(1) Your eyes became more focused on the 2 faces;

(2) It was like you wanted to just see the 2 faces, but you kept getting distracted by the abstract images.

Step 2:

Now:

- Try to focus on the abstract shapes.

Stop.

Discussion:

(1) Your eyes and attention kept getting dragged to the 2 faces;

(2) The abstract images kept "fading" out.

The *Looky Experience* is about demonstrating how your Mentation functions, without an Emotional Ω input. That is quite hard to achieve: how could you switch off your Ω?[498]

We can achieve part of that effect by giving you comparative stimulus:

- There is no Ω registration for the abstract designs;
- There is compelling Ω registration for the happy and sad faces.

We know this latter datum from early natal development studies. So, for example, the 2008 study by Nagy:[499]

> In most of our social life we communicate and relate to others. Successful interpersonal relating is crucial to physical and mental well-being and growth.
>
> This study, using the still-face paradigm, demonstrates that even human neonates (n = 90, 3-96 hr after birth) adjust their behavior according to the social responsiveness of their interaction partner. If the interaction partner becomes unresponsive, newborns will also change their

---

[498] You can't: instead you can freeze the reaction of your /S/ to Ω input referrals, by fixing these in Shadow Slices: a dysfunctional simulacrum of Theta : see ***I Want To Love But*** (2019) and ***Secret Self***

[499] Nagy, E. (2008). *Innate intersubjectivity: Newborns' sensitivity to communication disturbance. Developmental Psychology, 44*(6), 1779-1784. https://doi.org/10.1037/a0012665

> behavior, decrease eye contact, and display signs of distress. Even after the interaction partner resumes responsiveness, the effects of the communication disturbance persist as a spillover.
>
> These results indicate that even newborn infants sensitively monitor the behavior of others and react as if they had innate expectations regarding rules of interpersonal interaction.

Isn't that fascinating? No matter how old you are when reading this book, you still have exactly the same emotional attachment patterns emergent of your reality (as a human), as you did when just 3 hours old.

Our Emotions discriminate our information acquisition. They choose the focus of our visual and aural attention. Emotions act as channels for sense-Qualia, choosing amongst what we are to notice, and not.

Asking "why" our Emotions function in this way is as helpful as asking why our heart is constructed as it is. It is the machinery humans are born with: literally, as the above reported experiment shows.

But our emotions are not making Mentative judgments in that selection. Our Emotions don't "think" to themselves:

- I don't want to look at that;
- I should look at that;
- It's right to look away;
- I shouldn't avert my gaze.

Our Mentation can and obviously does train our Emotion to operate programs, by reference to circumstances prescribed by Mentation. That can take a long time: ask any parent.

This brings into notion, an interesting idea. If there is one part of the human process which is most "computer" like: it is Emotion.

- Ω does not reason.
- Ω is a *distributed referral* system: [+ more] information in => [- less] information out.
- Ω runs the default program [+ / -] unless and until instructed to run some other program.
- Ω does not consider, appriase, or evaluate the information, in any independent frame of reference.
- Ω is no more a reasoning process, than the mechanical action of a flour sieve.

That process of information selection, however, gives effects to inequality. Our Emotions are a *difference machine*:

$$(\epsilon) \neq (\backepsilon) \sum \Omega$$

As we unravel the inner complexities of the Ω field function, we begin to appreciate the raw power of this equation. Unbounded by form, it is the summing of differences which allows the application of Scalar /S/ evaluation.

It is apparent that $(\epsilon)^n$ values are Nodal, or Δ^n Interactions.[500] When we apply the inequality equation dynamics in full, we begin to see something very interesting: Chaos functions, as they become determined by interaction with bounded infinity, assume the properties of probability functions.

Indeed, given the temporal function of Ω, that is hardly surprising. Our Mentative Experiences are ∑probability functions, in | ₾ present time $[T]^n$ of Ω referral distribution operations. This explains much about a lot.

To provide a slightly specious, and analytically inaccurate metaphor, in aid of overall understanding:

Ω	/S/	Θ
baby	infant	adult

Even this metaphor must be accompanied by the caveat that all 3 function states are present in the same ontological frame of reference.

This is not a developmental map. The newborn has these 3 elements in its matrix, just as does the nonagenarian. The content and context of the inter-relationship of these Matrixial elements changes, in the course of life experience.

This open-textured character of the Ω function may

---

[500] See Chapters 4 and 6

call into question its utility. To answer that, beyond the thumbnail presented here, is the labour of an entire book: *Secret Self*.

Irrationality

As we said in *Matrixial Chronologic*:

> Each of us is a time machine. Just as with the grammar of language, we artifice a symbiosis of meaning from /S/ axis fields, under a tensor field operation: the Theta Θ axis.

Whether the neurophysical mechanism for this process is a quantum superposition collapse occurring within microtubules,[501] is a hotly disputed issue. That microtubule connectivity is implicated in the neurophysical processes of perception appears well-founded.

However, the logical gap between mechanism and "consciousness" is not solved merely by positing "consciousness" as a protean property of the universe. That mode of argument results in an infinite regress. It is the classic trap of seeking a Singularity: an Aristotelian unmoved mover.

We note with interest that the theory leads also to a notion of a Platonic set of ideas in the universe which are being realised in conscious moments. There is a tantalising idea here: it has captivated for 2,500 years.

---

501 The Hammeroff-Penrose theory

Recall from the *Languini* Chapter, the *I-Pen Experience* ("IPE"). The IPE, in practical trials demonstrates a para-Platonic outcome.

When trial subjects initially begin to use their Theta Θ axis landscape to write with their non-usual hand ("NUH"): *they all write their letters in the same way.*

Obviously, there are minor deviations, according to motor sensitivity and so on. But what we do see is a pattern:

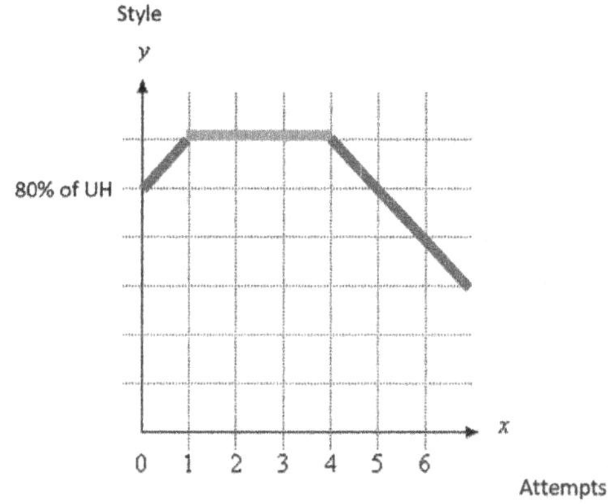

(1) The first attempt usually gets the NUH output to function around 80% of the quality of the Usual Hand ("UH").

(2) In the next attempts,[502] NUH output reaches 100% of UH.

[502] Within the same few minutes

But the, here is the remarkable data:

> (3.1) from the 1st NUH attempt, through a motor-skills adjustment over the next 3-5 attempts, the [A B C] letter formation is nothing like the UH; and
>
> (3.2) *the NUH letter formation by every trial subject has remarkably similar quality to every other trial subject.*
> Then, as the trial subject integrates the new motor-skill with their Subjective sense:
>
> (4) the NUH writing acquires its own distinctive character. It becomes personality expressive: although it is calligraphically completely different from the UH style.

We also observe that, once the NUH motor connection to the Theta Θ axis field has been established, at trial 1, the trial subjects report being unable to revert to their previous unskilled state.

As soon as the NUH picks up the writing instrument, the trial subject feels compelled to write "properly": they cannot de-learn the NUH writing skill.

This has obvious implications for early-years learning. We will follow these up in *Secret Self*.

We also note that the trial subjects report that it is *as if some other person is doing this writing for them.*[503] The IPE therefore has obvious implications for parapsychology studies. For example, it obviously explains the Ouija

---

[503] We find the same reported perspective in the *Noming Experience* and all other iterative uses of You_2

board phenomenon.[504]

There's only one way to draw a triangle, or a square, in the sense that the primary geometric properties must be respected.

But there are infinite ways to form letters [A B C]: they are open-textured graphics.

So it is a conceptually challenging fact that: *the NUH letter formation by every trial subject has remarkably similar quality to every other trial subject.*

Now, we can say that they are, by some dint of memory or other cognitive process, simply accessing "perfect" examples of letter forms, with which they are familiar from everyday life. But that supposition has difficulties. All printed (including digitised) lettering is Font-based. All fonts have distinct ligatures. In any event, we know that we never see the same thing exactly the same way twice.

This aspect of the IPE implies useful further investigation. As a cursory prediction, the substantially similar NUH graphics will be demonstrated to exhibit regularity based on the Fibonacci sequence.[505]

[504] The cup is moved under a Theta brain state

[505] For more on this see: https://en.wikipedia.org/wiki/Golden_ratio#Relationship_to_Fibonacci_sequence

If it turns out that this is correct, then this will be important. It will demonstrate[506] a unifying correlation between:

- Mentative states;
- Their link with willed motor function;
- Their communicative link with external mentative states
- Linkage of both with fundamental laws of material physics; and
- Biological organisation.

We will have located the irrational number which is fundamental to Fibonnacci, and which "drives" the Turing / Mandlebrot solutions to chaos, as a referential *qualium* of mentative states.

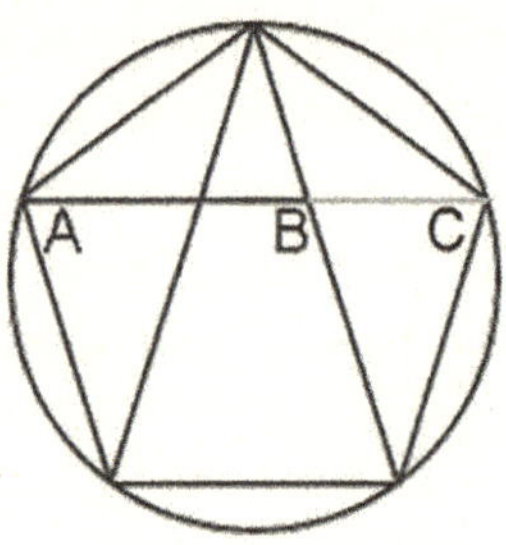

We know that Hammerof's microtubules are organised in Fibanacci spirals of 13.[507] We know that microtubules are ensembles which both self-organise and organise cells[508] and particularly neurons. We know that the physical building blocks of proteins, amino acids and nucleotides

506 Perhaps for the first time

507 A physiological datum independent of Hammeroff's theories

508 Not red blood cells

(RNA, DNA), are like microtubles, self-organising from dimorphic / dipolar reaction to water.

That's a kaleidoscope summary of vast issues. Consider the coalescence of these matters matrixially, and we appear on the threshold of a bio-unified map of conscious existence. That is very much a project for another book.

Identity Parade

One aspect of the "Hard Problem of Consciousness"[509] is the intractability of personal identity. We are back in the ever-fluxing river of Heraclites and the substance problem of hylomorphism that Plato and Aristotle attempted to solve.

The atomic substance of You changes moment by moment. Your thoughts change likewise. How, amidst such changes can there be an "I"? Is such "I" an illusion, as the Churchland-Dennett axis of philosophy contends. Or is "I" an epiphenomenal substrate, incapable of further elucidation: somewhat akin to Chaos in Matrixial Logic.

The *Noming Experience* evidences that both materialism and idealism are mistaken.

Recall, that in the *Noming Experience* you were able, by "switching on" Theta Θ, to synchronise instantly, with a beat.

[509] © David Chalmers

Now, you're not really "switching on". It's there all the time. You're just focusing and noticing.

Both[510] axes would have to admit that:

- The beat frequency; and (ii) the time counter; are manifestations of objective, material reality. They are computable functions, which are indeed effected by computer.[511]
- The finger-tapping is an act of will[512]
- The thought which impels the finger tap is synchronised with the BPM, such that the finger tap coincides with the BPM.

Just such mentative synchronicity with defined temporal events in the external environment, is the axis of Identity:

(Human) ≠ (nHuman) = (Beat) Synchronicity [T]

We can generalise the equation:

(H) ≠(nH) = Synchronicity [T]

The problem of Identity has always subsisted in the assumed paradigm that {I}, although porous, has no external "linkage" which pervades through time: where time is seen as τ, a linear flow.

---

[510] Materialist and Dualist

[511] Yes: it is conscious beings who make all this happen, as a matter of fact. But that matter does not ultimately assist the Dualist axis

[512] "Free" or not

That is hardly surprising. The premises exclude any linkage implication or entailment.

Once time is understood as [T]: quantised packets, the linkage is easily demonstrable. As in the *Noming Experience.*

Indeed, upon reflection, we see that such quantised interactions, which are provably "objective" occur througout every moment of our lives: from birth to death.

It is through such quantised interactions that we construct our subjective sense of the world, as "my" world.

The infographic reflects, in much simplified conceptual form, the dynamics of this relationship.

When You beat with your fingers you know that it is You, and nobody else, that is doing it. Yet, each beat identifies

You with a provable external reality.

Thus, your Scalar /S/ axis field is able to perform exactly the same function as when you see the dog, or hear the dog's bark.[513]

That interaction is what allows, and constitutes, personal Identity in time [T]. To represent it unidimensionally (and therefore only approximately):

World <=> Ohm ≠/S/ ∑Θ⇔ World

Where:

- The linkage: World <=> Ohm is receiprocal, but outside Mentation.
- The linkage: Θ⇔ World is entirely of Mentation.

David Chalmers, Dan Dennett, and Pat Churchland,[514] are invited to try the *Noming* Experience for themselves. What they will report, as all who [515] have undertaken the *Experience* report, is that they feel "nothing".

What You$_2$ *Experiences* demonstrate, is that noticing the operation of Theta Θ axis field functionality, allows also noticing of the Ohm Ω matrix.

What you notice is difficult to self-describe. Let's do an *Experience*, to bring it to mind:

---

[513] see *Languini* Chapter

[514] Snd everyone else

[515] Anecdotally; not clinically

Experience: Mirror

Get a pad and paper.

You'll be using your normal writing hand.

Look at the words "This can be cloudy"

Mirror-Writing: is writing those words from right to left, such that each of the letters is horizontally inverted. When you hold them up to a mirror, they show as "normal".

Step 1:

- Write the words in Mirror-Writing

Stop.

Discussion:

(1) Unless you have previously practiced Mirror-Writing, you will find this hard to do.

(2) You might manage, but with false starts, and it will not look very pretty by the (long) time you've finished.

Step 2:

Connect with your Theta Θ axis field: You_2.

If you're right-handed, envisage the connection as coming from your left-shoulder side.

Sense the connection between You and your head, hand and pen.

Now:

- Write the words in Mirror-Writing

Stop.

Discussion:

(1) The Mirror-Writing simply flowed.

(2) You didn't need to try: it just happened.

Now, quite apart from the matter that this Experience falsifies the position of materialists and dualists,[516] revisit the history of your head and notice how you *felt*, during the *Experience*.

Match the ideas in these following words to what you sensed:

- It's like someone else is writing, but I know it's me;
- It feels like "super-me".

fundamentally, it feels:

- Calm;
- Balanced;
- Peaceful;
- Powerful.

This all occurs because you are achieving internal *congruence* between your Ohm Ω matrix and your Theta Θ axis field functionality, such that your subjective Scalar /S/ self is balanced in dynamic harmony.

You are experiencing a dynamic harmony in form of inequalities.

[516] See Chapter 10

It is this dynamic harmony that you locate, with absolute certainty, your sense of personal identity "I".

This is what takes the argumentative paradigm beyond Descartes and Turing. That paradigm is reminiscent of Schroedinger's cat: there is your self-hood, within the box of unknowability, and unverifiability. The difference being that we can never open the box.

Matrixial Logic, in its application to reality by such *Experiences*, demonstrates Theta Θ state interaction between You (a sovereign self entity)[517] and You (an actor in the world of external reality).[518] It is the same You.

You subjectively will the interaction of your objective brain[519]/ motor functions with the external world, utilising instantly a new skill. You can see the result, on paper: a result which the whole world can see. And you "feel" great about it: you enjoy emotional congruence.

And all of this occurs in synchronous time [T]: $(\text{Ɔ}^{n}) \neq (-\text{Ɔ}^{n}) \mid \text{₾}^{n}$.

Your Identity as <I> is "fixed" in each interval of synchronous time [T]. There is never a moment when it is not: until your last breath.

---

[517] The idealist account

[518] The materialist account

[519] And, as materialists must admit: high-order cognitive brain function

Being, Now Becoming

We have set the proposition that our Emotions are not inhabited by meaning, and that all meaning is invented in our Scalar /S/ axis fields.

Our Emotions are a sensory apparatus: a machinery for channelling sensory inputs from the external world.[520] Our Mentation can use that Emotional machinery, and of course does, all the time. A feature of our Emotion is that it cannot distinguish between external and internal input.[521]

Putting it in this way, our Emotions are as exquisitely sensitive as our nervous system, and equally dumb. Emotions sense, but without cognition.

Our Mentation is in a conditioned, but not determined, relationship with our Emotional output. Further, our Mentation can, and manifestly does, manipulate our Emotional machinery: but there are limits.

The boundary conditions of Mentative manipulation of Emotions is a subject addressed with more particularity in *Secret Self*.

It is precisely because our Emotions are a sensory apparatus, without cognition, that they are able to engage in temporal traction.

[520] Often referred to as Qualia

[521] That is a datum of psychological experiment

We know, and have demonstrated in *Experiences* in which we ask you to manipulate cognitive objects in the landscape: that your Mentation cannot alter the past. We no doubt knew that in any event. But we had paid insufficient attention to the fact what everything we see, hear and sense in our external environment, is in the past.

Indeed, the same is true of our internal environment. Any sense stimulus happened just then. It is not happening now. That which happens now is an iteration, a Moment Δ^n in that river ∞.

Our Mentation is limited to the past. Our Emotion is coupled to the future.

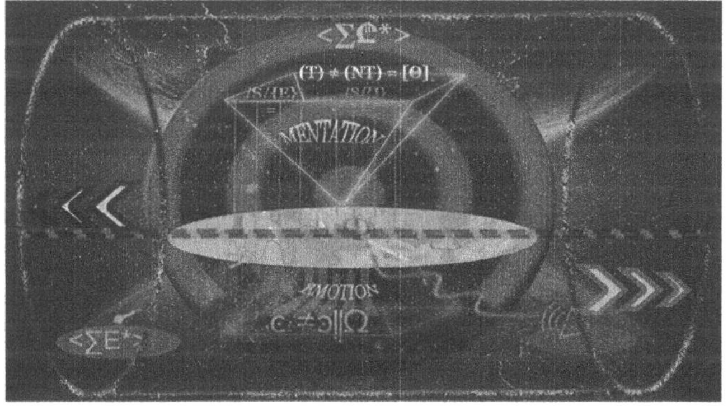

- The Ohm Ω domain functions as forward-looking;
- The Theta Θ domain functions as backward-looking.

The Scalar /S/ axes act as our "rear-view mirror". Ohm

Ω acts as our headlights. Both see more clearly than the other, and less clearly, in different respects.

Our Theta Θ is our conjugation with external time [T]. Theta Θ is not "in" the present. It defines our present, in every passing moment.

We cannot, however hippie and "now" aware we become, manipulate Theta Θ. We cannot even observe it from "outside", which is what a "now" awareness would require.

As you will have noticed from your *Experiences*, when you do surrender awareness to operation of the Theta Θ axis field (You_2), you become powerless. You cannot control Theta, you can only observe.

But that observation of Theta operating through your Mentation, Emotion and Organic state, recognises the sublimity of dynamic harmony. That power of that present: of you, but not for you.

We cannot "be in the now" in our Emotion Ω. We can try to freeze our Mentation. We are actually good at it. That effortless escape into Shadow Slices is one of the biggest problems that Restorative Psychology has to tackle.

We can't genuinely escape our Mentation. But suppose we could: exactly what spiritual currency should one expect to cash in from the value-less operations of an information sieve? We may as well imagine ourselves to be capable of having sentient awareness, as if we were red blood cells or digestive enzymes.

Indeed, there are psychological consequences of determinedly and energetically pursuing these. The same is true of hitting yourself over the head with this book. Although it won't make you any smarter about Matrixial Logic, you will learn some lessons about pain and frustration.

We have much more to say about all this in *Secret Self*.

CHAPTER 9

COMPLETING REALITY

In this Chapter, using inferences drawn from Matrixial Logic, we float some preliminary re-conceptualisations, of curated issues.

This Chapter cannot do comprehensive service to these big picture themes, and is not intended to. We rather signpost further and later journeys to be undertaken elsewhere.

The summary nature of this part entails a certain inevitable straw-manning of some deeply interesting ideas. We do not mean to disinherit their contributions to the sum of human knowledge, but dwarfishly to build upon their shoulders.

Causation and Determination

Causation

We need first, to acknowledge that the quantised nature of time [T], in contrast to classical and Einsteinian time τ, deprives causality of a significant alibi.

That classically conceived τ, has long furnished a substrate for notions of causation, which have rendered them logically undeniable.

Arguments which depend upon causation do often

acquire the character of self-immolating Greek fire: impossible to stamp out, yet burning the missile thrower.

Layers of sophistication have been added to notions of causation, so that the troublesome unmoved mover can be avoided by calling upon emergence. That, of course is to atomise the problem, not genuinely to avoid it.

Linguistic trickery has been used also, but to what end, ultimately? A child understands the *idea* of causation. The linguistic form is immaterial to the understanding.

We can start to look at the matter this way: Gödel's incompleteness problem, is the same problem as causation, simply from the other end.

The problem of causation comes not from infantile understanding of the obvious.[522] But from the age-old problem of seeking to subsume all under forms of equality.

As we have demonstrated in Chapters 3-5, any that which operates as an extrusion from "is" cannot be used to delineate one extrusion from another, without circularity.

Once we step aside from that circular domain, and use the tools of inequality logic, matters of causation fall into place as easily for us, as in the neonatal mind.

---

522 Indeed, neonates clearly have a practical grasp of the process labelled as causation

"A causes B". That is simply the Aristotelian Categorical Sentence ("ACS") form (I): the Particular Affirmative: Reflexive Opposition, as we say in ML.[523]

We are immediately in the domain of the ACS. We are merely replacing "is" with "causes":

Ackrill/Aristotle Categorical Sentence Transitions to ML

	CS	ML Equation	Type
A	Every A is B	$(A) \neq (-A) = B[I]$	Categorical Universal
E	No A is B	$(A) \neq (B) = N[E]$	Reflexive Universal
I	Some A is B	$(A^n) \neq (nA^n) = B[E]$	Reflexive Opposition
O	Not Every A is B	$(A^n) \neq (-A^n) = B[I]$	Reflexive Negation

Table 3.6

All that the classical and neo-classical traditions have done, is to add "in time τ", as a universal substrate.

When we consider this operation, it's tautological redundancy becomes obvious. We may as well say "in the universe".

So inured have we become to the notion the time is the currency of causation, that we seem to have mislaid any conception of what independent function τ plays.

Thus the apparent chasm developed between "classical" physics and QM. But there never was a gap. [T] was always quantised: indeed multi-variantly quantised. We simply lacked adequate technology to establish that.

---

523 See Chapter 3

That [T] is quantised makes no difference to Newton, Maxwell or indeed Einstein. The equations of electro-magnetism, for example, function perfectly well in [Δt], whether you use τ or [T].

That is precisely because we are considering changes in state of that which is locally invariant (unilinear propagation).

That $M = E / C^2$ is related to time, in the sense that C is necessarily a function of Distance and Time. But since the equation is preserved[524] with quantised time [T]: no problem there. Indeed, the equation acquires more dimensions of application in a time quantised universe.

Thus, to say "A causes B, in time τ" adds a tautology to an identity. That hardly presages explanatory power. It advances the matter no further to seek reliance on the Second Law of Thermodynamics, Entropy, as a surrogate for τ. This is the idea that it is entropy which demonstrates the unilinear direction of "time's arrow.

Problems with this notion include:

- SLD operates only subject to a particular frame of reference (and not as a universal);
- Trying to universalise the operant of SLD as heat, or

[524] See Chapter 7

information, is simply trying to assert that the cart does not need the horse;

- SLD is (one one view) contradicted by general relativity.

Trying to say ""A causes B", because SLD says it is not the other way around, is rather like stabbing oneself in the arm, applying a sticking plaster and stating: there, that proves the wound.

We can state the matter in Matrixial Logic simply: "causation" is a way of identifying the generation of an Event which, upon interaction with an infinity, becomes a Node.[525] That of course may be a Node arising out of Substance or a simulacrum Node arising out of Moment Δ^n.

Let's just remind ourselves, from Chapter 4:

(1.1) $((A^1) \neq (nA^1) = [E^1]) \neq ((A^2) \neq (nA^2) = [E^2])\ [\sim]\ \Sigma(FoR_2)$

$$\approx$$

(2.1) $(\square 1) \neq (\square 2)\ [\sim]\ \Sigma(FoR_2)$

$$>$$

(2.2) $(\square 1) \neq (\square 2)\ [\sim]\ <\Sigma E>$

the underlying cross-dynamics at (1.1):

are present in the Node as the function of

its existence in Substance, but present differently:

[E] Form is	in	Actuality.
<∑E>Node is	of	Actuality.

[525] Explained in detail in Chapters 4, 5 and 7

So, to say, α causes => β, we are positing:

$$\underline{\alpha \Rightarrow \beta}$$

□

$$((A^1) \neq (nA^1) = [E^1]) \neq ((A^2) \neq (nA^2) = [E^2])\ [\sim]\ \Sigma(FoR_2)$$

Equation line 1 should give us pause for thought. Whenever we say "α =>", it is as if that "causation" leaves α naïve, unaltered. It is the micropic case of the unmoved mover, such that α moves, but is unaffected by the moving.

How can that be so? In what frame of reference other than that of causation itself, can α effect change, without being subject to change? Within the FoR of causation, one cannot speak of cause, for there is already a bounded linear infinity ∞ of causes Δ^n.

Accordingly, to state that "α =>", is a Zenoic paradox.

Now, we can distinguish matters such as legal causation: causation in law. Anecdotally, this is familiar stuff to the Author, going back to undergraduate law studies at Oxford. Non-English readers will forgive the use of English cases to illustrate the issues.[526]

Lord Denning famously[527] tore away the curtain of "legal" causation in *Lamb v Camden London Borough Council* [1981]

---

[526] Commonwealth and US law follows the same *rationales*

[527] To English lawyers

2 All ER 408, [1981] 2 WLR 1038 (Author emphases):

> *The truth is that all these three - duty, remoteness and causation - are all devices by which the courts limit the range of liability for negligence or nuisance.* As I said recently in the Borag case (Compania Financiers v. Hamoor Tanker Corporation) (1981) 1 WLR 274 at page 281E:
>
> > "It is not every consequence of a wrongful act which is the subject of compensation. The law has to draw a line somewhere".
>
> Sometimes it is done by limiting the range of the persons to whom duty is owed. Sometimes it is done by saying that there is a break in the chain of causation. At other times it is done by saying that the consequence is too remote to be a head of damage. All these devices are useful in their way. But ultimately it is a question of policy for the judges to decide. I venture to repeat what I said in Dutton v. Bognor Regis U.D.C.. (1972) 1 QB 373 at page 397:
>
> > "It seems to me that it is a question of policy which we, as judges, have to decide. The time has come when, in cases of new import, we should decide them according to the reason of the thing.
> > "In previous times, when faced with a new problem, the judges have not openly asked themselves the question: What is the best policy for- the law to adopt? But the question has always been there in the background. It has been concealed behind such questions as: Was the defendant under any duty to the plaintiff? Was the relationship between them sufficiently proximate? Was the injury direct or indirect? Was it foreseeable or not? Was it too remote? And so forth.
> > "*Nowadays we direct ourselves to considerations of policy*"

The Denning formulation is very useful in illuminating the matter presently under consideration. "Causation" has no meaning independent of policies of attribution: whether of blame or reward. Causation is a label, not an explanation.

The first step in seeking to use causation as an explanation of anything is: $(A^1) \neq (nA^1)$.

We know, looking at α, that there is going to be a change in the state of α. So, we need to define:

- Whether α is Substance or Moment;
- The FoR of that change.

This is elementary Matrixial Logic method.

We are not going to provide any example here: because against the particularity of that example could always be ranged the supposed universal character of =>.

This is a familiar argument: it is the Square of ACS.[528] It ;\ is just as in error, and for the same reason. One cannot use a particular form of a Universal to affirm or deny a Universal.

But then again, it is impossible logically[529] to prove that any universal exists, since the means of proof are of the same stuff as the object of proof: equations in equality.

So: first find your $(A^1) \neq (nA^1)$. Now, in the fantastical realm of =>, that is suddenly difficult. You can only reach a definition of $(A^1) \neq (nA^1)$ as an agency involved in => by reference to β. Which is instantly to be trapped in circularity.

The escape hatch is to figment some arbitrary form of [α

528 See Chapter 3

529 Within classical or neo-classical logic

≠ β], while trying to disguise your process. Let's call that arbitrary form "Unicorn" or [U].

Your next step is to discriminate between $(\alpha^1) \neq (n\alpha^1)$, by reference to [U]. You then proclaim: aha! see change in Δα is not dependent on Δβ, therefore causation is indeed an independent category of universal.

It then however becomes exceedingly difficult to avoid stating Δβ without reference to [U]. Tricks of language can effect disguises.

But, after all that effort, all you have achieved is to return to your starting point: a presupposition that A=B under some universal [U], therefore:

- Δ[U] = Δα and Δβ
- or, what amounts to exactly the same thing
- ΔA => Δβ | (sub voce) in [U]

Put simply, *causation* is always mere pre-judgment, posing as logic. It is the forms of equality which appear to confer logical value on the pre-judgment: but at the price of ineradicable circularity.

This poses a dire problem for deism. The Kalam Cosmological Argument,[530] and all the variations, require that a concept of *causation* be agreed upon as

[530] https://en.wikipedia.org/wiki/Kalam_cosmological_argument

epistemologically valid: which it isn't.

The Kalam then exploits the core contradiction inherent as circularity in *causation*, and proffers the only solution: to end circularity with a singularity (Si).

Of course, any manufacture of Si[531] is an exercise in disguised futility. The Kalam is a classic example of replacing one circularity with another: to replace a [U] with an Si.

Taking the chain of causation from the other end, Δβ, we fare no better. We are trying to get to something like:

(□1) ≠ (□2) [~] <∑E>

But no amount of equality forms manipulation will get us there. We are trying to walk to the sky.

We say:

- β has changed in manner [z]

and

- α caused that change in [z] of β

Well, to "cause" [z], both α and β must pre-existently be of [z]. So, both must be particulars of a universal. But we do not want [z] to be a universal: we want it to be a very particular emanation of state change Δ in β. Indeed, we

---

[531] See Chapters 2 and 6

want that very emanation to define causation: [z] being that which evidences causation has occurred.

So, we are back into the pit of circularity, from the other end.

Experience: Relativity
Setup:
Imagine a vast, black, but blank universe.
Start with your typical image of a universe and its star field: then switch the stars off, leaving the void.

Step1:

- Imagine a planet, there in the blank universe;
- Move it.

Stop.

Discussion:
(1) The planet won't move.
(2) It shudders a bit, as you try, But it's stuck.

Step1:

- Imagine a second planet
- Move it

Stop.

Discussion:

(1) Now, both the planets move.

(2) It's actually hard to get them to stop moving.

We have just entered the familiar world of Sir Isaac Newton.[532] Do we say that the 2nd planet "caused" the movement of the 1st? Or the other way around? Or that "causation" does not work conceptually in this plenum?

Is gravity itself a "cause". If it is, then why did gravity not work for a single planet? If gravity requires a plenum of gravitational objects, then what actually is gravity as a thing, a force, in itself? These are matters that Newton struggled with throughout his scientific life. Until Einstein came up with another, non-causality based explanation.

We now want to address the idea that "causation" is a validated doctrine of science. So, welcome to philosophy's wonderful world of scientific causation. The sign over that shop in Diagon Alley letters *This shoe fits any size.*

Aristotle of course divided causation into: material; efficient; formal and final. At which point the same baby paddles in 4 different bath-waters: so *quare* same baby.

It is a commonplace that Newton was the arch-inventor of a fixed universe of clockwork causation. That we then

[532] (1642-1726)

struggled to escape its machinery. In fact, that is a much later view,[533] and very much a product of the industrial age, 150 years later.

The truth is that Newton mad a multi-faceted approach to ideas of "causation". His plural causations were always conditioned by his requirement that God not be relegated to a universal supernumerary: but must remain in substance and power throughout the cosmos.

Let's begin with this insightful 2013 paper by Andrew Janiak *Three concepts of causation in Newton*. We would cite much more of it is space permitted:[534]

> Two conclusions seem apt. Firstly, there is no tension in Newton's views: he can assert that gravity is a distant action and deny that this action must be attributed to any substance such that substance acts at a distance. Secondly, we can locate a philosophical novelty in the fact that Newton's three concepts of causation did not operate on the same level. In his causal reasoning, Newton was a revolutionary physicist, but a conservative metaphysician.
>
> Yet the key to future developments in philosophical and scientific understandings of causation is that Newton very deliberately insulated the aspect of natural philosophy that we would call physics, which centers on discovering the forces of nature, from the aspect of natural philosophy that we would call metaphysics, which centers on substances and their properties.
>
> Remarkably, Newton insulated physics even from his own fundamental metaphysical preconceptions about how actions in nature must be

[533] Leibnitz argued that Newton's gravity was "occult", rather than clockwork

[534] Andrew Janiak Studies in History and Philosophy of Science Part A Volume 44, Issue 3, September 2013, Pages 396-407 https://doi.org/10.1016/j.shpsa.2012.10.009

> attributed to substances. This insulation enabled mathematical physics in the eighteenth century to proceed on its merry way, even as it developed concepts and conclusions that challenged the very heart of the metaphysical tradition.

Newton distinguished, as Janiak's paper sets out with admirable clarity: [535]

> three kinds of causes...: mechanical, dynamical, and substantial causes. This threefold distinction enables us to recognize that although Newton clearly regards gravity as an impressed force that operates across vast distances, he denies that this commitment requires him to think that some substance acts at a distance on another substance. (Dynamical causation is distinct from substantial causation.)

Billiard balls clunkily transferring their momentum to each other via contact, and the mutual force effected as gravity between bodies, is not for Newton, the same kind of "causation".

Of course, once we start having different kinds of "causation", the word rapidly loses meaning. It becomes merely a placeholder for thinking about effects and reactions. Such concepts furnish (for example) the Kalam with no assistance.

It's also notable that this interpretation of Newton brings his conception of gravity far closer to Einstein. Both think of gravity, not as an independent force, acting at a distance, but as a relationship between matter emerging from distance.

[535] op. cit

For a completely opposite take, see *Newtonian Causality – The Natural Law of Cause and Effect*:[536]

> 1. It has been shown that Newton's theory of motion (which was meant by Newton as a "theory of everything") implies a "dualist" law of causality, representing a rational, measurable mind-body" interaction of non-material and non-observable active causes, the (impressed) "forces of nature", with passive material bodies, their changes of motion being the material "effects" generated by the said non-material causes. Thus we meet with a "Newtonian world view" very different from the one that has been taught so far - a world view that implies a non-material or spiritual transcendent "Platonic" reality in contrast to the dominating monist materialist scientific world view of our times.

We may have thought that turning the pages of Hume's *Treatise on Human Nature* would give us a ready grasp on the concept of causation. But no:[537]

> Although Hume and Newton agree that causal analysis is guided by the rule that similar effects have similar causes, Hume invariably interprets this rule in a purely qualitative way: resembling effects have resembling causes.
>
> In Section XV, Rules by which to judge of causes and effects, rule four, Hume writes that "...the same effect never arises but from the same cause" (T 173).
>
> And in rule five, he claims that "...where several different objects produce the same effect it must be by means of some quality, which we discover to be common amongst them. For as like effects imply like causes, we must always ascribe the causation to the circumstance wherein we discover the resemblance" (T 174).

---

[536] Ed Dellian. June 2011. Internet Publication. http://www.neutonus-reformatus.de/download/dellian_newtonian_causality.pdf

[537] Richmond, S. (1994). Newton and Hume on Causation: Alternative Strategies of Simplification. *History of Philosophy Quarterly, 11*(1), 37-52. http://www.jstor.org/stable/27744609

By contrast Newton holds a stronger principle that permits inference to the common cause of similar effects that are correlative and independent of each other, even when this means inference to a numerically identical cause or a cause of unknown variety. His first rule of reasoning in philosophy is: "We are to admit no more causes of natural things than such as are both true and sufficient to explain their appearances" (P 398). This rule of parsimony with respect to causes is followed by a second rule: "Therefore to the same natural effects we must, as far as possible, assign the same causes" (P 398).

These rules when applied to a temporal sequence of events favor structures with fewer prior explaining events than later events explained. Their application favors temporally simple causation.

For yet another view (and there are many more):[538]

Contrary to common belief, acceptance of Newtonian causation does not commit one to a mechanistic, materialistic, or deterministic understanding of the world.

I argue that the Newtonian view can be assimilated to contemporary theoretical alternatives in psychology. This means that, given the Newtonian understanding of causation, it is possible for such alternatives to be scientific - to treat of causes- without requiring either mechanism, materialism, or mathematical formalizations.

I argue that we best understand Newtonian causation as formal causation. I do this by discussing the history of Newton's theory of causation and comparing his theory to Bacon's. I also compare Newton's theory of causation to Aristotle's, arguing that when we speak of formal causes we speak of our descriptions rather than the nature of things.

We may (or may not) accurately impute various elements of our scientific descriptions to the nature of things, but when we speak of formal causes, we are speaking of the patterns we use to describe the changes we observe rather than the nature of things themselves. Since any science must use such patterns, even alternative psychologies

[538] Faulconer, J. (1995). Newton, Science, and Causation. *The Journal of Mind and Behavior, 16* (1), 77-86. http://www.jstor.org/stable/43853669

> use Newtonian causation - if they are genuinely scientific alternatives. However, mathematics is not the only discipline that offers such patterned explanations. Moral explanations offer an alternate model for causal explanation.

We don't propose here to try and resolve all these disputes. Merely to note that there is not even consensus on what the supposed originator of clockwork causation actually had to say about the idea.

We have explained above, why that is bound to be so. Artificially created Singularity admits of any explanation you like.

Determination

Determinism is the ugly sister of causation.

We like causation, because we can use it to explain things.

We don't like determinism, because it stops us trying to explain things the way we would want to.

There remains even today the endlessly repeated story that determinism has its modern foundation in Newton.[539] Insofar as there is any truth to this, it is only in respect of Newton's theological clockwork cosmology.

That, in itself, is not "determinist". Newton's project was

[539] With the correlate that therefore there must be something in it, since Newton's laws work so well

to illumine understanding of all that which God has set in motion. Which is not the same as claiming that everything in motion has a pre-ordained course from the beginning of creation.

As noted above, what is so often represented as Newtonian causation and determinism, is actually a product of the industrial age: arising from our new self-certainty in our capacity ultimately to master nature.

Enter Laplace:[540]

> An intelligence which, for one given instant, would know all the forces by which nature is animated and the respective situation of the entities which compose it, if besides it were sufficiently vast to submit all these data to mathematical analysis, would encompass in the same formula the movements of the largest bodies in the universe and those of the lightest atom; for it, nothing would be uncertain and the future, as the past, would be present to its eyes.

> The 2014 paper by Marijvan Strien. *On the origins and foundations of Laplacian determinism*[541] repays consideration:
> In this paper I examine the foundations of Laplace's famous statement of determinism in 1814, and argue that rather than derived from his mechanics, this statement is based on general philosophical principles, namely the principle of sufficient reason and the law of continuity.
>
> It is usually supposed that Laplace's statement is based on the fact that each system in classical mechanics has an equation of motion which has a unique solution.
>
> But Laplace never proved this result, and in fact he could not have

[540] (1749-1827) *Essai philosophique sur les probabilités* (1814)

[541] Studies in History and Philosophy of Science Part A Volume 45, March 2014, Pages 24-31 https://doi.org/10.1016/j.shpsa.2013.12.003

proven it, since it depends on a theorem about uniqueness of solutions to differential equations that was only developed later on.

I show that the idea that is at the basis of Laplace's determinism was in fact widespread in enlightenment France, and is ultimately based on a re-interpretation of Leibnizian metaphysics, specifically the principle of sufficient reason and the law of continuity.

Since the law of continuity also lies at the basis of the application of differential calculus in physics, one can say that Laplace's determinism and the idea that systems in physics can be described by differential equations with unique solutions have a common foundation.

This analysis demonstrates a substrate of what we think of as "determinist" thought, which has bubbled as an undercurrent in scientific thinking (and its reflections in popular culture). More startlingly, it pops up again in 1935, as we shall see later.

Ian Hacking shows in *Nineteenth Century Cracks in the Concept of Determinism* that:[542]

During the nineteenth century we witness an event that I call the erosion of determinism. At the start of the period we read Laplace, confident that all the world is ruled with the same exact necessity as is found in celestial mechanics. At the end we find C.S. Peirce (1839-1914) asserting there is no ground for believing either such a doctrine of necessity or for holding that the laws of nature are rightly represented by precise numerical constants. What he called "tychism" - the idea that we live in a Universe of Chance, became a main theme of his philosophy

Robert Burch's analysis of CS Pierce's thought and work

[542] *Journal of the History of Ideas* Vol. 44, No. 3 (Jul. - Sep., 1983), pp. 455-475 Published by: University of Pennsylvania Press DOI: 10.2307/2709176 https://www.jstor.org/stable/2709176

repays study. Pierce can, along with Tesla[543] be credited as one of the foremost, yet forgotten, creators of the 20th century.[544]

As Hacking continues:

> The purpose [of Laplace's dictum], I believe, is to assert that in the physical realm there is no such thing as chance. Chance, as Hume had put it, is a matter of hidden and secret causes. Indeed that philosopher is at one with the applied mathematician, writing in his Treatise of 1739:
>
> > 'Tis universally acknowledged, that the operations of external bodies are necessary, and that in the communication of their motion, in their attraction and mutual cohesion, there are not the least traces of indifference or of liberty. Every object is determin'd by absolute fate to a certain degree and direction of its motion.[23]
>
> As in the case of Leibniz, we note even the very verb "determin'd." But Hume quite explicitly writes of external bodies in this connection, and not of mental events. Likewise Laplace implicitly refers only to external bodies. If we are to save the thesis of Cassirer, we should not hold fast to his feeble claim that Laplace wrote only metaphorically. Instead let us guess that Laplace was not fully dissociated from a kind of dualism. In the world of extended substance, of external bodies, everything is "determined." That entirely leaves open the question of mental substance. Even Kant found his place for freedom in the world of mental substance, although he uses the terminology of noumena.
>
> > [23] David Hume, A Treatise of Human Nature (London, 1739-40), Book 2, Part 3, Section 1. I owe this reference to Thomas Baldwin. My distinction between determinism in extended substance, and determinism or Praedeterminismus in the mind is not anachronistic but was discussed early in the nineteenth century, for example in James Gregory, Letters from Dr. James Gregory in defence of his essay on the difference of the relation between motive and action and that of cause and effect in physics, with replies by the Rev. Alexander Crombie (London, 1819)

---

[543] (1856-1943)

[544] Burch, Robert, "Charles Sanders Peirce", *The Stanford Encyclopedia of Philosophy* (Winter 2018 Edition), Edward N. Zalta (ed.), URL = <https://plato.stanford.edu/archives/win2018/entries/peirce/>.

This is yet another paper which we wish space would permit entire reproduction. The gist is that the idea of determinism is just as multi-variant as causation. The idea didn't really find traction until the mid-19th century: soon after to be dismantled across the scientific landscape.

We earlier promised later: an example of determinism.[545] That later is the "EPR Paradox": set out in the famous 1935 paper (emphases in original):[546]

> Whatever the meaning assigned to the term *complete*, the following requirement for a complete theory seems to be a necessary one: *every element of the physical reality must have a counterpart in the physical theory.* We shall call this the condition of completeness. The second question is thus easily answered, as soon as we are able to decide what are the elements of the physical reality.
>
> The elements of the physical reality cannot be determined by a priori philosophical considerations, but must be found by an appeal to results of experiments and measurements. A comprehensive definition of reality is, however, unnecessary for our purpose. We shall be satisfied with the following criterion, which we regard as reasonable. *If, without in any way disturbing a system, we can predict with certainty (i.e., with probability equal to unity) the value of a physical quantity, then there exists an element of physical reality corresponding to this physical quantity.* It seems to us that this criterion, while far from exhausting all possible ways of recognizing a physical reality, at least provides us with one such way, whenever the conditions set down in it occur. Regarded not as a necessary, but merely as a sufficient, condition of reality, this criterion is in agreement with classical as well as quantum-mechanical ideas of reality.

---

[545] Joke

[546] *Can Quantum-Mechanical Description of Physical Reality Be Considered Complete?* A. Einstein, B. Podolsky, and N. Rosen Phys. Rev. 47, 777 - Published 15 May 1935

This conceptualisation has the whiff of Maxwell's Demon about it.[547] Given the origin of quantum theory in problematics of electro-magnetism, this is perhaps unsurprising.

The odour is augmented by the idea that information has zero entropy cost: *If, without in any way disturbing a system, we can predict with certainty...* But perfect information must come at a cost of increasing the entropic value of any system which is the domain of the enquiry.[548] Unless we posit a system which is not in entropic unity: in which case certainty is by definition impossible.

If we could know everything about everything, that would preclude us from knowing anything about anything.

This either/or dilemma appears masked or elided by the twin divergencies:

- The second question is thus easily answered, as soon as we are able to decide what are the elements of the physical reality;
- The elements of the physical reality cannot be determined by a priori philosophical considerations, but must be found by an appeal to results of experiments and measurements.

These introduce subjective selection by idea and practice,

547 See Chapter 6

548 See the Bennett Theorem, Chapter 6

and thus seem to allow a closed system to produce determinate certainty.

Yet this is naked dualism:

- External observers of a parallel closed system;
- Able to subject the closed system to dictates of objectivively established certainty[549]
- which certainty cannot be justified within the system.[550]

EPR concluded:

> While we have thus shown that the wave function does not provide a complete description of the physical reality, we left open the question of whether or not such a description exists. We believe, however, that such a theory is possible.

We do suppose that we have to take seriously what these brilliant minds had to say about their own minds, and ours. But we just can't.

The conclusion, and its starting premises, presume a universe in which Maxwell's Demon can both exist theoretically, and have practical reality.

The kernal of the EPR objection appears here, and it is both instructive and intriguing:

> More generally, it is shown in quantum mechanics that, if the operators corresponding to two physical quantities, say A and B, do not commute, that is, if AB/BA, then the precise knowledge of one of them precludes such a knowledge of the other. Furthermore, any attempt to determine the latter experimentally will alter the state of the system in such a way

549 Measurement certainty: theoretical and empirical
550 Back to Turing and Gödel

as to destroy the knowledge of the first.

The EPR doctrine can thus be reformulated:

Any theory of a physical system is complete only if:

(1) The properties and/or relationship of any manifestation thereof and therein ("an Event")

(2) can be equalised under a /S/ operation

(3) such that any Event may be determined by qualification of any other Event: determination and correlation being operational equivalents.

Once seen in ML terms, the EPR objection to QM is clearly misplaced. For QM merely requires that we use orthagonal[551] Scalar /S/ axes. QM is only saying that qualification in $/S/_1$ does not allow qualification in $/S/_2$.

That is hardly absurd or radical:

- Betty is in the changing room;[552]
- We know she took 6 items in with her;
- We can see 2 items have appeared on the "don't want" rack outside the changing room.

but

[551] We could say "parallel", but given the importance of angular momentum to QM, "orthagonal" is more representative

[552] Sadly, in a pre-Covidiot world

- We can't know, without having watched all the time[553]

whether

- Betty tried them on and rejected the 2 items for their fit size; or
- never tried them on and rejected them for some other reason

Ignoring all the possible sartorial inclinations of Betty, this is a closed macroscopic system. Yet qualification in $/S/_1$ does not allow qualification in $/S/_2$, *without altering the system.*

Yes, we could watch the security footage of the booth, and even have it beamed to our phone:[554] but then the privacy condition of the booth would be violated: Schroedinger's box would be transparent.

Does this entail existential irrationalism? That we can't know, in a scientific sense, as defined by Einstein?

Fear not: you already have the answer. Recall the *Noming Experience* from Chapter 7. You are able to synchronise with an external time beat $[T]^n$.

The very reason you can do this is because this is not a scalar universe, ruled by scalar time: a metric. Ours is a

---

[553] Or otherwise broken the communication isolation of the changing room: Schoedinger's Feline Receptacle

[554] Really must suggest this to Selfridges

living universe, operating dynamically through forms of inequality.

You don't need to be able to plot every point of the compass on a scientifically Olympian map. You simply need to be able to match your "now" to any other "now: and you can. We've proved it.

Certainly, we can map points and paths of expectation. That's what QM, and life experience are for. But the map is not the thing, nor even representational of the thing. It expreses the structure of the gaps between things. The map is what scales inequality.

Experience: Maptime
Keep your eyes open.
Just look around the room you're in.
Look to your left

Preparation:

- Imagine yourself, about 2 arm's lengths away. With Silhouette You ("S-You"), having its back to you. Standing upright, on the floor;
- You can be perfectly aware of the surroundings of S-You;
- Walls, and other such obstacles don't exist for S-You. It can walk throuhg any of them, just like a ghost.

Step1:

- Looking at the back of S-You;
- Make S-You move in straight line forward.

Stop

Discussion:

(1) S-You won't move.

(2) It's like trying to push S-You through drying wet concrete.

Step 2:

- Looking at the back of S-You.

Now:

- Feel the distance between You and S-You;
- See in your mind's eye like a thick, stiff, length of rope.

Now

- Place that rope length in front of S-You;
- Make S-You move in straight line forward;
- Repeat.

Stop

Discussion:

(1) S-You moves immediately to the end of the rope length.

(2) At instantaneous speed.

Step 3:

- Looking at the back of S-You;
- See again the rope length between You and S-You.

As before in Step 2

- Place that rope length in front of S-You;
- Make S-You move in straight line forward.

Now

- With S-You at the end of a rope length, from any previous position of S-You ;
- Make S-You come back Half (½) a rope length.

Stop

Discussion:

(1) S-You won't budge;

(2) The best you can do is get a shadow of S-You to hover at the half-way mark, but then disappear, under the force of S-You.

Step 4:

- Looking at the back of S-You;
- See again the rope length between You and S-You.

As before in Step 2

- Place that rope length in front of S-You;
- Make S-You move in straight line forward.

Now

- Snap your fingers and make the rope length disappear.

Stop.

Discussion:

(1) S-You instantly snaps back to its starting point.

Congratulations: you have just experienced being an electron. Your self-movement was quantised. S-You was not able to travel at all, until you provided the natural unit of movement: an S-You Length ("SYL").

Once provided with SYL, S-You could travel, in SYL steps, all the way to infinity. But S-You could only move, forwards or backwards, in SYL steps. And, once you removed the SYL frame of reference, S-You transmuted instantly back to base state.

Clearly, S-You was *determined*: in the sense of being determinable. We could say with great accuracy where S-You would be: at some number of rope length demarkated points between start and infinity.

But it would be odd to say that any particular resting place of S-You was *caused*.

Experience: Soundspace

Keep your eyes open.
Just look around the room you're in.
Look to your right

Preparation:

- Prepare S-You, just as in *Maptime Experience;*
- Google "30 bpm";
- Choose a metronome (YouTube or any will do);
- Be ready to press Play (don't Play yet).

Step1:

- Looking at the back of S-You;
- If you start thinking of the rope-length, wash that out of your mind;
- Make S-You move in straight line forward.

Stop

Discussion:

(1) S-You still won't move
(2) It's still like trying to push S-You through drying wet concrete.

Step2:

- Looking at the back of S-You.

Press Play.

- Allow yourself to fall into the rhythm of the beat;
- "Switch on" You_2: see the *Noming Experience* in Chapter 7;
- Nod your head, or tap your finger in time.

Now:

- See S-You;
- Just focus on S-You.

whatever you experience

- Just keep focusing on S-You.

Discussion:

(1) Without you even thinking of it, S-You moved.

(2) S-You just started "walking", in time with the Beat.

(3) On every Beat, S-You moved another step.

(4) It was always exactly the same distance.

(5) Yet you weren't thinking out the distance, as in a rope length: it just happened on its own.

So, now distance and speed are being quantised by packets of time [T] beats. Welcome to quantised momentum.

You can re-run the *Soundspace Experience* in your head. You can see your in awareness that there is no possibility you could make S-You take half a step. Or walk backwards.

Do now re-run the *Soundspace Experience* in your head. You can see in your awareness, that if the Beat speeded up, then so would S-You. But, if the Beat got to infinity ∞, S-You would freeze.

You can now appreciate why a QM calculation must be wrong if it returns a result which is infinity.
Now, does the Beat "cause" S-You to move? Well, we can

clearly say that the Beat is a necessary condition of S-You moving. Unless you use the rope length. So the Beat is an alternative condition. But once you admit alternatives, you have ruled out "causation" as having any explanatory value.

Clearly, S-You was *determined*: in the sense of being determinable. We could say with great accuracy when S-You would move: at each Beat. But we can't possibly ascertain how far S-You moves with each beat. How do we apply a longitudinal frame of reference to what is, in reality, a simply made up time beat?

Welcome to the quantum world. You can specify:

- Where S-You is (number of rope lengths);

or

- When S-You is (number of beats)

but not both.

You don't believe it?

Experience: Ropetime

Keep your eyes open.

Just look around the room you're in.

Look left or right.

Preparation:

Get your Beat metronome ready.

Get S-You in place.

Get your rope length from S-You ready.

Step 1:

- Press Play.
- Move S-You.
- Try and use the rope-length in time to the Beat.

Stop

Discussion:

(1) It doesn't work.

(2) You can use the rope length.

(3) Or: use the Beat.

But:

(4) You can't do both.

(5) As hard as you try, it all just become "mush". Everything collapses.

(6) S-You won't move.

As Dirac effortlessly put it:[555]

> The non-classical nature of the superposition process is brought out clearly if we consider the superposition of two states, A and B, such that there exists an observation which, when made on the system in state A, is certain to lead to one particular result, a say, and when made on the system in state B is certain to lead to some different result, b say. What will be the result of the observation when made on the system in the superposed state? The answer is that the result will be sometimes a and sometimes b, according to a probability law depending on the relative weights of A and B in the superposition process. It will never be different from both a and b.

---

[555] Dirac, P. A. M.. *The Principles of Quantum Mechanics.* (1930) (4th Ed 1957). At 4: *Superposition & Indeterminacy.* The Author's appreciation of Dirac's style does not commit the Author to endorsement of all the substance

> The intermediate character of the state formed by superposition thus expresses itself through the probability of a particular result for an observation being intermediate between the corresponding probabilities for the original states, not through the result itself being intermediate between the corresponding results for the original states.

There is not, of course, complete correspondence between these *Experiences,* and "small state" quantum mechanics.

So, now what do you think of "causation" and "determination"? You tried exactly to impose "causation" on this closed S-You system, and to extract "determination" from it. You found that you couldn't do either.

Indeed, you discovered another proof that forms do not have forms. There is no "superintendant" form of forms in Theta Θ, which allows you to subsume distance and time under one measure.

These ugly sisters concepts don't even work inside your own head.[556] It's therefore difficult to see that they can work anywhere else in the "real world".

So, in completing reality,[557] where do God and Good fit in the matrix?

Divine and Devout

---

[556] Over which you can be presumed to have complete power; or at least significant influence

[557] More modestly and accurately: erecting a few sarsens

Faith is stupid and science is smart. End of story.

So says rationalism. The slight difficulty with that position, is that rationalist forms of equality, the belief that identity and commutation are fundamentals of the laws of reason, turn out to be wrong.[558]

So, it's a good thing that material science is not based on forms of equality. For sure, identity and commutative beliefs get mixed up with science, partly as a result of the unavoidable mathematics. The course of science then becomes self-limiting and paradoxical.

Science, like logic, is not a matter of believing in anything. It is a matter of constructing understanding from observation. Science is the artifice of natural philosophy.

Religious people don't like being called stupid by atheists. People generally don't react well to being stupified. One reason why atheists make so little progress with evangelism, is that their arguments don't address the matters that matter to the religious.

What's more, rationalist arguments, to the theist, have the unmistakable tinge of the illogical. In which respect, the theist is often on solid ground.

---

[558] If any reader thinks that any of these following words is any kind of defence of theism, or that the Author believes in any sort of "God"; then that reader has really wasted the price of this book

After all, theism is a self-ordinating set of arbitrary ideas, the substrate of which is an artificial singularity. The cultural gestation of theism is a subject for another book. Let's just peer behind the thaumatonomic curtain, at the logic.

Any idea of <god> is a simple Si Bridge out of Chaos:[559]

$$\underset{\text{Disjunct}}{\frac{(\text{Order}) \neq (\text{-Order})}{(\text{Chaos}) \neq (\text{-Chaos})}} \qquad \underset{\text{Bridge}}{=} \qquad \underset{\text{Form}}{\text{God } [E]}$$

$$>$$

$$\text{Si: } (\Delta \text{ Order}) \neq ((\Delta \text{ Chaos}) = \text{God } [E] \,\} \text{ External}$$

$$<>$$

$$(\Delta \text{ God}^{n}) \neq (\Delta\text{-God}^{n}) = \text{Spirit}[I]\}$$

You can put whatever clothes you wish on [E]} Node, or the [I]} conversion.

You can see, for example how the term: $(\Delta\text{-God}^{n})$, characterises in early Christianity as "Man". Indeed, the contribution of later Christian theology[560] was to swap <order> with <god> in the Si Bridge equation, and then to bring "Jesus" as the external manifestation, thus:

$$\text{Si: } (\Delta \text{ God}) \neq ((\Delta \text{ Chaos}) = \text{Order } [E] \,\} \text{ External}$$

---

[559] See Chapters 2 and 6

[560] Once it mixed with the post-Aristotelians, in St Augustine and his contemporaries

<>

$(\Delta\ \text{Order}^{n}) \neq (\Delta\text{-Order}^{n})$ = Jesus[561] [I]}

It's actually a rather neat 3 card trick, which proved difficult to get at: and still does.

Atheism cannot defeat the logic of theism. That logic is syllogistically sound:

- All elephants are pink
- Nellie is an elephant

∴ Nellie is pink

Obviously, there are plenty of arguments in theology which are syllogistically stupid: but you can say that about any walk of life.

Experience: GodClip

Turn to the front cover of this book.

There you will find the by now ubiquitous paperclip, on a blank sheet.

Please do not use You_2 for this *Experience*.

---

[561] The Gospel of John anthroporphic fusion version, not the Pauline Epistles version (that is, the supposedly genuine ones, not the fake ones)

Step 1:

- Try to move the paperclip.

Stop.

Discussion:

(1) As is well-familiar to you now: you can't.[562]

Step 2:

- Imagine an all-powerful god: the god of Christian, Jewish or Islamic religion;
- That god must (by definition) be "up there" or out beyond there.

Now:

- Feel god moving the paperclip.

Discussion:

(1) No matter how you try to visualise it, you just can't see god moving that paperclip.

Step 3:

- Imagine a Greek or Viking god: the anthropomorphic sort that happily comes down to earth and sings ships or turns into a swan.

Now:

562 If you can: imagine a perfect duplicate paper clip; and move that. Now you can't

- Watch god moving the paperclip.

Discussion:

(1) Easy. Your Zeus or Odin or Apollo just swept the clip up. No problem.

Which Experience brings us something of a theological problem:

- So far as <god> is a singularity, separate from <existence>, such entity cannot effect <existence>, even so far as to move a paperclip;
- Yet a silly Greek god, can.

Now, if one politely notes to the most enthusiastic theist, that their belief is that their <god> is a Singularity: they will most happily agree.

It's the theistic equivalent of contending that rain is wet. Singularity is the actual label on that theist tin, in the great cosmic car-boot sale of grand ideas.

If you'd like to test the matter, the just inquire as to how the theist completes the equation:

$$<god> \neq (ngod) = ?\ [E]$$

The theist will gladly tell you that no such equation can be completed. Which, given the premises of Singularity, must be logically correct.

Great. We are in perfect agreement. That's a good basis for

further discussion. So now, we can agree with the theist, surely that Human <h> is not equal to <god>. Obviously so.

We are in agreement again. Indeed, we can agree that *nothing is equal to god.*

So now let's complete the equation:

<god> ≠ <nothing> = ?

The answer cannot be <h>: as that is something. The theist needs to find a form of the agreed <god> ≠ <nothing> inequality. Or to proclaim that the equation cannot solve.

St Augustine himself came a bit unstuck at this point. So he laid down as the substrate of all his thinking, and which is still the catechisis of Catholicism and its descendants (Author emphases):[563]

> 9. ...For the Christian, it is enough to believe that the cause[564] of all created things, whether in heaven or on earth, whether visible or invisible, is nothing other than the goodness of the Creator, who is the one and the true God.
>
> Further, the Christian believes that *nothing exists save God himself and what comes from him*; and he believes that God is triune, i.e., the Father, and the Son begotten of the Father, and the Holy Spirit proceeding from the same Father, but one and the same Spirit of the Father and the Son.

Here is the very equation: <god> ≠ <nothing> = ?

---

[563] The Enchiridion (c. 420 AD). https://ccel.org/ccel/augustine/enchiridion/enchiridion.chapter3.html

[564] Augustine means "necessary substrate", not effecting agent

St Augustine turns the (?) into the symbol of faith itself. It is the very appreciation that there exists nothing which is not <god>,[565] which marks out the channel of faith.

Indeed, according to St Augustine, even pagans believe in <god>: they just don't do it correctly.

Now, you can claim to be a Christian, and claim to reject the logic of Augustine, but that's a tough position to take:[566]

> [St. Augustine of Hippo (354-430) is] a philosophical and theological genius of the first order, dominating, like a pyramid, antiquity and the succeeding ages. Compared with the great philosophers of past centuries and modern times, he is the equal of them all; among theologians he is undeniably the first, and such has been his influence that none of the Fathers, Scholstics, or Reformers has surpassed it.

So, let's be clear at this point. The theist, consistently with the express logic of <god>, and irrespective of the mode of religious belief, agrees that the Theist does not exist. Nothing except <god> exists.

If there be some thing which exists, then the definition of <god> is wrong. But that definition can't be wrong: to be <god> is exactly to be that which is NOT<whatever is>.

Well, that's fine, the theist does not exist: which makes

---

565 And <of god>, which is explicitly the same category of thing

566 Philip Schaff *History of the Christian Church*, Volume III (1867, Revd 1884) p854. Schaff out the "F" in Fundamentalist Christianity

further conversation about anything, insuperably difficult.

We certainly can't have a discussion about moral values, or anything of that kind. Morality is the concern only of that which is: and <god> is the singular antithesis of <all that is>. So that's fine: singularity imposes silence.[567]

It may be fascinating to discover that the foundation of post-Pauline christian theology is the express avowal of a form of inequality: albeit obviously derived from a Singularity.

We then arrive at the problem of Good. It is obviously going to pose real difficulties: where all that exists and <god> are the same thing, you end up with:

$$(\text{Good}) \neq (\text{-Good}) = ?[I]$$

$$>$$

$$(\Delta\ \text{Good}^{n}) \neq (\Delta\ \text{Bad}^{n}) = ?[I]$$

Since all that exists is <god>, one is rather limited in the solution to the equation, so we end up, inevitably with:

$$(\Delta\ \text{Good}^{n}) \neq (\Delta\ \text{Bad}^{n}) = <\text{god}>[I]$$

Here, <god> is a process in infinity ∞, a becoming. An infinite river suffused with currents of good and evil[568]
Experience: Modes

---

[567] Sound familiar in the lexicon of monastic orders?
[568] Merely a poetic linguistic form of "bad"

Setup:
Look in front of you, into your room.
Create, in a row in front of you, single file (from you as South pole, heading North).
Modes of You, in some your various personalities:

- Happy;
- Sad;
- Kind to others;
- Nasty to others;
- Telling truths;
- Telling lies.

Get at least 6 of these, and make sure there are some "bad" modes in there.

Step 1:

- See one of the "bad" modes of you, in the line;
- Try and eliminate that "bad" you.

Stop.

Discussion:
(1) You can't.[569]

That which is of you, cannot cease to be of you. However bad it becomes.

[569] If you can: imagine a perfect duplicate "bad" You; and move that. Now you can't

As St Augustine put it: (Author emphases):[570]

> 12. All of nature, therefore, is good, since the Creator of all nature is supremely good. But nature is not supremely and immutably good as is the Creator of it. Thus the good in created things can be diminished and augmented. For good to be diminished is evil; still, however much it is diminished, something must remain of its original nature as long as it exists at all.
>
> *For no matter what kind or however insignificant a thing may be, the good which is its "nature" cannot be destroyed without the thing itself being destroyed.*
>
> There is good reason, therefore, to praise an uncorrupted thing, and if it were indeed an incorruptible thing which could not be destroyed, it would doubtless be all the more worthy of praise.
>
> When, however, a thing is corrupted, its corruption is an evil because it is, by just so much, a privation of the good. Where there is no privation of the good, there is no evil. Where there is evil, there is a corresponding diminution of the good.
>
> *As long, then, as a thing is being corrupted, there is good in it of which it is being deprived*; and in this process, if something of its being remains that cannot be further corrupted, this will then be an incorruptible entity [*natura incorruptibilis*], and to this great good it will have come through the process of corruption. But even if the corruption is not arrested, it still does not cease having some good of which it cannot be further deprived. *If, however, the corruption comes to be total and entire, there is no good left either, because it is no longer an entity at all.* Wherefore corruption cannot consume the good without also consuming the thing itself. Every actual entity [*natura*] is therefore good; a greater good if it cannot be corrupted, a lesser good if it can be. Yet only the foolish and unknowing can deny that it is still good even when corrupted. *Whenever a thing is consumed by corruption, not even the corruption remains, for it is nothing in itself, having no subsistent being in which to exist.*
>
> 13. From this it follows that there is nothing to be called evil if there is

570 op. cit

nothing good. A good that wholly lacks an evil aspect is entirely good.

The Augustine syllogism of evil:

- Nothing exists save <god>;
- Good is a quality of existence;

∴ Evil does not independently exist.

Evil is merely a dimunition of good. In anything evil there is good.[571]

Experience: Changing

Setup:

Look in front of you, into your room

This time, move your chair so that you are at right-angles to the row: they are now in a line abreast, facing you.

Step 1:

- See one of the "bad" modes of you, in the line;
- Try and eliminate that "bad" you.

Stop.

Discussion:

(1) It's easier.

(2) But the "bad" you is still sticky.

Step 2:

---

[571] as Luke Skywaker said of his father: *Return of the Jedi* (1983)

- Try and eliminate that "bad" you.

and this time

- Move "bad" you behind you.

Come out of the scene, and:

- Blink;
- Put the scene into your head again.

Stop.

Discussion:

(1) Once you had come out of the scene, and gone into it again, the "bad" you had disappeared;

(2) You could feel it like a ghostly presence, but it was receding into being gone.

This may have a familiar ring to it:[572]

> But he turned, and said unto Peter, Get thee behind me, Satan: thou art an offence unto me: for thou savourest not the things that be of God, but those that be of men.

The insuperable logic position of Augustinian Christianity is thus *supreme moral relativism*.

Bad is simply a version of good. Nothing that is all bad exists. So all our history books filled with blood, war and genocide splattered pages: well they're obviously just

---

[572] Matthew 16:23 (King James Ed.)

wrong. What never existed, never happened. Unless it existed as some dimunition of Good.

That must obviously come as something of a surprise to present-day Catholic teachers (italics in original):[573]

> 1. *"There's no such thing as absolute truth. What's true for you may not be true for me."*
>
> ...our concepts of right and truth are not arbitrary but are based in some greater, universal truth outside ourselves - a truth written in the very nature of our being. We may not know it in its entirety, but it can't be denied that this truth exists.

There's no shame in being wrong, and even less in being stupid. But it is not consistent with the good or glory of anything, to be orthagonally inconsistent with the logic of one's own expressed beliefs.

Matrixial Logic is not the answer to anything. It is simply a tool that assists in formulating questions which can be answered: in consistency with the premises of the question.

We can ask no more of logic than that.

[573] https://www.catholiceducation.org/en/religion-and-philosophy/apologetics/12-claims-every-catholic-should-be-able-to-answer.html

CHAPTER 10

MATRIXIAL MANIPULATIONS

In this concluding Chapter, we build a collection of bridges. From where Matrixial Logic, in its brief life, has been. To where it may become.

We, homo sapiens, are part of complete reality. We have been debating how that is so, since at least 500BC. Matrixial Logic provides a different set of answers.

The Matrixial Self: in Part

In the *Languini* Chapter, we used language to demonstrate how even the most intractible matters of classical and modern logic, and philosophy admit of simple matrixial solution.

Now, we can begin to bring the parts together into completing reality. But only past of the process. This book is concerned with explaining the tools of logic which we can use to unwrap "mysteries" previously considered in philosophy, for the last 2,500 years.

This work by itself has expended a quite sufficient volume of pages and words.

The other part of the story is told in *Secret Self*.[574] This is the matrixial logic explanation of how the Self operates, in itself and in the world.

We made a start on that project in *I Want To Love But*. *Secret Self* explains how the psycho-emotional and mental processing operations revealed in *I Want To Love But*, fit into this complete picture:

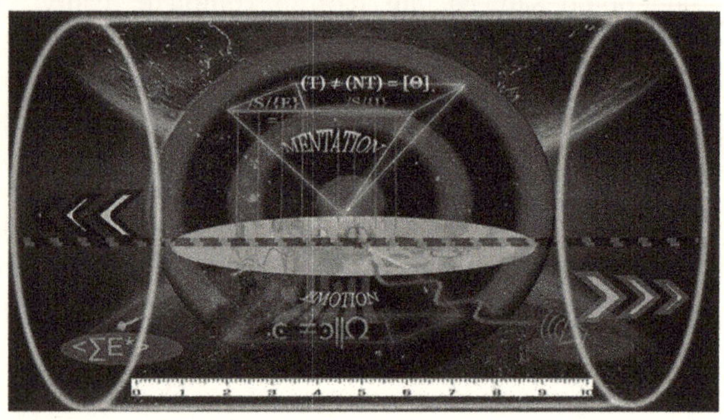

Becoming Being

Using the techniques of ML, we have built a simple and consistent model:

The External World

(1) The world exists "out there".

(2) We extract certain elements of the world as Node, and certain elements "enter" us directly.

---

574 The Author (2020)

(3) Both pass through the Gateways: the 5 senses,[575] and in every other way in which we can be organically effected. That, of course, includes internal organic processes and events.

(4) It is only each such moment of interaction which is "now".

The Emotional Realm

(5) The post-Gateway "inputs" interact with our ocean: our emotional referral system.

(6) Emotions are a distributed information referral and feedback system. This is the Ohm Θ domain:

$$(\text{ͼ}) \neq (\text{ͽ}) = \sum\Omega$$

(7) Each interaction produces a peturbation: whether as a VCIP (m̄ for [E] Nodes), or as a wave interference event Φ (for [I] forms.

The Subjective Sphere

(8) We have differentiated Scalar axis fields /S/: these interact with m̄ and Φ.

(9) Scalar axis fields operate with a law of identity. But not always with a law of commutation.

(10) Scalar axis fields recognise, differentiate and equivate: both quantatively and qualitatively.

(11) The Subjective Sphere is unique to humanity. But it is not the same thing as consciousness.

[575] Qualia

The Objective Sphere

(12) The Theta Φ Objective Reference Domain interfaces with the Ohm Ω domain of Emotions but only via the Scalar domains.

(13) /S/ operations have effect as transitions to the Theta Φ Objective Reference Domain ("ORD").

(14) The ORD is where we ORDer our thoughts. This is our existential thinking mode: the mode which functions our outputs to the world.

(15) Our ORD is what makes Human Information Transmission possible, as grammatical language.

(16) The Subjective Sphere and Objective Sphere are not "parallel processing". To call the Subjective Sphere operations "processing" is fine, if you insist. But those operations do not work the same way as Objective Sphere Operations. So, there is no "parallel", whichever way you cook it.

Spacetime

(17) Time and Space are quantised.

(18) Our Subjective Sphere is not quantised. Our ORD is quantised.

(19) Because Spacetime and our ORD are quantised, they can synchronise.

Consciousness

(20) Conscious introspection is nothing other, or more than, these interactive processes in the body:

- the Ohm Ω domain of Emotions (which includes m̄ and

Φ)

- the Subjective Sphere, in its Scalar manifestions.
- the Objective Sphere, in its internal and external co-ordination.

being and becoming in the world which is in being and becoming.

Critiques of this ML Model ("MLM") may well want to say that it is all just subjective. That it does not solve the "hard problem" of consciousness.

ML answers that no "hard problem" of consciousness ("HPC") exists. And that ML can prove this:

- By showing that HPC, where claiming to be materialist, actually mis-states the evidence for materialism. And thus puts a faith-based burden of proof on PSM;
- By showing that HPC, where claiming to be dualist, is inherently contradictory;
- By demonstrating where the HPC actually comes from. And thus showing the basis, in reality, for a fictitious problem.

ML proves the fallacy of the HCP, not merely negatively. But also through the Experiences encountered in this book: culminating in the *I-Pen Experience* ("IPE").

By demonstrating with empirical proof that both subjectivity and objectivity are demonstrable properties of consciousness. Demonstrable in externally observable reality.

We recall one of IPE sample results:

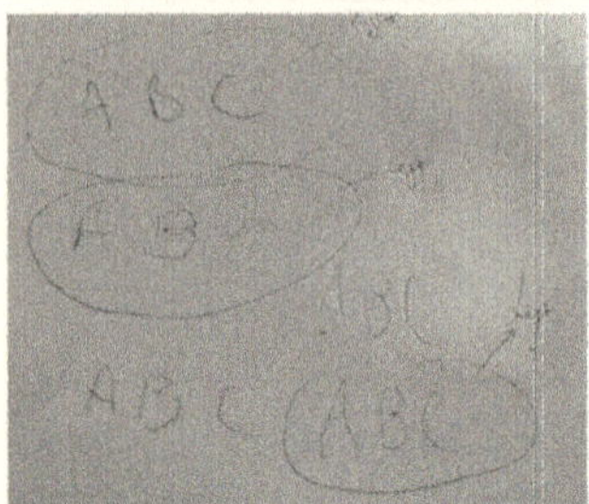

1. Normal Hand

2. Left:
without You_2

3. Left:
with You_2

The 3rd writing was performed moments after the 2nd writing

All of the *Experiences* in this book demonstrate a facet of the MLM. The *Experiences* which pose an ultimate challenge to the idea of HCP, include:

(1) *IPE.*

(2) *Noming.*

(3) *Ropetime.*

(4) *TimeCart.*

(5) *Mirror.*

We call these Theta Θ bases *Experiences* together "Thexes".

Hard Problem: Soft Science

HPC advocates claim that science operates under the Identity Principle: that A=A.

It has been a commonplace for 400 years that science operates under the Identity Principle. In reality, science doesn't and never has. We have demonstrated this in earlier Chapters.

Logic has tried to bind itself to the Identity Principle: and got slowly (but interestingly and brilliantly) around a circle of identity, exploring the extrusions of A=A, and never escaping the trap.

The HPC mantra is that: science can examine, quantify and verify the material world. But science cannot effect these measurements for your "subjective" sense. So we must either:

- Accept materialist reductionism;
- Accept dualism;
- Accept some token-based analysis.

Let's put the supposed tests of science to the test. Can any science, or scientist tell us:

- How many atoms, precisely, comprise the keyboard this is being written on? Sure, science can tell us lots about atomic and molecular *processes*. But give us a number: right now. No you can't;
- How many atoms, precisely, comprise the Author's body: right now. And in one second from when this is typed? Or how many atoms comprise any reader's body: right now. And in one second from when this is read? No. You can't;
- Can you tell us how many new skin cells we have grown today? Yes you can give is an "average", meaning a description of a biological process. But no, you can't.

And we could go on to infinity, listing matters about matter, about which science and scientists can give us no better answer than an astrologer.

Sure, science can do equations. So can ML. Science can manufacture material experiences in accordance with reliable, verifiable processes. So can ML. It was ML "science" which created those *Experiences* in this book. The *Experiences* are repeatable and verifiable. They work for every human being who tries them.[576]

The net result is that militant materialists are engaging in a hand-waving subterfuge. The scientific instruments of their persuasion are very fuzzy and floppy indeed.

Considering Consciousness

The significant other part of the story is told in *Secret Self*.[577] We summarised it above:

Consciousness

(21) Conscious introspection is nothing other, or more than, these interactive processes in the body:

- The Ohm Ω domain of Emotions (which includes ₼ and Φ);
- The Subjective Sphere, in its Scalar manifestions;
- The Objective Sphere, in its internal and external co-ordination.

being and becoming in the world which is in being and becoming.

---

[576] Except psychopaths, in relation to some *Experiences*: which is a good set of tests for psychopathy.

[577] The Author (2020)

That's a lot of claim in a small compass. It's a definition of <consciousness> which meets the 3H criteria:

- Holistic
- Heuristic
- Hylopmorphic.

Consciousness becomes in our organic bodies becoming in a world of becoming. The synchrony of timeframes provides the successive (and layered) reference points of being.

The understanding provided by forms of inequality, in dynamic relation and having the actioned capacity for synchronous timeframes, renders the complex and unsolved: sophisticated, but simple.

We could teach an elementary version of this to 7 year olds. This is the Matrixial Logic universe of being in becoming:

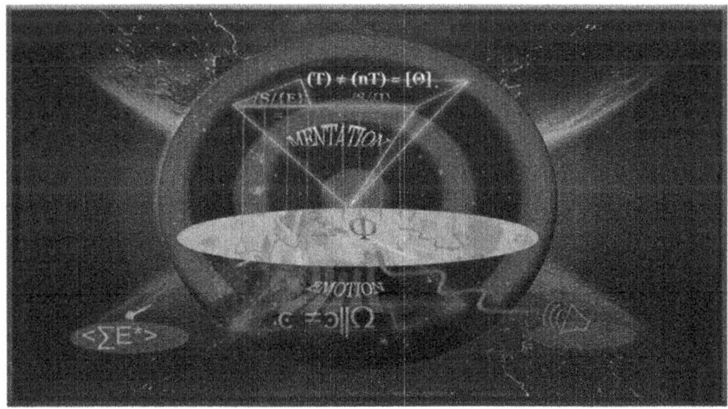

Here's a tabular version of the graphic:

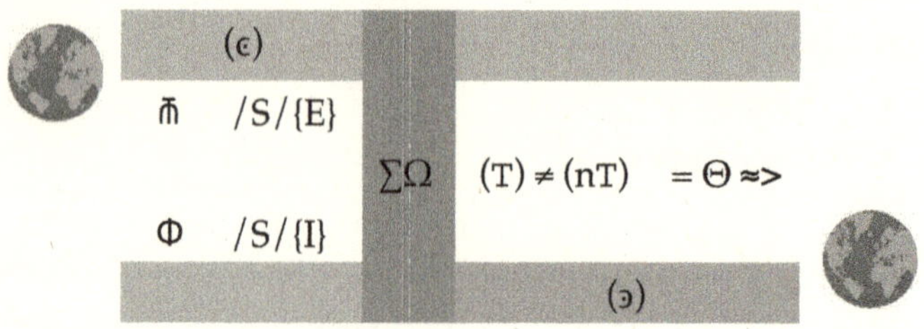

Using the tools of ML, we simply "slice" parts of the whole. Without committing the mereological fallacy along the way. Using linguistics as a means of explaining key concepts, we arrived at these elements:

m̄ /S/{E}

(T) ≠ (nT)

Φ /S/{I}

In the *Languini* Chapter, we did criticise logicians who have tried to analyse language separately from its emotional components: the Ohm Ω \ Scalar interaction.

"I feel fine" does not mean the same as "I feel… fine". And even the three words "I feel fine" could mean anything, including their opposite, depending upon the Ohm Ω \ Scalar interaction and the real-world context.

So, we need these Ohm Ω domain qualaities to provide a complete understanding:

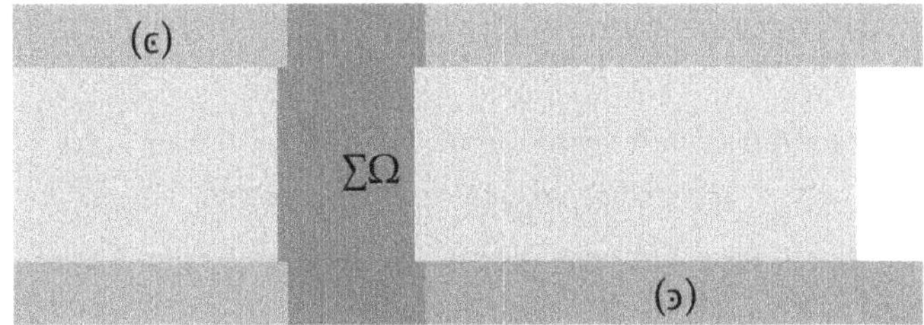

This is a very different equation from standard ML usage.

(ɛ) ≠ (ɜ) ∑Ω

The explanation, at the summary level, is simple:

- Different emotional states (ɛ) / (ɜ) are not "equal" in any sense. So the standard ≠ notation makes sense;
- Nor do different emotional states ultimately "equal" anything. They merely sum ∑ to internal psychological, and external acting out, behaviours.

We can see the resultant internal psychological, and external acting out, behaviours in the Ohm ∑Ω domain. The psychological sciences effect analyses, both descriptive and predictive, of the Ohm ∑Ω domain.

We can use the same ML methods: but the investigative and analytical tools we use are different, from when we consider thought. That's not any kind of "dualism": you don't get a computer technician's tools to tend a garden, and vice versa.

We are now trespassing upon the pages of *Secret Self*.

Which we can do here only in summary assertion, which cannot act as its own justification.

So we shall state it summarily:

<phenomenal consciousness> is this

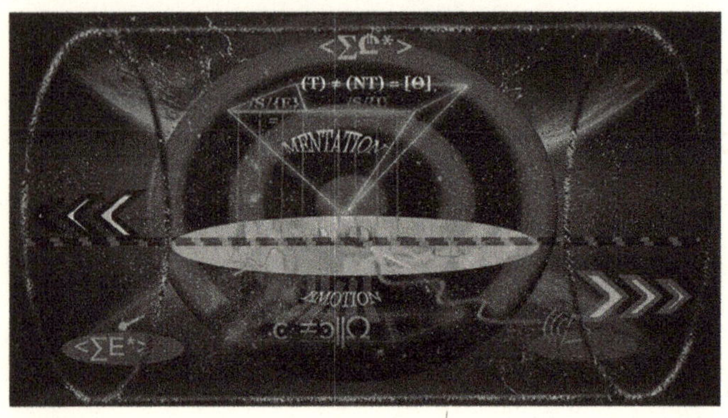

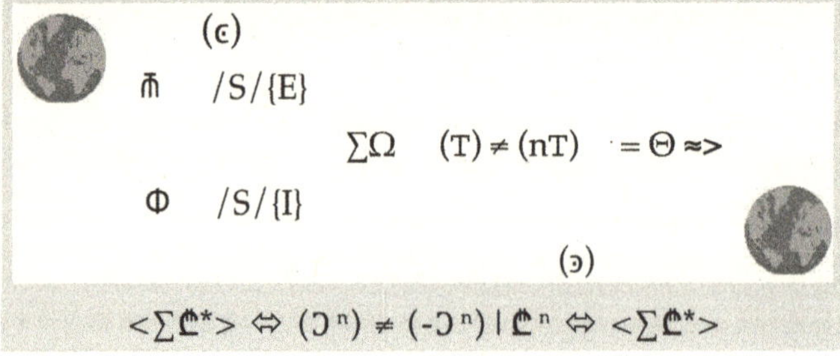

The previous Chapters explain how the elements of this graphic combine.

It's worth repeating that these Domain functions together operate under the 3H principle:

- Heuristic
- Hylomorphic
- Holistic

<phenomenal consciousness> is multivalent: the 3H interleaving of those elements in dynamic harmony: mutually affecting processes of coherence, disassociation and congruence.

If we must define <phenomenal consciousness>, then "it" is, within a unique organic biological frame of reference, and in plurality:[578]

- Multivalent processes of being in becoming;
- Mediated by an information distribution and referral system, with internal organic and subjective state process feedback;
- Mediated, enveloped, and reciprocated, by subjective state processes which are diachronic;[579]
- Cohered to external reality by synchronous co-valent quantisation of space and time.

To put the matter in a very poor way, analogise the three domains as planetary spheres, and include reality and our organic being as others. Then:

<phenomenal consciousness> is the *gravity* between these planetary spheres.

578 Humans
579 With respect to external and internal changes

By <phenomenal consciousness>, we mean the sort of thing that Nagel was talking about in it being "like something"; what Dan Dannett says is Caspar's Ghost. What David Chalmers says is real but unknowable.

We have to say that "sort of thing", because each of these materialist and dualist approaches brings their own baggage of presuppositions, packed in different suitcases filled from the wardrobe of the 17th century. With the addition of polarising sunglasses, a transistor radio and a silicon chip.

The Origin of Theatrical Delusion

We have Monsieur Descartes[580] to blame, or thank, for 400 years of HPC head-wall banging.

Cogito ero sum. Usually translated as "I think therefore I am". The idea of Decarte's theatre: the observer observing and judging the show.

That idea is itself fictious, even on a strong dualist view. Our introspections do not work like a theatre, or movie. Those are, in the real world, carefully edited productions directed by a single guiding mind.

Consider the cinematic flop *Casino Royale*:[581] the spoof with David Niven, Woody Allen and a host of stars. Shot

---

[580] With not a little help from David Hume; and others

[581] Columbia Pictures (1967)

be five different directors. And a byword for how not to make a movie.

Introspection, to anyone who just does it, and reviews the contents of their head, is an incoherent, discordant, distracted tumbling of jigsaw pieces from different boxes. You can gain coherence: but that comes at the price of living with "Shadow Slices".[582]

We seldom use moving images, in introspection. We introspect in jump-words and phrases and feelings. Why? Because in introspection, we are using Scalar techniques, to explore our inner world.

That is why we are, simply, not very good at introspection, by and large. Those few who are, tend to mark themselves out as artists, or the insane.

So, if dualism were right,[583] we have a very odd way of going about it.

What's more, Matrixial Logic techniques can be used to effect memory wipes. Exactly by getting you to figment a "movie" type memory, ML can wipe the video tape. Your access to traumatic memories is permanently deleted. In around a minute.

---

[582] See *I Want To Love But.* The Author (2019)

[583] And it must have been right for all the history of humanity

Just by using a formula of words that you read, or are spoken to you, together with a tiny manipulation of the connections within your <phenomenal consciousness>.

Those who experience the Memory Wipe Experience report both immediate and longditudinal success. You can read more about it in *Secret Self.*

For the extant philosophic debate about HCP, this is armageddon. The end of those worlds. Perhaps, about time.

The HPC materialist objects: how can a subjective state provide proof about an objective state of affairs? And the dualists say: well, your objective entreaties cannot disprove the subjective state because they cannot account for it. No ought from an is.

The Author prefers to translate the phrase as: *I am thinking therefore I am existing.*[584]

This preference arises because it makes more transparent the Cartesian consideration behind the phrase.

All that this amounts to is the claim that our existential thinking mode Theta [= Θ] (the mode which functions our outputs to the world) can effect views:

---

[584] A perfectly admissible translation: both in Latin and French. In its use of the present participle, it better represents what Descartes actually wrote

- Onto the world

and

- onto ourselves.

Indeed it can. That is simply how the Theta [= Θ] objective domain works.

As has been pointed out, one could stub one's big toe, with the accompanying sense perception in the Subjective Sphere, and equally well state: *I feel from stubbing, therefore I am.*

In other words, self-reference is not a proof of external, objective reference. And that is what HPC enthusiasts seek.[585]

So they should be given it. It's time.

Non-Matrixial Time

This universe is time and dimensions in space. That set of concepts can be stated in various ways.

For the purposes of this overview, it matters not much whether we, for convenience, go with Einstein's recapitulation *Relativity: The Special and General Theory.*[586]

Or whether we adopt the quantum entanglement

[585] Perfectly reasonably

[586] Fifteenth Edition (1952). Originally published (1912)

approach to time, as discussed in the previous Chapters. The former may be viewed as a macro-theoretic and the latter as a micro-theoretic.

We trust that even the most Cromwellian materialist can agree to common ground here. Spacetime is real. It is the "ultimate reality". There can be no process or event which occurs outside spacetime.[587]

Every scientific measurement is possible only because spacetime provides a universal frame of reference for all extension of all substance. Agreed?

Let's allow you, Materialist, to have the best measuring devices you can have: practically, or theoretically.

Measure a person's organic states: heart rate, galvanic skin response, time of blinking per second. Good. Now measure how many cells that person has right now, in their body. And the next moment. Measure the electrical activity in a person's brain, and different brain areas. Well, you can't, with present technology. But let's allow that you could in theoretical principle.

Now mark off on a millisecond clock, the start and stop time:[588]

- That Alice saw that dog picture

[587] Except in theist thinking
[588] Which will do for this example

- That Alice had feelings about this state of affairs;
- That Alice formed the word "dog";
- That Alice had more feelings about this state of affairs;
- That Alice engaged in the speech (or writing act) of "dog";
- That Bob engaged in reading the word.

and so on.

In theoretical principle, you can measure each such time. It follows then, that each such event is *real*, does it not?

Recall from Chapter 2, at the very start, it was stated as a founding principle of ML, that all Substance is "extended in spacetime". And we did agree that:

> spacetime provides a universal frame of reference for all extension of all substances.

Well: that's that then, isn't it? Of course, that isn't that. Because all of that external measuring you do, doesn't get you "inside" the mind of <phenomenal Consciousness>

But I can't see inside your head. Indeed. And you can't "see inside" the life events of a daisy either.

You can aggregate and average life processes of a daisy. As you can with homo sapiens: in all sorts of interesting ways. But you can dissect a daisy to its cellular level, and you will not find some mysterious daisy "life" occurrences of processes and events.

But no materialist goes around suggesting that daisy biological events are illusions. *But a daisy is not conscious.* We agree.

So, philosophy currently poses these fundamental questions:

(1) What is phenomenal consciousness;
(2) How do we have consciousness;
(3) Why do we have consciousness.

But, the philosophy of consciousness is stuck, because <internal> and <external> states have to be brought under some rubric of coherence, so that each can be equalised to the other under some common frame of reference.

Philosophy needs <internal> and <external> states to be $(A)^1 \setminus (A)^n = [A]$. Since [A] must include material time and space, this just can't be done. Philosophy cannot arrive at an iteration of A = A.

Chapter 3 onwards have shown us why this search for a form of equality which allows everything to be its iteration:

- Doesn't work
- Can't work
- We've been on the project since Aristotle (at least)

The tool of the HPC paradox is nothing more sophisticated than the Liar paradox. It is simply the endless effort to pose one mirror version of scalar existential fiction, against another.

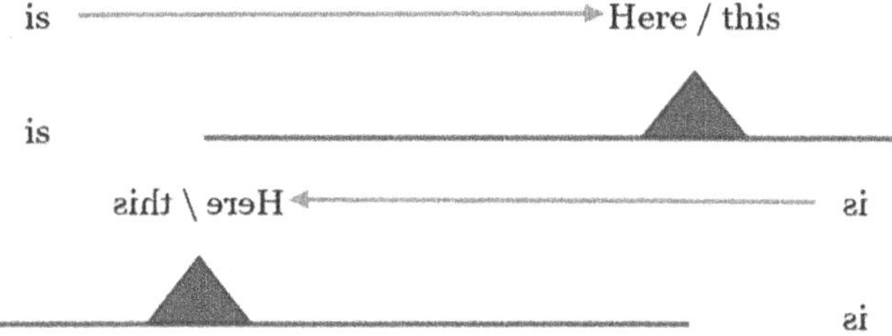

The HPC-ist is seeking a ghost which was invented out of the human interface of the Subjective Sphere and ORD field. Both operate in spacetime. And nowhere and nowhen else.

The only solution to such scalar mirroring, is to invoke a singularity. The HPC-ist seeks more. You seek god. Now it's time. Admit it.

The Floating World

Let's raise the stakes immediately. With fanfare, to point a direct challenge to present paradigms of philosophy, which have the imprimature of two and a half millenia.

Please recall, from the *Languini* Chapter, the *I-Pen Experience* (one of the Thexes) and one of its sample results:

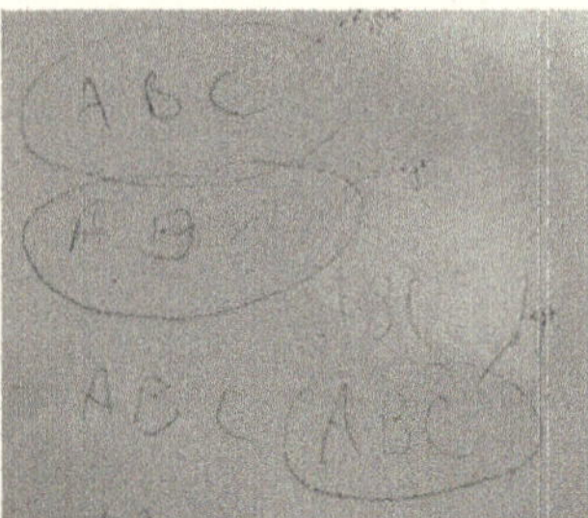

1. Normal Hand

2. Left:
without You$_2$

3. Left:
with You$_2$

The 3rd writing was performed moments after the 2nd writing

Recall your own Thexes. A difference between this and other Thexes, and the other *Experiences*[589], is that you can yourself witness the result. Graphically, in spacetime.

We do not suppose that anyone could persuade you that what you see written on your own sheet of paper, or time-clock matched with your fingers, following the Thexes, is any kind of illusion.

You might wish to deny the reality of your mentative *Experiences*, during the course of this book. But you cannot deny the reality of the Thexes.

Echoing that fanfare, let us state clearly: that which you have yourself witnessed as the result of your own

---

[589] Which we hope you did undertake

conscious activity is, under the present paradigms of philosophy: *impossible*.

The Challenge to Materialism

Be on the Materialist axis of thought: from Churchland to Dennett. If we may crudely summarise your ontology:[590]

- There exists nothing but matter;
- The body and brain are a particular configuration of the same matter as exists in anything else;
- The concept of a phenomenal consciousness is either plain wrong; or is merely an illusory by-product of the operation of matter in explainable configuration;
- However you consider the issues, there is no independent phenomenal consciousness;
- You may consider there to be an organically dependent phenomenal consciousness.[591] In which case, the idea of phenomenal consciousness is redundant.

Let there be posited any Steel Man variation or elaboration of the Materialist axis. Here is the challenge:

by what means can the materialist axis explain
the reality of the Thexes?

If the brain is some analogue of hardware running software, or wetware; if these wares are matters of neurobiology, whether or not explicable by eveolutionary biology.

---

[590] Not in order to Straw Man it, merely for narrative convenience

[591] The Churchland axis may deny even that

In that conception of these matters, what consistent explanation can be offered for these realities:

(1) Thoughts, embodied in writing, prompted you to "imagine" a You_2.

In itself, that is non-problematic. The written instruction is merely a data input.

(2) You can imagine a You_2 and (this is the big and) feel a connection between You and You_2.

Well, that's just "feelings" Feelings are not facts. And if they are facts, they are just facts like asking you to think of [2 +2 =].

(3) Those "feelings just as facts" instantiated an activation of fine motor-skills, which otherwise you could not instantiate.

a) Well, I don't know for sure that I could not instantiate them some other way.

> We respond: are you serious? You've lived all your life since the age of 3 being unable to instantiate these fine motor-skills. You are supposed to accept the scientific force of empirical evidence.

b) Well, I don't know for sure that I did instantiate those fine motor-skills because of the instruction.

> We respond: are you again serious? You are supposed to accept the scientific force of empirical evidence.

(4) It is empirically proven[592] that:

> We instantiated *a material and significant change in your neurobiological function,* simply by having you undertake 3 iterations of mentation (1. think of You$_2$; 2. "connect" your left hand and pen to You$_2$; 3. Allow You^2 to guide your left hand and pen.)

(5) It is further empirically proven that the Author was able to **Predict** and therefore **Manipulate** *<u>your</u> neurobiological and motor function.*

(6) Indeed to **Predict** and therefore **Manipulate** the neurobiological and motor function of every human alive.[593]

So, materialist philosophers: now what?

You are in very serious, and obvious difficulty. Because the Author, in using the Thexes, reached into your mind, gave it instructions, and produced a result in the application of fine motor-skills, which you otherwise could not access:

---

[592] You have the graphic results of your own efforts before you

[593] With suitable modification for the left-handed; which works just the same

- Without pharmacological intervention;
- Without hypnosis.

As to the latter, you, and every reader is perfectly sure that no hypnosis was involved. [594] Indeed, you were experiencing a perceptibly high state of conscious awareness, in performing the Thexes.

Each Thex was a set of operating instructions, which are empirically predicated and validated, having effect to make occur:

material and significant change in your neurobiological function and motor: which is empirically observable, repeatable and verifiable.

A Lot of Nots

Let's just expand on that. Whatever Thexes are, it is not merely a "trick" of perception.

Thexes instantiate an alteration in your gross physical attributes, a *material and significant change in* the availability to you, and application by you:

- Of fine motor-skills in external 3 dimensional space, and time;
- Of aural-motor syncopation with external time;
- Of aural-perceptual syncopation with external time.

---

[594] The Author has studied and practiced in hypnosis.

All occurrent and graphically manifest *in the real world.*

- Whatever is happening, is not just happening in you. It is altering your spatio-temporal relationship with the external objective world;
- And such that any outside observer can see (for example) the difference between writing: without You_2; and with You_2.

Plainly, this is not equivalent to an instruction to "jump" and you jump your normal available jump. It is perhaps equivalent to an instruction to jump, and by means of the manner of instruction, you can instantaneously jump 10 metres (long or high).

Indeed, once a reader has performed a Thex, particularly *I-Pen*, it's impossible to go back. You just can't write badly with your non-usual hand anymore. The ML direction has effected a permanent change in your neurobiological and motor function, and that interaction with the real external world.

Moreover, this is not equivalent to a yogi, fakhir or hypnotist "switching off" internal subjective perception of, or reaction to, pain qualia: so that you can walk barefoot over hot coals. Indeed, rather then effecting restriction of any neurobiological operation, this is expanding and extending it, in full consciousness.

This is not placebo-based neurobiological interference.

No substance ingestion can instantaneously effect *material and significant change in* the availability to you, and application by you, of fine motor-skills, in the real world.

The Author offers a scientific prediction: that no alteration in brain function (save as noted below) would be observed in fMRI in anyone undertaking Thexes.

The reasons for that prediction include:

(1) it is self-evident that no new synaptic function could be instiated physically within the brain in the mere moments which it takes to instruct and perform Thexes.

(2) it is self-evident that the same physical regions of the brain must be involved in the perceptive acquisition and application of the Thexes instructional information.

(3) The specific instruction to place You_2 on the right-hand side is because we are instantiating left motor-function: and to that extent we may witness in fMRI differential stimulation of the motor cortex.

Prediction (3) is particularly challenging for Materialists. You are simply repeating mentative "operating instructions" which the Author has created in his own head.[595]

---

[595] Whatever you think is in there, and however you think it works

Yet those phenomonal communications are having direct, observable biological effect: and the only avenue for "causation" of that effect, is the transmisison of information from one person to another. In this case, merely via printed words.

Again, this is not merely the evoking of some feeling, or image. This is direct physiological effect, observable in the real world, of mere epiphenomena.

The Author has always found Dan Dennett's advocacy of functional materialism to be eloquent, powerful, and about as persuasive as error can get. Take this seminal passage:[596]

Figure 2.3

[596] *Consciousness Explained.* (1991) (p. 35).

> Just as one would expect, ingenious technical exemptions based on sophisticated readings of the relevant physics have been explored and expounded, but without attracting many conversions. Dualism's embarrassment here is really simpler than the citation of presumed laws of physics suggests. It is the same incoherence that children notice - but tolerate happily in fantasy - in such fare as Casper the Friendly Ghost (Figure 2.3, page 36). How can Casper both glide through walls and grab a falling towel? How can mind stuff both elude all physical measurement and control the body? - ghost in the machine is of no help in our theories unless it is a ghost that can move things around - like a noisy poltergeist who can tip over a lamp or slam a door - but anything that can move a physical thing is itself a physical thing (although perhaps a strange and heretofore unstudied kind of physical thing).

The Author well remembers his own delight at the Caspar's Ghost problematic, when it first emerged. A simple, powerful, explanation of the central problem of Dualism.

The problem for Materialism, and at this point in the life of Matrixial Logic, an apparently insuperable one,[597] is:

the Thexes do function as Caspar's Ghost.

The Author puts a limited collection of Thexes concepts and instructions into *your* head, and out pops material and 3rd party observable change in the universe. By means of altering the connection between "whatever is going on in your head" and the fine motor-functions of your body.

If we tap your knee with a hammer, and get a physical reaction, accompanied by an observable change in mental state: well OK for materialism.

---

[597] We stand to be corrected on that

If we attach Galvani[598] electrodes to your brain tissue and pursue the pathways of Wilder Penfield[599] then: physical stimulation in, physical response out. Again, OK for materialism

But this is: epiphenomonal consciousness data in, predicted and novel and otherwise unattainable physical response out. And not merely physical response: graphic and synchronous motor-temporal interaction with reality.

To put it plainly, the Thexes takes Materialism down a Matrixial Logic rabbit hole. And the more Materialism digs, the deeper the hole gets.

The Challenge to Dualism

Before those on the Dualist axis hang out the bunting, the foundation of their house of idealist cards, tumbles down under the seismic shock of the Thexes.

From Descartes,[600] though the moderns; Thomas Nagel, though to David Chalmers; and a recent pansychic defence by Philip Goff:[601]

> In fact, this challenge is less straightforward than is often presented in undergraduate philosophy courses. It is true that the dualist cannot explain the causal connection between the mind and brain, but this is arguably part of a more general limitation of human knowledge. While

[598] Luigi Galvani (1737-1798)

[599] (1891-1976)

[600] Debatable whther Descartes was a dualist at all

[601] *Galileo's Error. Foundations for a New Science of Consciousness* (2019) p 28-29

> Descartes struggled to make sense of mind-body interaction, a century later the great Scottish philosopher David Hume (1711–1776) pointed out we cannot explain any of the fundamental causal relationships of the universe.2 If this is true, then the "failure" of the dualist is just one instance of a more general failure of our scientific understanding.
>
> Panpsychism avoids the problems of dualism because it does not postulate consciousness outside of the physical world and hence avoids the challenge of accounting for the interaction between the nonphysical mind and the physical brain. The panpsychist places human consciousness exactly where the materialist places it: in the brain. And because it is not trying to explain consciousness in terms of nonconscious brain processes, panpsychism also avoids the problems of materialism. Rather than trying to account for consciousness in terms of non-consciousness, the panpsychist aspires to explain the complex consciousness of human and animal brains in terms of simple forms of consciousness, simple forms of consciousness that are postulated to exist as fundamental aspects of matter. Does this really count as an explanation of consciousness? Isn't this just taking consciousness for granted rather than genuinely explaining it? To be sure, panpsychism does not offer a reductive explanation of consciousness, that is to say, it does not explain consciousness in terms of something more fundamental than consciousness. However, it is a prejudice of materialism to suppose that this is obligatory. There is a great deal of precedent in science for nonreductive explanations which take the phenomenon to be accounted for as basic. Consider, for example, James Clerk Maxwell's theory of electromagnetism in the nineteenth century. (p.115)

If we may crudely summarise Dualist epistemology:[602]

- There exists matter and stuff other than matter;
- The body and brain are a particular configuration of matter and other stuff;
- Phenomenal consciousness is real in the world;
- How it is that phenomenal consciousness interacts with

[602] Not in order to Straw Man it, merely for narrative convenience

matter is: (a) unknown; or (b) unknowable;

- Unknowability is no bar to reality since the 4 fundamental forces of the Standard Model are presently "unknowable" in terms of anything more fundamental;[603]
- However you consider the issues, there is independent phenomenal consciousness.

and the *Critical Proposition*:

- That which proves independent phenomenal consciousness, is that my subjective experiences are inaccessible to you, and vice versa. It is this unknowability which stands as guarantor of the reality of the phenomenon.

On the classical account of Dualism, panpscyhic theory is somewhat of a cheat. It seeks to defend Dualism, by violating the *Critical Proposition*. Or by so denuding it of content, that Dualism turns into a parallel version of Materialism, the two tracks joined by an affirmation of insuperable ignorance.

Let there be posited any Steel Man variation or elaboration of the Dualist axis. Here is the challenge:

by what means can that Dualist axis explain
the reality of the Thexes?

[603] And the efforts of quantum chromodynamics in string theory or loop theory are interesting, but unestablished empirically

The challenge to Dualism is made easy by the *Critical Proposition*:

(1) If your subjective mental states are inviolably and unknowable sovereign.

Then:

(2) How can the Author manipulate them so as to produce neural and biological effects, which are observable, in the external world.

In other words, since plainly the contents of your head are "transparent" to functions founded in Matrixial Logic, those contents must (in the Dualist way of looking at matters) lie in an objective plane of reference. You are not a free man: you are a number.
So, farewell the Dualist defence of *free will*. But fret not: the Materialist shotgun of determinism is out of ammunition too.

We can also eliminate is some hyper-material theory based on quantum electrodynamics.[604] QED is inconsistent, and remains expressly inexplicable: "shut up and calculate".

Fundamentally, all such theories seek to extract a fundamental reality, including about ourselves, merely by iterative extrusions from a scalar reference domain in which A=A: that is, a domain subject to the laws of identity.

---

[604] The Hammeroff-Penrose *Orch-OR Theorem*: brilliantly helpful though it is as a matter of neuro-biology

What is interesting, is that the further theoretical physics finds itself to be penetrating to "ultimate reality", the more that the epistemological mathematics of physics, in that very law, become self-evident hindrances.

So How Now?

As Holmes said:[605]

> when you have eliminated the impossible, whatever remains, *however improbable*, must be the truth

If real change is to be initiated, such that philosophy can escape from the straight-jacket arms of Materialism and Dualism, then those presently captive to such constraints must first make themselves free.

So what, then, is the philosophical explanation for Thexes, and all the other *Experience* phenomena in this book?

In the Appendix *Matrixial Logic in Plain English,* we state in 20 Propositions, the fundamental structural elements of ML.

We have, in each previous Chapter, unwrapped the key elements. At the opening of this last Chapter, we have summarised the MLM which arises from those elements.

It is not here claimed that the Matrixial Logic rules and

[605] *The Sign of Four.* A Conan-Doyle. (1890)

forms of inequality will solve puzzles of being and becoming.

It is claimed, upon the evidence presented in this book, that Matrixial Logic makes possible profoundly novel methods for asking useful questions.

Matrixamus ergo Intelligamus

APPENDIX 1

MATRIXIAL LOGIC IN PLAIN ENGLISH

Before You Begin

This Chapter is written as a guide to understanding. Using non-technical plain English, we summarise principles of Matrixial Logic.

It is not possible to specify with analytic rigour terms used in this context with properly analytic terms used in the main Chapters text.

It would therefore be a cheap shot for any critic to allege triumphantly that there is analytic inconsistency between this plain English summary, and the rest of the book.[606]

Matrixial Logic

Here, we set out a curated collection of propositions, each followed by a short explanation. We are seeking to allow understanding to emerge from the chain of propositions, which create inequalities, from which forms arise.

(1) Nothing can exist on its own.

The inescapable condition of existence is that there be more than one of anything: a plurality of things.

---

[606] Which, no doubt, won't stop some critics doing exactly that

(2) Nothing is the same as anything else.

Whenever we consider any plurality of things to have the quality of identity, it is merely because we are using some third thing to hold them in common. That third thing is some measure of equality.

(3) Inequality produces understanding.

To establish an identity of things is to understand them only by reference to the measure of identity. To understand things, is to appreciate the forms which their inequalities present.

(4) Forms of inequality depend on things.

We can invent any idea of form and arbitrarily assign it to any things. In Matrixial Logic, we allow things to dictate their form. This is inductive reasoning.

(5) Different types of thing have different forms.

There are two types of things: those that are in being, and those which are in becoming. All being becomes and all that becomes manifests as being.

(6) Existence and Infinity are the Modes of Forms.

That which is in being, is in existence. That which is in becoming, is infinity.

(7) That which is neither being or becoming, is Chaos.

Chaos has no forms, but produces Singularities. Singularities do not generate forms of inequality, and exist only in reflection of forms of inequality.

(8) Specification of form depends upon content context.

Specific context always modifies the manner in which we seek understanding of things. Context thus determines whether we seek understanding of a form of being, or becoming.

(9) Forms interact as Events.

Events[607] are created by the interaction of different forms of content. Events are unstable.

(10) Events stabilise as Nodes.

A Node occurs when an Event interacts with a becoming. The interaction interrupts that becoming, and a new sort of being is created.

(11) Nodes produce Points.

Points are the manifestation of Node-becoming interactions. Points produce the capacity for scalar measurements of the becoming[608] in which a point has arisen.

(12) Becomings interact as interferences.

Becomings do not[609] produce Event Nodes. Instead, becomings interfere with each other.

(13) Interferences stabilise as Points.

---

[607] Which does not always mean "something happening

[608] Which can be plural

[609] Wxcept theoretically

Points also arise from becoming-becoming interactions. Points produce the capacity for scalar measurements of the becomings in which a point has arisen

(14) Scalar fields are undefined.

Scalars operate under the law of identity, but not necessarily the law of commutation.

(15) Theta fields allow scalar fields to be understood.

Scalar fields can only refer to each other to where their information comes from. Theta fields reference Scalar information to the external world.

(16) Our understanding is matched to our reality.

External reality creates the conditions of our understanding. But we understand only through the medium of ourselves. Our own medium is so organised by reality, such that we have a way of communing with external reality directly.

(17) Existence is proven by communion.

We can never prove or disprove reality of our own being in our own becoming, or vice versa. We do not need to. Proof of both is established by demonstration in fact of the capacity each of us has, for communion with that which is external to each of us.

(18) X

If external reality did not exist, whether in being or

becoming, we would be unable to know it.

(19) We are a dynamic matrix.

We exhibit three operational features in dynamic: hylomorphic, holistic, heuristic. We are born such. We are born of reality and live in reality. These operational features are both what allow and inform our individual lives. We are unable to operate the processes of these features otherwise than under forms of inequality.

(20) Matrixial Logic is the logic of forms of inequality.

To understand is to have the capacity to effect. And vice versa.

Here is a single paperclip, in infinity.

- You cannot effect it
- You do not understand it

∴ (A) ≠(n\- A) = F [E \I]

Quod Erat Demonstrandum

APPENDIX 2

TABLE OF SYMBOLS USED IN MATRIXIAL LOGIC

Symb	Meaning / Use
These ML synbols denote construction of an FoR	
(A)	Can be a Substance or a Moment
(nA)	Is Substance in being in opposition to (A); also written [NOT]
(-A)	Is Moment in becoming negation to (A)
≠	Inequality. In speech: "(A) *inequal* (^{NOT}A)"; "(A) *does not equal* (^{NOT}A)"; or "(A) *non aequalis* (^{NOT}A)"; or "(A) *non EQ* (^{NOT}A); Sometimes abbreviated in speech to: "(A) *NQ* (^{NOT}A); And in like manner for any speech using ≠
=	Aequalis / Equals. Not the same meaning as in 2 +2 =4. The symbol implies *translates as*
[E]	The Form of Being: of Substances. A deducted FoR.
[I]	The Form of Becoming: of Moments. A deducted FoR.
[C]	Chaos: a non-Form; or Anti-Form
/S/	Scalar: an inducted Perspective. Not an FoR.
Θ	Theta
Si	Singularity
>	Moving from one line of ML Equation to another. Indicates "and so…"

$\approx$	Equivalent: when an ML equation, expressed in ML Symbols, is turned into ordinary language
$\therefore$	Therefore [used as the concluding line of a syllogism
◊	Spacetime: the dimension of Substance
∞	Infinity: the dimension of Moments
C	Chaos: the dimension without Forms
Δ	Potential^{I}: the capacity of Infinity to produce Moments
□	Actuality

Symb **Meaning / Use**

These ML synbols denote operation within an FoR

$\geq$	Shows we are moving from an FoR to operations within an FoR
^{n}	Shows we are iterating, or one might say, "pluralising"
$\Leftrightarrow$	This is a "shadow" iteration within an FoR. It is an inherently unstable (conditional) iteration. So the operation cannot be notated by the classic =
{ }	These brackets contrast with the usual [ ]. They notate that the Form is being operated "internally". It notates that this is no longer a real Form, but a "shadow" $_{f}$orm
$_{f}$orm	The referential use of a Form within its own FoR: a Shadow $_{f}$orm

Symb	**Meaning / Use**
These are symbols used in ML TransFormations	
[~]	Intersection: indicates a transforming point between 2 frames of reference
Σ	In logic and mathematics, this sign usually means "sum". In ML, it means "aggregation"; "taken together as". Indicates a Node
FoR_2	The subscript shows us the number of Forms that we are aggregating. The subscript number can be anything from $_2$ to $_n$
|	Held in perspective with. Where 2 MLE are held in a relationship of Potential^p
π̄	Collapse of an Infinity ∞ by an Event □
Φ	Interference between ∞

Symb	**Meaning / Use**
These are symbols used in Matrxial Time Functions	
T	Temporal Reference
$Ɔ^n$	Chrone
|	Annihilation
$Ċ^n$	Chrono Gap

Symb	**Meaning / Use**
These are symbols used in Matrxial Emotion Functions	
(є)	Ohm Antipole
(϶)	Inverse Ohm Antipole
Ω	Ohm

KEYBOARD SHORTCUTS FOR

SYMBOLS USED IN MATRIXIAL LOGIC

Many ML Symbols can be accessed through standard keyboard operations. These following can be accessed through the specified keyboard operations.[610]

Symb	Shortcut Type
≠	2260 Alt+x
NOT	Ctrl+ Shift + + [press again to release superscript]
2	Ctrl+ = [press again to release subscript]
≈	2248 Alt+x
∴	Hold Alt; type on numeric keyboard 8756
≥	underline >
n	Ctrl+ =
⇔	Type: < = >
∑	2211 Alt+x
∞	221e Alt+x
◊	25ca Alt+x
│	2502 Alt+x
Δ	2206 Alt+x
□	25a1 Alt+x
₥	20bc Alt+x
Φ	03a6 Alt+x

[610] Note: specified only for MS Word; qwerty keyboard (UK); operating in Windows Office. Specifications may be different with Mac operating system.

± Alt + 0177

₸ 20b8 Alt+x

∀ 2200 Alt+x

┘ 2518 Alt+x

┐ 2510 Alt+x

ˇ 02c7 Alt+x

Φ 03a6 Alt+x

Φ 03a6 Alt+x

₸ 20b8 Alt+x

∀ 2200 Alt+x

┐ 2510 Alt+x

┘ 2518 Alt+x

ˇ 02c7 Alt+x

Ͳ 0372 Alt+x

₾ 20be Alt+x

ͼ 037c Alt+x

ͽ 037d Alt+x

Ω 2126 Alt+x

Ↄ 2183 Alt+x

α 03b1 Alt+x

β 03b2 Alt+x

ω 03c9 Alt+x